LNG: Fuel for a Changing World—A Nontechnical Guide

LNG
FUEL FOR A CHANGING WORLD
A NONTECHNICAL GUIDE
2ND EDITION

MICHAEL D. TUSIANI

GORDON SHEARER

Copyright© 2016 by
PennWell Corporation
1421 South Sheridan Road
Tulsa, Oklahoma 74112-6600 USA

800.752.9764
+1.918.831.9421
sales@pennwell.com
www.pennwellbooks.com
www.pennwell.com

Marketing Manager: Sarah De Vos
National Account Executive: Barbara McGee Coons

Director: Matthew Dresher
Managing Editor: Stephen Hill
Production Manager: Sheila Brock
Production Editor: Tony Quinn
Book Designer: Susan E. Ormston
Cover Designer: Elizabeth Wollmershauser

Library of Congress Cataloging-in-Publication Data

Names: Tusiani, Michael D., author. | Shearer, Gordon, author.
Title: LNG : fuel for a changing world : a nontechnical guide / Michael D. Tusiani, Gordon Shearer.
Description: Second edition. | Tulsa, Oklahoma : PennWell Corporation, [2016]
  | Includes bibliographical references and index.
Identifiers: LCCN 2016006622 | ISBN 9781593703691
Subjects: LCSH: Liquefied natural gas industry.
Classification: LCC HD9581.2.L572 T87 2016 | DDC 338.2/7285--dc23
LC record available at http://lccn.loc.gov/2016006622

Printed in the United States of America

1 2 3 4 5   20 19 18 17 16

Contents

Illustrations

Figures

Tables

Acknowledgments

The LNG industry has changed considerably since the original publication of *LNG: A Nontechnical Guide* in 2007. As a result, this revision is now a necessity. For example, at that time, the industry expected the United States to be a large importer of LNG; today it is exactly the opposite—it is envisioned to become one of the world's largest exporters. Plus, the Fukushima nuclear accident in 2011 made LNG the favored substitute fuel, and the significant decrease in the price of oil has made LNG more attractive to buyers, thereby increasing its demand.

Gordon Shearer, a friend and colleague, once again coauthored this new book. His expertise and many years of experience were invaluable in this effort. Moreover, we are both indebted to Frank Spadine, without whom this book would not have been possible. Because of his tremendous contribution, it is more his than ours.

Our thanks also go to colleagues who have generously provided their expertise when needed: Fred Adamchak, Jim Briggs, Captain Doug Brown, Nicholas Buttacavoli, Majed Limam, Lawrence Noto, and Mike Reimers.

Finally, several individuals stand out as worthy of our deepest appreciation: Melody Lawrence, Lucio A. Noto, Anita Odedra, and Liz Van Houten. They were kind enough to edit the book and provide valuable input.

I thank my wife, Beatrice, and my children, Paula and Michael and their families, for their support and encouragement. Lastly, my brother, Joseph, who continues to inspire and guide me through life. This book is for him.

Michael D. Tusiani
May 6, 2015
New York City

Michael has acknowledged and recognized the efforts and support of my former colleagues at Poten & Partners and the others who gave so much of their time and experience to support the writing of this book. Like Michael, I would specifically recognize Frank Spadine, who undertook the lion's share of the effort.

I thank my wife, Aleksandra, and our children for their support and understanding during this effort. Most of all, I would once again give my deepest thanks to Michael, my friend, collaborator, and colleague of many years, and one of the great figures of our industry. Writing this second edition has provided further testimony to our friendship and testament to Michael's vision and desire to ensure that we have reported the state of the industry as comprehensively and accurately as we can.

Gordon Shearer
May 6, 2015
New York City

Introduction:
The Liquefied Natural Gas
Industry—An Overview

A "perfect storm" of $100 oil (at least until the middle of 2014), coupled with technological developments such as horizontal drilling and hydraulic fracturing, fast-track floating regasification terminals, and a dimming outlook for nuclear power in the wake of the Fukushima disaster, has created a boom in the LNG business. At the same time, the commercial structure of the industry has evolved. Spot and short-term sales now represent more than 25% of the market, while new business models have opened up opportunities for a multitude of new industry participants. As of December 2014, there were far more LNG projects under construction and in the planning stage than ever before. As the number of exporters and importers increases, LNG will continue to be the fastest growing segment of the hydrocarbon industry for the foreseeable future. Although the industry has impressive potential, it also faces major challenges. Companies may have to take greater commercial risks in the hope of reaping larger financial rewards.

In an increasingly environmentally conscious world, natural gas, because of its clean-burning characteristics, is consolidating its position as the fossil fuel of choice. It promises to be the only hydrocarbon that will continue to increase its share of global primary energy supply. Once described as a bridge to a clean energy future, natural gas now seems to be a strategic long-term component of it. This increasing role for natural gas is being enhanced by the development of unconventional gas from shale in the United States and coalbed methane in Australia, contributing to an ever-expanding natural gas resource base. Even as abundant and relatively cheap unconventional gas brings down prices and spurs demand, much of the world's gas reserves will remain surplus to local needs and will need to access distant markets. Developers will turn to LNG as the way to commercialize these resources.

Given the scale and likely rate of development, Poten & Partners projects the global LNG market will surpass 360 million tons per year (Mmt/y) or 490 billion cubic meters per year (Bcm/y) by 2020 and 415 million tons per year (565 Bcm/y) by 2025. The industry has already more than doubled from 103 million tons per year (140 Bcm/y) at the turn of the century to 243 million tons (330 Bcm) in 2014. The consensus view holds that the rate of growth in the LNG industry will be about twice that of global natural gas. But this will only occur if hundreds of billions of dollars are invested to make the additional LNG supply available and to build the ships and infrastructure to deliver it to consumers. Much of this investment is being made in Australia and in North America. These two regions are likely to rival Qatar as the world's largest LNG producer and exporter in the next decade. Longer term, East Africa (Mozambique and Tanzania) may become a key LNG supply region, with potential for multiple projects to monetize more than 120 Tcf of proven reserves, but these projects face greenfield and geopolitical challenges. On the market side, new outlets are opening up in both developed and developing countries, spurred by the advantages of natural gas compared to other fuels and at times by the availability of floating regasification terminals.

This promising future should not be taken for granted. If history has taught us anything, especially recently, it is that industry dynamics can change quickly, and with unpredictable consequences. When the first edition of this book was published in 2007, the consensus among industry experts was that the United States would be the next major LNG import market. Export projects in Africa, the Middle East, and beyond were being built to supply a wave of new import terminals in the United States (by far the largest natural gas market in the world) and Europe—including the United Kingdom. Among the projects developed primarily to supply the United States and the United Kingdom were six so-called megatrains in Qatar, each with a capacity of 7.8 MMt/y (10.6 Bcm/y)—by far the largest ever built. These megatrains made Qatar the largest LNG producer in the world.

At the time, shale gas was in its infancy and few foresaw the impact it would have on the natural gas market and the LNG industry. In short order, US shale gas turned the world's gas market on its head. The development of shale gas reversed the expected decline in US natural gas production, halving the need for Canadian pipeline imports and almost eliminating the need for LNG imports. In the United Kingdom and Europe, weak economic conditions, coupled with heavily subsidized renewable energy, a broken carbon trading scheme and an onslaught of cheap coal (ironically displaced in the United States by shale gas), curbed natural gas and LNG demand. Prices fell sharply in both regions. Export projects built to supply these markets struggled. For example, Qatar Petroleum and its foreign partners had built two massive LNG import terminals, one in the United States and the other in the United Kingdom, each capable of receiving the production of two megatrains, nearly 15 MMt/y (20 Bcm/y) each. By 2010, these investments were looking less than stellar. Similarly, the other new LNG terminals built to serve the North American and European markets were largely empty.

The export ventures were largely saved by another unforeseen event—the March 2011 tsunami and the resulting Fukushima nuclear disaster in Japan, which resulted in the eventual closure of the nation's 54 nuclear units. This closing created a huge jump in LNG demand to fuel the nation's gas-fired power plants. In 2012, the first full year without nuclear production, Japan's LNG imports climbed 17.5 MMt/y (24Bcm/y) to 87.5 MMt (119Bcm/y). This increase amounted to nearly 8% of global LNG production. In addition to imports into Japan, imports into Latin America also surged during this period, driven by a number of factors.

The resulting "tight" LNG market—anticipated by neither buyers nor sellers—boosted spot LNG prices to over $20/MMBtu and encouraged buyers to scour the globe for LNG supplies. This contributed to a wave of investment in Australian LNG projects, where robust oil-linked sales prices and firm purchase commitments underwrote project viability in the face of overheated construction costs. Project sponsors there made extensive use of modular construction in an attempt to shift work overseas to avoid high-cost local labor. In addition, Australia witnessed the first sanction of a project based on a floating LNG production unit—Prelude LNG led by Shell—and the first LNG export projects based on unconventional natural gas (coalbed methane). As more projects moved through the development process, sponsors began offering equity in both the upstream and the liquefaction complex as an inducement to the buyers to commit to their projects. These arrangements were designed to share risk as well as financial returns with the buyers.

With high construction costs and increasing overruns, spurred on by severe labor constraints, Australian LNG project development slowed down and essentially halted by 2012. Buyers began to turn their attention to North America, where the flood of shale gas was hitting the market, driving down prices and virtually eliminating the appetite for LNG imports. With falling North American prices, high prices and high demand in Asia opened up the possibility of capturing the widening arbitrage value between the two regions. Growing short- and medium-term volumes of LNG, increased trading, and the emergence of LNG "portfolio" aggregators all helped to challenge and change the conventional wisdom of the LNG market. Entrepreneurial US terminal owners, with their import facilities lying fallow, saw the opportunity to add liquefaction trains at their sites—at far lower unit capital costs than new greenfield projects—that would transform North American gas into LNG.

Terminal owners adopted a capacity tolling model to commercialize their underutilized import facilities (Cheniere's model involves the free on board [FOB] sale of LNG, placing responsibility for the supply of feed gas on the terminal owner, but is otherwise commercially indistinguishable from tolling). In this model, capacity holders, who include utility companies, LNG aggregators, and trading firms, lease liquefaction capacity, buy their own feed gas from the grid, and take the resulting LNG from the plant on an FOB basis. Terminal costs, including capital charges, are guaranteed through fixed charges, with the "tolling" customers taking both price and volume risks. For Asian utilities, these projects offer the attraction of Henry Hub–linked prices, which diversify the pricing in their LNG purchase portfolio away from traditional oil-linked terms. The LNG also provides destination flexibility absent from most of the utilities' other supplies. For the aggregators and traders, LNG supply linked to Henry Hub opened up the opportunity to benefit from price arbitrage by reselling volumes to markets where oil-linked LNG prices set a higher benchmark. But, while margin plays between Henry Hub and oil-linked prices could have generated huge arbitrage profits in the conditions prevailing in early 2014, they could also result in significant losses when the price relationship between Henry Hub and crude oil tightens, as it did in early 2015.

The technologies that drove the North American shale gas revolution were first viewed as a threat to the LNG industry because they largely eliminated the need for LNG imports. But as a benefit, they spawned an LNG export business, which was entirely new. However, these same technologies have been driving US shale oil growth as well, leading to rapid declines in oil imports. Just as the United States has become the global leader in natural gas production, the country is on track to replace Saudi Arabia as the world's largest oil producer. Thanks to shale oil, by the end of 2014, US crude oil production climbed to nearly 9 MMb/d from 5 MMb/d in 2005. Saudi Arabia produces 10 MMb/d. Coupled with slowing demand for oil in the world economy, this growth contributed to a collapse in crude oil prices. As 2015 opened, Brent was trading at under $55/bbl and WTI (West Texas Intermediate) at less than $50/bbl. Saudi Arabia has declared that it will defend its oil market share without regard to the price consequences. In the process, it apparently hopes to curb the growth in US oil production.

If the fall in crude oil prices proves to be more than a temporary event, it will have significant consequences in the LNG industry and could undermine the financial viability of new liquefaction projects—including those grassroots projects based on unconventional natural gas resources. For example, each project in Queensland, in eastern Australia, where coalbed methane supplies the feed gas, will require an ongoing A$1 to A$2 billion (US$0.75 to US$1.5 billion) annual investment after the project is commissioned, in order to maintain feed gas levels. This will result in a higher marginal cost of production than any other integrated LNG project. In the United States, customers who are not utilities buying for their supply needs could find the trading margin between the Henry Hub–linked price and other regional pricing has disappeared, or worse, turned negative. With crude oil prices between $60 and $70/bbl—equivalent to about $9–$10/MMBtu for LNG delivered in Asia and $7–$8/MMBtu in Europe—margins will shrink to unsustainable levels. To break even at $9/MMBtu in Asia, Henry Hub would have to fall to about $2.50/MMBtu. New US liquefaction ventures, especially ones based on grassroots development, may have a difficult time finding the customers required to proceed, let alone the $50–$100 million up-front investment to secure the necessary permits. Potentially adding to the problem, lower prices for associated liquids could boost the break-even production cost for shale gas in the United States and Canada.

LNG demand could present another challenge to the industry. Just as Fukushima helped spur the current generation of LNG export projects, the restart of nuclear plants in Japan could contribute to a supply surplus. The Abe government has indicated a determination to restart many of the nuclear plants. However, local opposition is very strong and could slow the pace of restarts and in turn, slow the decline in Japanese LNG imports.

These developments will increase the pressure on sellers to find other outlets. While lower LNG prices should spur demand, the United States, the largest natural gas market, will be an exporter and not a net importer. Meanwhile, continuing tough economic conditions threaten to undermine the LNG import outlook in Europe, and previous peak import levels may not be reached before the end of this decade. Of course, unforeseen events could change this, particularly given the political climate in Russia, the largest pipeline gas supplier to Europe.

Lower prices could spur demand in China, India, and other emerging markets. Since the first edition of this book was published, China and India have become major LNG importing markets, and lower prices will further enhance their natural gas demand. Slowing economic growth in China, however, coupled with rapidly increasing pipeline imports from Turkmenistan, Myanmar, and Russia, could temper China's appetite for LNG. Full realization of their import potential, however, awaits progress on downstream market reforms. Concurrently, new markets have flourished (primarily by implementing fast-track offshore floating regasification terminals), particularly in Southeast Asia, the Middle East, and South America.

Another positive development, not foreseen in the earlier book, is the maturation of floating regasification terminals. These regasification ships have opened new markets in many locations, and under a variety of new commercial approaches to the LNG market. The new markets would benefit immensely from competitively priced and ample LNG supplies. However, many of them are characterized by less-than-steady baseload demand, and often less-than-creditworthy buyers.

While it is impossible to identify the circumstances that will impact the LNG industry in the future, this book aims to provide an understanding of the commercial and technical underpinnings of the industry, and a method for analyzing unfolding industry-impacting events. One thing is certain—there will be unforeseen events or "wild cards" that will cause the LNG industry to develop in new and unanticipated ways. Just as the past few years have witnessed vicissitudes such as the "shale gale," a sustained economic slump in Europe, the Japanese nuclear crisis, a sudden and steep drop in oil prices, and the maturation of new technologies, the next decade holds the promise of more to come. The forecast may call for a few strong winds and thunderstorms, but the future of natural gas, and especially LNG, remains bright.

1

The Liquefied Natural Gas Industry

Introduction

Liquefied natural gas (LNG) promises to be a growing strategic component of the global clean energy future. Long considered the ugly stepsister of oil, natural gas's clean-burning characteristics are increasingly prized in an environmentally conscious world just as the resource base has been boosted dramatically by new technologies, which are unlocking unconventional gas resources such as shale gas. Some environmentalists have touted natural gas as a bridge to a clean energy future, but it will be more than that. Given its abundance and lower emissions than other fossil fuels, natural gas seems destined to become the dominant fossil fuel source over the next few decades. But natural gas reserves located far from markets must overcome investment costs and transport hurdles. Today, LNG is the most established and effective way to clear these hurdles. This book addresses all aspects of the LNG industry, particularly the striking transformations since the first edition of this book was published in 2007. In these pages, the reader can learn how the LNG industry has evolved and why—to steal a phrase from the International Energy Agency (IEA)—the "golden age" of LNG is at hand. Since 2007, shale gas has emerged as a dominant supply, floating regasification has entered the mainstream, floating liquefaction is under construction, and, perhaps the biggest game changer of all, North America has shifted from potentially becoming the world's largest LNG importer to becoming its largest LNG exporter. Any one of these changes might justify a new edition; together they compel it.

Energy demand grows inexorably with expansion of the world economy. Economic growth has slowed dramatically in the industrialized world, but China, India, and other emerging economies are now creating new LNG demand. Concerns over global warming have not abated as renewable energy is developing more slowly than its supporters would wish, even with favorable tax and policy initiatives to support it in many countries. At the same time, prospects for future nuclear developments were devastated by the March 11, 2011, Fukushima nuclear accident following an earthquake and tsunami in Japan.

Remaining reserves of conventional oil are concentrated in increasingly fewer countries, not all favorable to the major consuming nations. Yet technological advances are unlocking new oil and gas resources, upsetting the global producer–consumer balance; the US shale oil and gas revolution is a prime example. National oil companies (NOCs) have grown in importance as higher prices combined with ready access to technology have enabled them to operate independently of the large international oil companies (IOCs) on which they traditionally depended. Despite having access to ample cash flows, the IOCs are struggling to grow profitably as their access to conventional oil reserves becomes more restricted. In addition, IOCs face growing competition from other producers, including the emerging NOCs. Given a broader geographical footprint, the complexity of large-scale projects, the faster growth in demand, and the more benign environmental characteristics, it is no surprise that natural gas, especially as LNG, has become a growing objective for the IOCs. Long gone are the days when the geologist returned home to report the bad and the good news about recent exploration endeavors—the bad news that they had failed to find oil, and the good news that they did not find natural gas.

In the 1970s, an era of worldwide concern over limitations to commodity production, natural gas was seen as a limited resource, too precious to consume in many applications. In fact, some producers described natural gas as the noble fuel, worthy of being sold at a premium to both oil and coal. Government policies among the developed nations encouraged this perception by restricting the use of gas as a matter of policy, and controlling prices. But in the early

1980s, natural gas prices were falling in the large US market, and a glut of the fuel emerged following decontrol of wellhead natural gas prices. Low prices, coupled with its clean-burning characteristics, made natural gas a favored fuel. This trend was spurred on by the development of highly efficient combined cycle gas turbine (CCGT) technology to generate electricity in the United States and globally.

In the 21st century, natural gas has emerged as one of the fastest growing sources of primary energy and the fastest growing fossil fuel. This trend is expected to continue for the foreseeable future because the world's natural gas resource base has been transformed by the exploitation of unconventional gas. Spearheaded by the unexpected boom in shale gas production in North America, this game-changing development may spread elsewhere, significantly adding to global gas supplies. In June 2011, the IEA released a report titled "Are We Entering a Golden Age of Gas?" lauding the potential of unconventional gas. The IEA concluded that "natural gas is poised to enter a golden age, but will do so only if a significant proportion of the world's vast resources of unconventional gas—shale gas, tight gas, and coalbed methane—can be developed profitably and in an environmentally acceptable manner."[1] The report made the general public aware of what the energy industry had already concluded: gas is not merely a bridge to a clean energy future; it is almost certainly an integral part of it.

According to the IEA, natural gas is the only fossil fuel that will increase its share of the global primary energy mix over time, a trend that is already under way in many key markets. In the IEA base case, natural gas will increase from 21% of primary energy in 2010 to 25% by 2035.[2] Although oil remains the largest component of primary energy, accounting for about 32% of total world energy demand in 2010, its share has slipped from over 40% in the early 1980s, while the share of natural gas has increased from 19% to nearly 21% in the same time frame (fig. 1–1). Oil will maintain its top ranking through 2035, but its share will slide to 27%, only slightly ahead of natural gas. Coal will slip to third place, dropping from 28% to 24% as natural gas displaces coal as the preferred fuel for power generation.

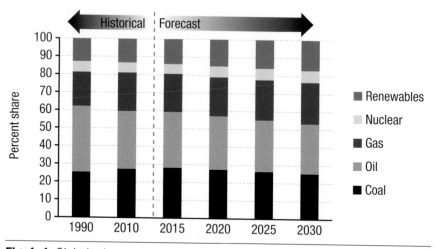

Fig. 1–1. Global primary energy sources. (*Source:* IEA, Poten & Partners.)

The IEA forecasts gas demand to grow by more than half, from 3.4 trillion cubic meters/year (Tcm/y) today to 5 Tcm/y by 2040. In the IEA's view, such growth will primarily be driven by the power sector, which accounts for about 40% of incremental gas demand.[3] Lower prices in some key markets, coupled with the advent of competition in traditional gas and power markets, have promoted increased gas usage (fig. 1–2). Advances in CCGT technology in particular have enhanced the economics of generating electricity from gas, as lower capital costs and markedly higher operating efficiencies of CCGT plants, along with lower emissions, have given natural gas a competitive edge over coal. More competitive natural gas prices will also spur industrial use, particularly in the United States. Natural gas is even showing growing promise in the transportation sector, though this will remain a function of the relative prices of oil and gas, the building of fueling infrastructure, and the rate at which owners of vehicles and vessels will adopt new technologies.

Despite the strong growth in consumption over the past two decades, worldwide reserves of natural gas have grown faster than consumption. Even conservative estimates show gas supplies ample enough to cover global demand for many decades. According to BP, technically proven remaining reserves totaled 185.7 trillion cubic meters (Tcm), or 6,557.8 trillion cubic feet (Tcf), at year-end 2013, representing a reserve-to-production (R/P) ratio of 55 years.[4]

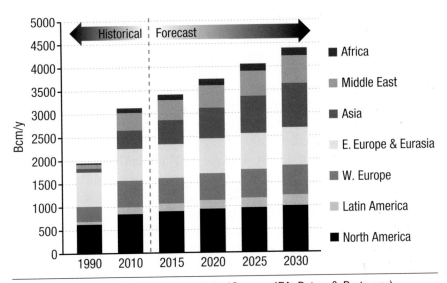

Fig. 1–2. Natural gas demand by region. (*Source:* IEA, Poten & Partners.)

According to the IEA, ultimately recoverable resources, the measure of long-term production potential, are dramatically higher, at 790 Tcm (27,900 Tcf) for an R/P ratio of 241 years.[5] This broader classification includes large volumes of unconventional gas resources expected to eventually be discovered and included in the proven category.

As new gas basins are exploited, they are often surplus to local needs and far from demand centers. Compared to other fossil fuels, natural gas is more expensive and technically more challenging to transport over long distances, whether by pipeline or by ship as LNG. As a result, the global natural gas industry has remained divided into a series of regional and even local markets. Nevertheless, there is a growing international natural gas trade, which Cedigaz estimated to be 1,048 billion cubic meters (Bcm; 37.0 Tcf) in 2013, accounting for about 30% of global gas supply.[6] This trade was split with about 70% by pipeline and 30% in the form of LNG, compared to 24% just five years earlier. Most industry experts expect the LNG share to continue to grow. The IEA forecasts the global LNG market will grow from 300 Bcm in 2012 to almost 600 Bcm in 2040.[7]

As of February 2015, Brent crude oil had fallen to approximately $60 per barrel (WTI to about $50/bbl) from a two-year run at or near

$100/bbl, and natural gas prices had reached record highs in many markets (though they are now falling with oil prices). The price of coal has fallen as coal has been pushed out of the US generating mix. Oil, a globally traded commodity, is priced accordingly. Natural gas is still traded mostly regionally. Natural gas prices in some markets (Asia and partially in continental Europe) are determined under long-term sale and purchase agreements (SPAs) at prices linked to oil, and in other markets (the United States, the United Kingdom, and increasingly northwest Europe) by gas-on-gas competition. Where gas trades on its own fundamentals, the trend has been for gas prices to fall below oil-linked equivalents (particularly in the United States). As a result, arbitrage opportunities and spot/short-term trades now account for about 30% of the global LNG trade, encouraging existing players to adopt new business models and new firms to enter the industry. North America is emerging as a potentially large LNG export region. Bringing US LNG to the global market is creating price linkages to Henry Hub rather than oil and offering buyers supply and price diversity, which carry their own benefits and risks. The dynamics of global gas pricing appear likely to shift under the combination of US supplies and lower oil prices.

The Global LNG Industry

LNG consumption has more than doubled, from 103.3 million tons per year (140.5 Bcm/y) at the turn of the century to 243 million tons (330 Bcm) in 2014, according to Poten & Partners.[8] A combination of growing environmental pressures, new LNG production capacity, future economic growth, and competitive pricing promise a substantial expansion of LNG demand in the future (fig. 1–3). Poten & Partners projects LNG demand to surpass 360 million tons per year (490 Bcm/y) by 2020 and 415 million tons per year (565 Bcm/y) by 2025.[9] Growth in LNG demand is anticipated in every major region except North America, which has been impacted by domestic shale gas production. The United States, once expected to become a major LNG importer, now promises to become a leading LNG exporter, even though lower domestic gas prices in the United States may revitalize segments of the nation's industrial base and promote domestic natural gas demand.

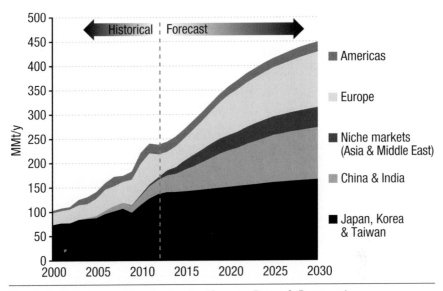

Fig. 1–3. LNG demand by geography. (*Source:* Poten & Partners.)

Hundreds of billions of dollars will need to be invested in order to make this additional LNG supply available. Much of this investment is underway in Australia and the United States, with East Africa and Canada poised to follow. However, dramatic cost escalations and environmental issues in Australia may threaten further growth there, and similar dynamics may be appearing in western Canada, threatening the emergence of an LNG export industry in that region. In contrast to the oil industry, LNG is a more technically, financially, and commercially challenging energy delivery system, well suited to the strengths and competencies of major oil- and gas-producing companies. There is no world price for natural gas or LNG, and some end-markets remain under the control of regulated utilities and essentially closed to competition. Simply having access to large untapped reserves of natural gas is no assurance that these reserves will be monetized easily or quickly or that their development will be financially viable.

In simple terms, the LNG industry involves identifying large reserves of natural gas with little or no prospect of securing local markets, liquefying the natural gas at very low temperatures (–163° Centigrade), shipping the LNG in specially designed tankers to markets, and storing and regasifying (or vaporizing) it before injecting it into a pipeline grid, at which point it becomes indistinguishable from pipeline gas for the end user (fig. 1–4). In its liquid form, natural gas shrinks to less than 1/600th of its gaseous state, making its transportation and storage more efficient. However, little is simple about LNG. Bridging the natural gas gap between supply and demand via LNG is one of the energy industry's most challenging issues and also one of the most expensive.

| Upstream | Liquefaction | Transportation | Regasification | Market |

Fig. 1–4. The LNG value chain. (*Source:* Poten & Partners.)

The traditional LNG industry is based largely on a series of virtually self-contained projects made up of interlinking chains of large-scale facilities, requiring huge capital investments, bound together by complex, long-term contracts, and subject to intense oversight by host governments and international organizations at every stage of the process. Ironically, this effort is applied to the simplest hydrocarbon—methane—and involves no chemical or other changes to the commodity (except to its temperature and purity) from the time it is produced until it reaches the final consumer.

Even though LNG has represented a major source of natural gas and a significant share of primary energy for decades in Japan and other areas of the world, it was for many years considered a high-cost niche segment within the energy industry. Given the numerous hurdles and uncertainties that faced each project and the need to coordinate the multiple disparate elements involved, many projects failed to achieve realization or collapsed before they had run their course. Skeptics warned of impending disasters that would threaten the safety and economics of operations, especially import terminals, which are generally located much nearer to population centers. Still, the LNG industry has not only prevailed, it has flourished.

Each project took years to develop, so that decades had to elapse for the industry to reach a critical mass of well-functioning projects and prove that the technical and commercial model could be successful. For the first few decades, the industry had relatively few participants, loosely known as the "club." Yet club members were already cognizant of what the rest of the world would later realize—that well-executed LNG projects can generate solid profits and stable financial returns over many decades.

The industry has proved its reliability and stability under a variety of market and economic conditions, accompanied by a safety record second to none for an operation of its scale. Demand growth and high energy prices through mid-2014, coupled with advances in technology, are driving more planned and proposed LNG projects than at any point in history. The commercial structure of the LNG industry is changing, revitalizing the business and opening it to new entrants. LNG is now playing an important and growing role in meeting global natural gas demand. More and more, LNG is traded in flexible spot and short-term trades, demonstrating its ability to serve energy markets on relatively short notice. This was amply demonstrated by the dramatic increase in LNG deliveries into Japan to generate electricity when the nation's nuclear plants were closed after the Fukushima Daiichi nuclear accident in March 2011.

The Global Natural Gas Industry

Although the ancient Greeks, Romans, and Chinese made limited use of natural gas, it was only in the 20th century that it became a significant source of energy. Thanks to technological advances in long-distance, high-pressure pipeline transportation, natural gas became widely distributed, allowing it to substitute for town gas produced from coal, which had been introduced to many cities in the late nineteenth century. These technological advances contributed to the steady growth of world natural gas demand throughout the 20th century, with the pace of growth accelerating in the later decades of the century and into the 21st century. During the second half of the 20th century, gas traded in the form of LNG became an important component of international trade, particularly in Asia. According to BP, global natural gas demand has more than doubled since 1970 to 3.35 Tcm (325 Bcf/d) in 2013.[10]

Natural gas is used extensively for heating and cooking in residential and commercial settings, and as a process fuel and feedstock by industrial consumers. However, while the number of customers for these applications continues to grow, overall consumption growth is muted as equipment efficiencies improve. The power sector has emerged as the main demand driver accounting for about 40% of future incremental gas demand in the IEA's view, because of environmental benefits and efficiency of natural gas compared with other energy sources.[11] Technological advances in CCGT power plants have shifted the economics of power generation away from coal in favor of natural gas, even with gas's higher prices. Historically, gas-fired power generation was primarily used for peak generation in simple cycle gas turbines, or as a direct substitute for residual fuel oil (resid), since the variable costs of generating baseload electricity using other fuels, such as coal and nuclear, were lower. Now CCGTs are increasingly being operated in mid- and baseload mode, particularly in North America where the shale gas boom has lowered gas prices, leading to direct competition with coal in power generation.

The age of the gas turbine

Until the 1970s, coal and oil were the principal fuels used to generate electricity. However, the use of oil for electricity generation slowed following the oil shocks of the mid-1970s. This, in turn, contributed to a rapid expansion of nuclear generation from the 1970s until the mid-1980s, when public opposition and escalating construction costs brought nuclear development to an abrupt halt in many countries. At the same time, environmental concerns associated with greenhouse gas emissions began to slow the development of coal-fired generation. Many countries also began a process of liberalizing their wholesale electricity markets, bringing an increased focus on the overall costs and efficiencies of power generation technologies. Costs were no longer guaranteed recoverable as they were under the classic utility regulatory environment. Lower capital cost technologies with shorter construction times began to win favor, setting the stage for the expansion of CCGT plants fueled with natural gas.

When the first gas-fired generating units were built in the United States during the 1960s, they were small and relatively inexpensive to build, but they were inefficient and depended on high-cost natural gas supplied through the classic utilities to run them. Moreover, natural gas supply was often curtailed during periods of peak demand. As a result, these peaking units were employed only to provide a rapid source of power during periods of peak electricity demand, which largely occurred during the summer months. With the exception of Japan, natural gas was rarely used outside the United States for power generation because it had to be imported, and these imports were expensive and often raised issues of supply security.

Moreover, many countries adopted policies prohibiting the use of natural gas in power generation, such as the Fuel Use Act of 1978 in the United States. An unusual case was that of Japan, where natural gas was used widely by the large power utilities in direct substitution for oil in classic steam turbine power plants. This allowed Japan to diversify its energy supply from an overdependence on Middle East oil to sources closer to home. The baseload nature of Japanese power utility demand helped to underwrite the success of the early LNG export projects by assuring a stable, secure, and financially appealing market with costs passed on to captive electricity customers.

With the introduction of CCGT units in the 1980s, the removal of prohibitions on the use of gas for power generation, and the introduction of competition in wholesale electricity markets, the boom in gas turbine generation was underway. In a combined cycle plant, a gas turbine is combined with a steam turbine, which uses the waste heat from the gas turbine to increase efficiency. Modern CCGT plants have thermal efficiencies of approximately 61%, compared to efficiencies of closer to 46% in modern coal-fired power plants. During most of the 1980s, annual sales of gas turbines were around 300 to 400 units, with an average size of only 30 megawatts (MW) reflecting their use in peaking applications. By 1990, annual sales exceeded 600 units, soaring to 900 units by mid-decade and 1,500 units in 2000. They remained at approximately this level through 2013. Turbine sizes increased dramatically as well. The largest gas turbines are currently from General Electric (470 MW in simple cycle) and Mitsubishi Heavy Industries (460 MW), while Siemens' largest is 375 MW.

Deregulation has also set in play factors favoring CCGTs. The introduction of wholesale competition in the power sector is increasing demand for these units. Gas turbines can be deployed rapidly, which reduces market risk, since developers can match their decisions more closely with market requirements. Unit capital costs are much lower, making financing easier and reducing the risk that the developer may be left with stranded assets. Finding a site for a gas-fired power plant is easier than for a coal plant, widely viewed as dirtier, or for a nuclear facility, widely perceived by the public as too dangerous.

These developments contributed to a major increase in the use of gas for power generation since the early 1970s. In the United States for example, the share of natural gas in the power generation mix has increased from 16% in 2000 to 29% in 2012.[12] Concurrently, coal's share has fallen from 52% to 39%. In the US Department of Energy's (DOE) view, by 2040, the coal share at 34% will match natural gas at a 33% share. Despite subsidies, the share for renewables increased slowly from 9% in 2000 to 12% by 2012 and is forecast to rise to 16% by 2040 according to the DOE.

The growing share of natural gas in the power sector has contributed to a slide in oil's share of primary energy in the United States. From 1980 to 2012, oil's share of the energy mix dropped from 46% to 36%, while natural gas's share rose from 19% to 27%.[13] By 2040, in the view of the DOE, oil's share of primary energy will still top gas, but not by much. The oil share will have declined to 31%, compared to the 30% contributed by gas. Yet, natural gas–fired power is not the answer to everything. The bankruptcy of companies such as Calpine and Mirant demonstrated that single-minded bets on natural gas carry their own risks, in this case associated with rapidly increasing fuel prices. This risk now appears to have been mitigated by the surge in unconventional gas production, which is contributing to more stable and predictable natural gas prices in the United States.

Internationally, similar patterns have emerged. The IEA is forecasting 7,000 gigawatts (GW) of new electricity supply capacity will be added between 2013 and 2040, compared to the 6,000 GW installed today. Of this, the largest contributions will come from renewables (a very ambitious and unlikely to be realized 4,000 GW) and natural gas (2,000 GW). The IEA sees coal, nuclear, and oil as only making small contributions (1,000 GW combined) to global electricity capacity additions in the next 35 years.

Competing against other fuels

While natural gas may continue to gain market share at the expense of other hydrocarbon resources, coal remains very much in the picture, nuclear was showing new signs of life until Fukushima, and renewables are helping to change the energy mix, dramatically, if the IEA forecast is realized. Addressing coal's environmental issues is not a trivial undertaking, expectations for renewable fuels and associated technologies may be overstated, and the nuclear industry is severely challenged by public opposition as well as by issues presented by the disposal of nuclear wastes and the high cost of building a nuclear plant. Meanwhile, the expansion of renewable energy supplies is dependent upon government subsidies that could be challenged by financially stretched treasuries.

Oil, long the king of energy consumption, comprises the biggest share of global primary energy and will continue to do so for the foreseeable future. Oil is relatively cheap to ship globally, and international markets are well established. Vehicle transportation markets, especially in India, China, and other emerging countries, will continue to expand. Natural gas still struggles to compete with oil for transportation. Until technological breakthroughs occur and/ or legislation passes favoring natural gas or other alternative transportation fuels, oil will continue to dominate this market in most areas of the globe, and transportation use will continue to drive oil demand, accounting for some 70% of the petroleum market. Even in the transportation sector, oil demand growth promises to be moderated by more fuel-efficient vehicles.

In the past, the lack of infrastructure to transport and distribute natural gas was a barrier to increasing natural gas consumption. However, the 1970s oil shocks provided an impetus for improving infrastructure, leading to the construction of major pipelines from the North Sea, Russia, and North Africa to Europe and significant expansion of existing pipeline networks in North America. With oil trading at over $100/bbl for several years until mid-2014, the competitiveness of the LNG delivery chain had been enhanced significantly, promising to make LNG the most rapidly growing segment of natural gas. Today, it constitutes the fastest growing hydrocarbon.

In regions where downstream natural gas pipeline infrastructure is limited, oil continues to be used for residential and commercial heating purposes. Natural gas is slowly making inroads in developing countries, especially Latin America (Mexico), South America (Venezuela, Brazil, Argentina, Chile), eastern Europe and Russia, and Asia (China, India, Thailand). The IEA concludes that natural gas demand growth will shift to more nations like China and India.

According to the agency, these two nations are expected to register the biggest gains, with Chinese demand growing by 7% per year to 545 Bcm (19.2 Tcf) and the Indian market expanding by nearly 5% per year to 180 Bcm (6.4 Tcf) in 2035.[14] The Middle East is also expected to experience significant growth, increasing from 404 Bcm (14.3 Tcf) to 640 Bcm (22.6 Tcf) in 2035 for power generation and gas-based industries. The two largest gas markets, the United States and Russia, will show steady expansion according to the IEA, but the outlook could be understating demand in the United States. According to BP, US demand reached 737.2 Bcm (26.0 Tcf) in 2013, already outstripping the IEA's 2020 outlook as power generators switched to cheap gas from coal. Demand in Europe continues to struggle because of economic recession and is only expected to regain its 2010 peak towards the end of the decade. Europe has seen significant growth in both coal consumption and renewable energy during the past several years, further threatening the role of gas in that region.

Another reason for the recent growth in popularity of natural gas is that it is the cleanest burning fossil fuel. Its low carbon content (compared to oil and coal) contributes to its attractiveness. Following the Kyoto Treaty, many developed countries have adopted targets to reduce carbon dioxide (CO_2) emissions and intend to enforce these targets through a combination of emission trading regimes and fines for noncompliance, further improving the comparative economics of natural gas.

Although coal, nuclear, and hydroelectricity have historically been the dominant fuels for power generation, the improved economics of CCGTs coupled with environmental benefits are allowing natural gas to capture new markets where delivery infrastructure can be built and natural gas can be delivered on a cost-effective basis. New gas-fired power plants have the lowest capital costs per megawatt of any power plant and are easier to site than most other types of plants.

The coal industry is moving to address the concerns associated with emissions. New technologies are more fuel-efficient and produce lower CO_2 emissions. Power companies are looking at the development of carbon sequestration as a way to address this issue. While these are expensive solutions, government policy in many countries could favor them to further energy supply diversity and security and to preserve jobs in their coal mining regions. The advantage natural gas has over nuclear energy has much to do with public perception. There are widespread safety concerns about nuclear power because of the perceived risks associated with potential radiation leaks, long-term spent fuel disposal, and nuclear proliferation; by contrast, public opinion of natural gas–fired generation is typically positive or neutral. The negative sentiment associated with nuclear power, heightened by Fukushima Daiichi, makes constructing new plants difficult and will encourage the retirement of aging ones. Nuclear facilities have enormous capital costs, and even though new technologies and standardization of designs could decrease the costs of constructing new plants, natural gas is unlikely to lose market share to nuclear.

With the exception of hydroelectricity, renewable sources have historically not been economically competitive with fossil fuels. However, hydroelectric power generation is restricted to certain areas and subject to variations in weather. Major hydroelectric power projects carry their own environmental risks and can force the migration of local populations, as was seen in the Three Rivers project in China. New construction is more likely in locations remote from local populations and can be very expensive. The cost of long-distance power transmission becomes another factor.

Solar power and wind power have clear environmental benefits over natural gas for power generation. The economics of wind energy have changed as costs decline, and wind increasingly appears competitive with other fuels, aided by favorable government tax treatment. Large-scale wind farms, virtually nonexistent before the 1990s, are becoming increasingly common in North America and Europe. Wind stands to be the fastest growing of all energy sources in the coming decades. However, the construction of massive wind farms, both onshore and offshore, is facing opposition in some communities. While wind will undoubtedly capture some of the power generation market, natural gas is considered more reliable and economic, and will be required as a backup when the wind does not blow. Solar power is also advancing but at a slower pace than wind; however, this could change as costs continue to fall. Indeed, some producers forecast that solar power in regions with favorable sunlight conditions may be a competitive source of electricity before 2020. While alternative energy has a rapidly expanding future, it will still account for less than 10% of US primary energy supply by 2040, according the DOE in its 2014 outlook.[15]

The IEA is far more bullish on renewables, forecasting them to outstrip all other forms of power generation capacity additions over the next three decades and become the world's largest source of electricity by 2040. The IEA forecasts hydropower to grow from 3,700 terawatt hours (TWh) in 2013 to 6,200 TWh by 2040, with wind and solar growing from 800 TWh to 4,700 TWh in the same time frame.

Natural Gas Resource Development

Natural gas supply

Historically, natural gas reserves remote from markets were viewed as a nuisance potentially capable of impeding the development of oil reserves. These remote gas reserves are often called "stranded." As the options for monetizing natural gas have expanded, some previously stranded reserves, discovered decades earlier, are now being developed. For many of the major energy companies, increasingly denied access to oil reserves or being forced to settle for onerous financial conditions to develop them, gas is becoming the "new oil." Coupled with technology improvements in exploration and production, the energy industry increased proven worldwide natural gas reserves by 2.6 times between 1980 and 2013 according to BP statistics, and reserves continue to climb as unconventional gas resources are exploited.

Natural gas resources

Proponents of the natural gas era point to the large size of the world's natural gas resource. At the end of 2013, world proven natural gas reserves totaled 185.7 Tcm (6,558 Tcf).[16] Prospects to increase the proven category are excellent as the energy industry increasingly focuses on natural gas, in particular unconventional gas, a relatively new and underexploited resource.

While natural gas resources are ample on a global basis, they are not evenly distributed. They are often far from major demand centers and have not been developed. About 61% of the world's reserves are located in the Russian Federation and the Middle East, though this share promises to be reduced as potential unconventional gas reserves (including shale gas) move into the proven category.[17] However, consumption is dominated by North America and Europe (including Russia), accounting for at least 60% of global gas consumption. The comparative R/P ratio, which indicates the number of years it would take to deplete proven reserves at current production levels, also points out the regional disparity. For example, proven reserves in the Middle East would last for 141 years. By contrast, the R/P ratio in North America is 13 years, demonstrating a more economically efficient use of this valuable resource. However, these R/P ratios alone do not tell the full story of the ultimate resource base, since the US proven gas resource base does not include vast unconventional gas potential. The United States' R/P ratio has increased from less than 10 in the 1970s to its current level as the industry has added proven reserves faster than consumption. In fact, the United States is now the top global natural gas producer, toppling Russia from this ranking.

The continuing imbalance between the location of gas resources and demand centers promises a major expansion in the international gas trade by long-distance pipelines and in the form of LNG. International gas trade totaled 1.0 Tcm (37 Tcf) in 2013: approximately 70% by pipeline and 30% in the form of LNG. But building the infrastructure to move natural gas long distances requires multibillion dollar investments, which raises the question of whether or not incremental gas supply can be delivered long distances to markets at competitive prices.

Pipelines

One of the first modern commercial applications for gas was in street lamps in the eastern United States and Europe in the nineteenth century. This application became obsolete when nations electrified, and natural gas did not play a major role in the energy picture until pipelines were constructed to transport the fuel from producing areas to consumers. Although natural gas had been transported in wooden pipelines much earlier, the first metal natural gas pipeline was built in 1872. Stretching five and a half miles, it brought gas from a producing well to the town of Titusville, Pennsylvania. These first metal pipelines were extremely inefficient. It was not until after World War II, when pipeline technology dramatically improved, that natural gas became a major part of the energy mix. In Europe, the foundation of the gas transmission system was laid between 1970 and 1975, fueled by a giant gas discovery in Groningen, the Netherlands.[18]

Natural gas pipeline systems are characterized by three applications: gathering systems, transmission pipelines or trunklines, and distribution grids. For the most part, in this book the term "pipelines" refers to trunklines, which transport natural gas from supply areas to markets. These long-haul pipelines are usually between 16 and 48 inches in diameter and can extend for thousands of miles, operating at high pressures maintained by compressor stations along the route. In mature markets such as the United States and Europe, a complex web of trunklines connects to other trunklines, storage facilities, large end users, and distribution grids.

Gathering systems generally comprise smaller diameter pipelines that take gas from the wellhead to central processing facilities where impurities are extracted. However, offshore gathering systems may be physically indistinguishable from transmission lines. Gathering line pressures can vary significantly, usually as a function of the wellhead pressure of the producing well. Distribution pipelines also tend to be smaller in diameter and disseminate natural gas to consumers in market areas. Distribution lines normally operate at medium-to-low pressures. While very large consumers such as a steel mill or an electric generator may be directly connected to a trunkline, the vast majority of medium-to-small consumers obtain their gas through a distribution grid, which is run by a local distribution company (LDC).

From an environmental and safety perspective, pipelines pass with flying colors. Aside from the initial impacts associated with construction, the pipelines themselves have very little environmental impact. Thanks to safety and security systems that detect corrosion or leaks, most problems can be identified and corrected before they become significant. Although accidents happen from time to time, these systems are considered to be among the safest modes of energy transportation.

Pipelines remain the major means of transporting the world's natural gas to consumers, accounting for over 90% of natural gas deliveries. The United States alone has over 305,000 miles of interstate and intrastate pipelines.[19] Interstate lines, which cross state borders, account for 71% of the total. The industry has overcome increasingly complex challenges of distance and terrain, and long, deep underwater crossings. Some existing and proposed pipeline projects are nearly 4,000 miles long and cost billions of dollars to build. Pipelines to deliver shale gas from fields in eastern British Columbia to planned liquefaction projects on Canada's Pacific coast could each cost billions of dollars to build. Generally speaking, an offshore pipeline is more expensive than a similarly sized onshore one.

It is not always possible or desirable to link supply to market by pipeline because of terrain, right-of-way issues, politics, and/or distance. Sometimes geographical features such as underwater faults or coral reefs may prevent construction, or a planned pipeline may interfere with an environmentally protected area or other public infrastructure already in place. Landowners may not allow (and may not be required to allow) the pipeline to be built on their land. A route may traverse politically unstable regions or countries that are on unfriendly terms with the supplying or consuming country, or it may simply not be a viable option to supply an island nation like Japan and Taiwan or small local demand centers in an archipelagic nation like Indonesia or Malaysia. LNG is often preferred because it is transported by ship, which minimizes the number of international borders that must be crossed and provides destination flexibility not available from a pipeline.

LNG Industry Development

Historical background

Petroleum products were first transported by tankers in the 1860s.[20] A century passed before natural gas was transported by ship. While it is relatively easy to store, load, and transport oil or other petroleum products, natural gas has to be turned into a liquid at very low temperatures, then stored and transported in this form.

Two seventeenth-century physicists, Robert Boyle and Edme Mariotte, are credited with discovering that air was compressible; this led to insights into how gas might be pressurized and condensed. Numerous subsequent experiments were conducted to establish the optimum method of reducing the volume of natural gas, and it was determined that increasing gas's density could be achieved by extreme pressure, extreme cooling, or a combination of the two. Compressed natural gas was generally considered too difficult and dangerous to transport around the world because of the lack of suitable materials to contain the high pressures and the risk of explosion in the case of a sudden release of gas. But with extreme cooling to below its boiling point (in this case, −163°C), natural gas could be reduced in volume by a factor of over 600 times, then stored and shipped at atmospheric pressure. In its liquid form, natural gas exhibits a property known as autorefrigeration, whereby continuous evaporation draws heat away from the liquid and does not require anything more than insulation to maintain the gas in liquid form as long as the evaporated gas, known as boil-off, is removed from the storage tank.

The process of cooling natural gas to extremely low temperatures began as a means of extracting helium from natural gas for use in US military balloons in the early 1900s. Shortly thereafter, advances in metallurgical techniques in Europe and the United States produced metals, notably aluminum and steel alloys, which would not become brittle at extremely low temperatures as most metals do. Storage facilities could thus be built for the supercooled liquid.

The first conceptual scheme for LNG was devised by Godfrey Cabot in 1914, when he patented a liquefaction plant and a barge design to prove that waterborne transportation of natural gas was technically feasible. However, he never pursued the idea. In 1939, the first commercial LNG peak shaving plant was built in West Virginia. Two years later, the East Ohio Gas Company built a second facility in Cleveland. This peak shaving plant operated without incident until 1944, when the facility was expanded to include a larger storage tank. However, World War II created a shortage of stainless steel alloys, resulting in a tank built from steel with inadequate nickel content. In 1944, this tank ruptured and natural gas leaked into the adjacent sewer system and residences. Sadly, it ignited and killed 128 people—the largest disaster in LNG history. Subsequent investigations resulted in new standards for the materials used for the handling of LNG, preventing this from happening again. This incident put LNG development on hold for a decade.

In the 1950s and 1960s, William Wood Prince, president of the Union Stockyards of Chicago, faced with escalating electricity rates, began to study the liquefaction of natural gas in Louisiana to barge it up the Mississippi River. The British Gas Council was also looking to transport natural gas to supplement supplies in areas stretched thin by manufacturing and household use. Union Stockyards subsequently joined forces with Continental Oil Company and the British Gas Council to turn an old World War II dry bulk carrier into an LNG ship, the *Methane Pioneer*. This vessel was used to transport LNG from Lake Charles, Louisiana, in the Gulf of Mexico, to Canvey Island in the United Kingdom in 1959. It was the first commercial shipment of LNG. After a major natural gas discovery in Algeria, the United Kingdom and France signed contracts with Algeria in 1961 and 1962, respectively, and the first commercial-scale liquefaction plant at Arzew, Algeria, became operational in 1964.

During the 1960s and 1970s, liquefaction plants were built in Alaska, Libya, Brunei, Abu Dhabi, and Indonesia, as well as Algeria. Import terminals were developed in Japan, France, the United States, and Italy, later joined by terminals in Belgium, Spain, Taiwan, and South Korea. However, owing to an oversupply of natural gas in Atlantic markets in the 1980s, only two new greenfield LNG export projects were put into service in that decade (one in Australia and the other in Malaysia), while expansions continued in Indonesia. Most of the capacity added in the 1980s was designed to serve Asian LNG markets, which were growing rapidly as they did not have access to domestic natural gas or pipeline imports as did Europe or North America. In the early 1990s, demand was catching up with supply in the Western world, causing a rebirth of LNG projects targeting those markets, and new LNG export projects were commissioned in Qatar, Nigeria, Oman, and Trinidad between 1996 and 2000.

By 2000, LNG trade had reached 103 million metric tons per year (MMt/y; 140 Bcm/y) (fig. 1–5).[21] Fewer than 20 projects had been commissioned since the start of the commercial LNG business in 1964. The pace of LNG development may have been hampered by an inflexible business structure including the following factors:

- LNG project sponsors were typically large international oil companies (IOCs) partnering with national oil companies (NOCs).

- The buyers, most often large, regulated gas or power utilities, signed purchase agreements that were relatively inflexible on volume and destination.

- Contract prices were linked to those of crude oil, ensuring acceptability in the end-market and providing comfort to lenders.

- Each project evaluated its shipping needs, and dedicated LNG tankers were built and owned by the project or backed by long-term charter arrangements.

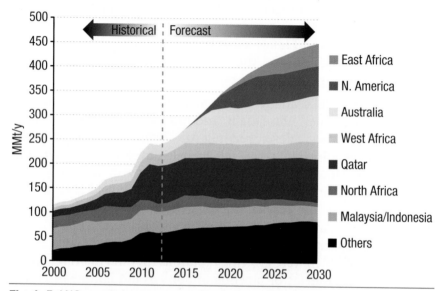

Fig. 1–5. LNG supply by geography. (*Source:* Poten & Partners.)

The pace of change accelerated as the LNG industry entered the 21st century. Downstream markets deregulated and countries such as China and India offered potentially very large markets. Successive trains (i.e., successive liquefaction units) at Trinidad LNG introduced new business models, including tolling arrangements for liquefaction trains with purchase contracts featuring greater destination flexibility. These contributed to the emergence of both spot and short-term trades, which helped balance the market. As a result, trading companies entered the business. Shipowners started ordering LNG tankers on a speculative basis since they could trade these vessels in spot and short-term trades until long-term charters could be secured. They also ordered new LNG ships to benefit from lower costs following the competition created by the entry of South Korean shipyards into the business.

Post-2000—the Qatari era

Qatar implemented a range of options to maximize the value of the North Field, the world's largest nonassociated natural gas field at approximately 900 Tcf, including LNG, domestic sales, petrochemicals, and gas-to-liquids (GTL). The national oil company, Qatar Petroleum (QP), through partnerships with foreign oil majors, delivered LNG results that were hardly conceivable at the turn of the century. The RasGas and Qatargas ventures built six 7.8 MMt/y (10.6 Bcm/y) trains. Dubbed megatrains, these new trains were by far the largest ever built, pushing the country's nameplate, or design, liquefaction capacity to 77 MMt/y (105 Bcm/y), including 30 MMt/y (41 Bcm/y) of existing capacity in eight conventionally sized trains.

Qatar, through its Nakilat shipping firm, stimulated shipping demand by ordering a fleet of giant LNG tankers consisting of 19 Q-Flex vessels sized between 210,000 and 216,000 m³ and 13 Q-Max vessels of 266,000 m³ cargo capacity to provide transportation services for the megatrains. The venture partners envisioned a giant "floating pipeline" to deliver cargoes to large purpose-built LNG import terminals developed by QP and its foreign partners in the United States (Golden Pass in Texas), the United Kingdom (South Hook in Wales), and Italy (Rovigo). When LNG was no longer needed in the United States following the surge in shale gas production, or in the United Kingdom as gas demand dropped in the face of a prolonged economic downturn, the cargoes were diverted to LNG-short markets in Asia, where demand was particularly strong after the Fukushima Daiichi nuclear disaster. The economies of scale generated by the Q-class LNG vessels under the point-to-point trade scheme were lost under the diversified trade patterns.

Qatar, which first exported LNG in the 1990s on a relatively modest basis, produced 77.9 MMt (105.9 Bcm) in 2013, dwarfing all other exporters by a wide margin. This contributed to the global LNG trade more than doubling from 103 MMt (140.1 Bcm) in 2000 to 243 MMt (330 Bcm) in 2014.[22] In just 14 years, the business had expanded more than it had in the previous four decades. Concurrently, the number of importing countries more than doubled to 26. While Japan remained the world's top importer, its share of global LNG trade dropped from 52% in 2000 to 36% in 2014, even though the closure of Japan's nuclear reactors boosted demand to a record high of 88 MMt (120 Bcm). The LNG shipping fleet climbed to 369 ships from around 110 in 2000 with another 113 new buildings on order at year-end 2014, a precursor of a further LNG trade expansion. The average ship size also increased dramatically, resulting in an even larger tonnage increase although no one else ordered Q-class vessels. A new class of LNG vessels with onboard regasification, called floating storage and regasification units (FSRUs), opened up small markets on a fast-track and cost-competitive basis. At the same time, floating liquefaction (FLNG) production units were being constructed.

Spot and short-term trades, which comprise single- and multi-cargo transactions as well as term deals up to four years in duration, were once considered a marginal LNG business but now are an established feature of the industry. In 2000, these trades accounted for less than 10 MMt/y (14 Bcm/y) of LNG and amounted to less than 10% of total trade. In 2013, around 70 MMt (95 Bcm) of LNG was moved under these deals and accounted for about 30% of global LNG trade. Many major LNG sellers once considered spot trades too small in volume and value to be worth their time, and were concerned that supplying LNG to the spot market at prices below the level of their long-term contracts would undermine their relationships with their long-term buyers. Furthermore, they added complexity to carefully crafted shipping logistics (fig. 1–6). At that time, buyers were only interested in spot trades if they helped meet an immediate supply shortfall. They were not a part of a long-term supply strategy, even to meet seasonal demand peaks. This has changed. The flexibility provided by these trades is now a critical component of the business for both buyers and sellers, as they adjust to unforeseen events such as nuclear shutdowns in Japan.

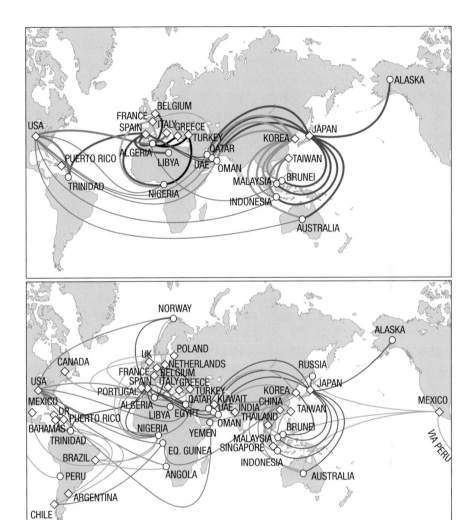

Fig. 1–6. The evolution of global trade of LNG. Top: 2000 World LNG trade routes. Bottom: 2012 World LNG trade routes. (*Source:* Poten & Partners.)

Post-2013—the Australian era

While projects in Qatar drove the last LNG export expansion, Australia is driving the next one. In 2014, Australian LNG production totaled 23 MMt (31 Bcm). By 2018, based upon projects that are already under construction, Australia will be able to produce 85 MMt/y (115 Bcm/y) of LNG at 10 complexes, challenging Qatar's top rank. According to cost estimates, sponsors will lay out upwards of $200 billion to build 15 liquefaction trains with costs inflated by a variety of factors, including a difficult labor market, remote construction sites, exchange rate fluctuations, and water issues for unconventional gas development. Faced with high unit production costs, Australian projects rely on revenues generated by oil-linked term sales contracts at relatively high oil prices for economic viability.

The high-cost construction environment, in large part due to labor constraints, is spurring innovation. Developers are turning to modular construction, extensively employed at Pluto LNG, which was commissioned in 2012, and FLNG production, which is the technology selected at the Prelude LNG project, in an attempt to shift construction to less labor-constrained locations. FLNG is now being promoted at other potential Australian LNG projects, with one project considering as many as three units. The modules and the FLNG units are built in shipyards or construction yards in Asia to benefit from lower construction costs. The Australian industry was also the first to advance liquefaction projects based upon unconventional gas resources. On Australia's eastern coast in the state of Queensland, LNG developers are constructing three projects to convert coalbed methane to LNG for export, with a combined capacity of 25 MMt/y (34.0 Bcm/y) across six trains that are scheduled to be commissioned during the middle of this decade. In contrast to "conventional" integrated LNG projects, where the majority of the capital is invested up front, the eastern Australian projects will need to make significant ongoing capital investments in gas drilling and production given the production profile of coal seam gas wells, which decline very rapidly.

2020 and beyond—the US game changer

US LNG projects could be the next industry game changer. Numerous liquefaction trains are being added at underused import terminals where they can benefit from the existing infrastructure, thereby lowering capital costs. The numbers are mind-boggling, with proposed projects totaling more than 240 MMt/y (326.4 Bcm/y), though it is very unlikely that they will all be built. Poten & Partners projects that US LNG exports will reach 60 MMt/y (82 Bcm/y) by 2025, with more to follow.

While subject to considerable debate, most analysts have concluded that there will be ample US natural gas resources to enable significant LNG exports with limited impact on US natural gas prices. This has encouraged the US DOE to process project applications for sales, including sales to countries without Free Trade Agreements (FTAs) with the United States. This non-FTA authorization is critical to advancing such projects because most of the large LNG-consuming nations do not have FTAs with the United States, with South Korea and Chile as notable exceptions. As of July 2014, the DOE had issued eight permits allowing the export of 77.8 MMt/y (105.9 Bcm/y) of LNG to non-FTA countries. This could make the United States one of the top three LNG exporters shortly after 2020. As of October 2014, four projects had reached a final investment decision (FID), with the first LNG scheduled to hit the market by late 2015.

What makes US liquefaction a game changer is not just its potential size, but the sharp departures from the commercial structure of traditional LNG schemes. To date, most of the US projects have adopted a tolling model under which firms lease liquefaction capacity at the plant for a fee (Cheniere's Sabine Pass and Corpus Christi projects are exceptions). The capacity holders will often arrange their own feed gas from the grid, rather than the traditional structure where dedicated upstream (and otherwise stranded) reserves are processed at a purpose-built LNG plant and then sold by the venture under destination- and volume-restricted long-term sale and purchase agreements (SPAs) to end users. While many of the capacity holders may deliver volumes into their home markets, their contracts contain no volume or destination restrictions. Many of the capacity holders have active trading operations and are already major players in spot and short-term trades.

The US liquefaction projects sanctioned to date are brownfield, piggybacking on storage, marine, and other infrastructure at existing, underutilized import terminals benefiting from considerable investment savings. (Cheniere's Corpus Christi LNG project on the

Gulf Coast promises to be the first greenfield project and should begin construction in 2015.) There are additional savings because none of the upstream capital costs of traditional LNG projects are required. For example, a three-train project on the US Gulf Coast typically costs around $10 billion, but a comparably sized greenfield project in Australia may cost multiples of this.

Capacity holders will still have to buy feed gas from the grid at US market prices. This final major departure from traditional projects, one that has proved attractive to buyers so far, is the linkage to gas hub prices (primarily at Henry Hub in Louisiana) rather than to crude oil. The lure of this Henry Hub price alternative to traditional oil linkage has proven to be quite effective, assuming that oil prices recover from early-2015 lows. International firms—for their own market use or for trading—have snapped up tolling capacities at these projects in the United States. Henry Hub is now not closely correlated to oil prices, giving the capacity holders the ability to diversify price risk in their supply portfolios. Unlike traditional LNG projects, these projects have no need to fund upstream development and most of the required pipeline infrastructure is quite modest. While this lowers the up-front investment, it also leaves these projects (or more precisely their offtakers) exposed to a supply price risk over which they have no control, other than the option to suspend taking LNG while continuing to make their capacity payments.

The essential ingredients of an LNG project

Unlike many other energy delivery systems, LNG has tended to be organized on a project basis, involving not only the development of the upstream infrastructure, the gas production, gathering, and liquefaction facilities, but also the shipping and import terminals, which are coordinated with buyers. This approach reflected the very high capital investment needed for each unit of energy delivered, as well as the historical separation between the LNG suppliers (large IOCs and NOCs) and the LNG buyers (principally gas and electric utilities, both private and state owned). Until the turn of the century, the industry grew at a measured pace with a relatively limited number of participants who had the ability to finance the projects.

In order to mitigate the participants' financial risks, each project weaves together a complex web of contracts and agreements. Managing a project in which buyer and seller have disparate interests and are both subject to scrutiny from and/or involvement of their host governments presents a substantial challenge. Other barriers to successful project execution are the many commercial, technical, and legal challenges that must be addressed. Meeting these challenges requires the coordination of dozens of experts, including lawyers, engineers, contractors, shipowners, bankers, government representatives, buyers, and consultants.

The following is a summary of some of the complex issues involved in developing an LNG project, which will be discussed in detail later in this book.

Sufficient gas reserves. By the time an LNG project is being considered, the existence of ample low-cost natural gas reserves has usually been confirmed. A traditional LNG project must have dedicated proven reserves that allow the project to operate at its design level for 20 to 30 years, with an additional reserve margin to protect against unexpected production declines or better-than-expected plant performance. This helps ensure financial viability and is typically a requirement of both the buyers and the funding institutions. The cost of producing the reserves must be fairly low in order to make the LNG project economic. The presence of liquid hydrocarbons along with the gas can improve project economics, as this can create an additional, high-value revenue stream.

Long-term commitments from buyers. The capital costs required to build a facility usually dictate that a downstream buyer or buyers, who will contract for the majority of the plant's output, must be secured via long-term "take-or-pay" contracts. The traditional structure of most contracts in the LNG industry involves the buyer taking most of the volume risk with limited flexibility, and the seller taking most of the price risk, with limited opportunity to revisit those terms. These traditional structures are now being modified as the industry matures and the options and opportunities for buyers and sellers expand.

Unlike the oil business, where the production profile tends to build gradually over time and the commodity may be shipped relatively inexpensively to any number of markets on tankers that can be contracted for on relatively short notice, liquefaction plants generally produce large quantities of LNG shortly after they come on line, and access to LNG tankers and market outlets is more limited and can be significantly more expensive. Under ideal conditions, production and consumption reach maximum levels as early as possible and maintain this level for the duration of the project. For this reason, it is preferable to market LNG to buyers with access to well-established natural gas markets. In countries with small or nonexistent markets, an LNG sales contract might be anchored by an electric-power generating plant or an industrial buyer that can consume large volumes of regasified LNG. Smaller customers would be added as a distribution network is built out from the terminal.

This commercial structure, the foundation of the LNG industry for many years, is changing. Sellers are moving downstream and buyers upstream. Destination flexibility is becoming widespread and some firms are adopting a portfolio supply strategy to balance uncertain downstream market needs, diversify risks (price and source), and capture arbitrage opportunities.

Access to capital. LNG projects usually require sizable shareholders with the ability to fund major capital investments, either on an equity basis or through borrowings. Project (or limited recourse) financing can play a major role in funding LNG projects, even those involving the largest oil and gas companies. Successful project financing requires a well-constructed project concept with robust and stable commercial arrangements, which can ensure the repayment of billions of dollars of financing over many years in a potentially volatile energy price environment. Financings have traditionally been underpinned by long-term SPAs at oil-linked prices.

Strong relationships. Although the hardest to define in concrete terms, the relationships, reputations, and experience of the various stakeholders are among the most important aspects of an LNG project. Strong relationships between project sponsors, customers, governments, lenders, and contractors are a key condition to long-term project success. The reason that trust and relationships are so important is simple: in an enterprise spanning many decades, subject to many changes unimagined at the outset, the underlying commercial and legal agreements may be inadequate to ensure the project's success, and the relationship between the parties is the only assurance that the business can address the problems that inevitably will arise. These relationships are especially important at the outset when the project is least defined and the parties are preparing to invest significant sums of money, often before any firm contracts are in place. It can take years to build these types of relationships and reputations, and breaking into the LNG industry can be more challenging as a result.

Technical details. All physical links in the LNG chain require technical analysis, including feasibility studies, engineering designs, project execution, and operational plans. These will determine the technologies that are used, the size, compatibility, and integration of each component of the facilities, and any additional required infrastructure.

Unique logistical challenges must be faced (and overcome) on each project. Planning for these sometimes tangential details may be projects in their own right. For example, when the Nigeria liquefaction plant was being constructed, it was necessary to house over 10,000 people at a remote site where there were no existing accommodations. Such related undertakings can cost the project millions of dollars. Canadian LNG export projects must overcome the cost of building multibillion dollar pipelines to deliver feed gas from shale gas fields in eastern British Columbia across two mountain chains to the coast where the liquefaction projects will be built. In eastern Australia, project joint ventures processing coalbed methane into LNG must manage the vast number of wells that must be drilled over the life of the project to maintain the necessary feed gas.

Commercial issues. At the outset of a proposed project, it is important for the participants to agree on the principal commercial terms. This involves the project structure and ownership of various components of the project, including shareholders' agreements and project development agreements; negotiating and structuring a series of end-to-end contracts (gas supply agreements [GSAs]) covering the production of natural gas at the wellhead all the way to sales to the end users (SPAs); defining the host government's role and revenue share (which may or may not be determined in advance); agreeing on and negotiating the contracts associated with the plant design and construction; and finally financing the project.

Safety and siting. Safety is of paramount importance in the LNG industry. The assets are so expensive that any repairs required by a major accident could amount to hundreds of millions of dollars. The lost production could also mean billions of dollars of lost revenues. A major accident could mar the industry's reputation and set back other proposed projects. Finally, insurance costs are a major expense for project investors, and an accident could drive them so high that profitability would be reduced. Lenders require an independent technical analysis of the facilities' design as well as confirmation of compliance with regulatory requirements and good industry practice. For all of these reasons, companies go to great lengths to design safety into the facilities from the outset, and to provide continual training and resources for employees working on and around LNG facilities and ships to ensure LNG's safe transportation and storage.

These considerations are most acute for regasification terminals since the most desirable locations are often near densely populated areas. There is widespread misunderstanding of the safety of these terminals. In an era when the threats of global terrorist activities have dramatically increased, siting new terminals or moving tankers near populated areas has become even harder due to local regulations and public perceptions. This was particularly evident in the United States and spawned the development of offshore terminals employing a new technology, FSRUs.

In the case of a greenfield project, both supplier and buyer must control sufficient land to accommodate their respective facilities and ensure legal compliance with safety and environmental conditions. These sites must have access to waterways which are deep enough to accommodate the LNG tankers or are capable of being dredged. Expansions of existing facilities generally do not create the same level of concerns.

External advisers. Legal issues are present from the onset of a proposed project. Lawyers with specialized industry knowledge and experience are instrumental in drafting contract and project terms. The contracts and agreements associated with an LNG project are complex, especially when it comes to allocating and mitigating the parties' risks and in addressing contingencies for unforeseen circumstances. Lawyers must also advise on, and in certain cases help decide, the applicable law that will govern each element of a project. Legal issues are a major component of every project.

Each project will also involve the input of a variety of other technical advisers who will provide expert support to the project sponsors. These will include consultants who will help prepare the necessary environmental impact statements and evaluate the gas reserves and markets, shipping alternatives, insurance, safety, and security, among other tasks. The project sponsors, buyers, and financiers will all seek independent evaluations of the various aspects of the project as part of their own due diligence exercise leading up to the final investment decision.

Government regulations. In most aspects of an LNG project, government regulations can play a significant role. Many liquefaction plants are constructed in relatively remote locations and in countries that often have no defined environmental, safety, or other regulatory guidelines. This can introduce significant project delays as host country governments wrestle with these issues. Often times in these cases, the project sponsors will utilize regulations from other countries, guidelines issued by the World Bank, or best practices promulgated by industry associations. Adopting these guidelines is often a prerequisite to securing international financial support and insurance for the project.

Once the project moves beyond liquefaction, the degree of government scrutiny and involvement remains high. LNG tankers are subject to a variety of regulations and conventions issued by the International Maritime Organization (IMO), the tanker's flag state, and the maritime regulatory bodies of the countries in which the tanker is expected to load and unload its cargo. Regasification terminals also are subject to stringent regulatory guidelines in their host countries that govern safety, security, and environmental aspects of the terminal and the vessels calling on it. In addition, the importing country may exercise economic regulation over the import terminals, requiring open access and the filing of tariffs governing the terminals' use. This will be particularly the case where the LNG terminal is owned by a public utility.

The US model

New LNG export projects in the United States present a different set of challenges. Feed gas will be purchased from the domestic natural gas market with the understanding that these gas supplies will be priced at market levels. Unlike the integrated projects, the emerging US model offers no access to liquids to boost project returns, as the heavier hydrocarbon components have generally been stripped from the gas supply before it enters the transmission grid. Resource adequacy and access are not the risks, but price certainty is another matter, since the feed gas buyer has limited ability to mitigate price risk.

The US LNG projects have adopted a tolling structure (Cheniere's projects are an exception) whereby creditworthy firms lease liquefaction capacity. Under the typical lease, the capacity holders pay a fixed-cost component, which is sufficient to cover all expenses including loan payments and the owners' financial returns but not feed gas. The lease payments are in a form that could be characterized as "hell or high water," and obligate the capacity holder to make the payments irrespective of events, unless for defined (and usually narrow) force majeure events. The tolling agreements (LTAs) bear some similarities to terminal use agreements (TUAs) at regasification facilities. The acquisition and delivery of feed gas to the plant is the capacity holders' responsibility (Cheniere being the exception as it undertakes this obligation itself, though the basic commercial outcome is the same).

Unlike the integrated projects, the US projects rarely involve any of the IOCs in the ownership or operation, with the principal ownership residing with smaller independent companies who have little or no role upstream or downstream of the liquefaction plant (Shell at Elba Island and ExxonMobil/QP at Golden Pass are exceptions). This pattern emerged because the early US liquefaction plants are based on conversions of existing import terminals to export plants, and with a few exceptions, the IOCs took no ownership role in the import terminals. This means the financing is more complicated. While the projects look to the debt markets for up to 70% of the funding on a nonrecourse basis, the equity share requires the project sponsors to raise equity financing through equity issuance (in either the public or private market), bring in partners with financing capability, or even consider initial public offerings to raise the funding.

US safety and regulatory codes present another set of challenges, as the United States has generally more complex regulations governing these plants and permits broad public participation in the siting and approval process. This can add time and cost, as much as $100 million to bring a liquefaction project through authorization to final investment decision (FID).

Finally, the US model has much less of a relationship component. Generally, the LNG purchased from the project is not subject to destination restriction, and many of the buyers are LNG aggregators or traders with no assured access to import terminals. As a result, any failure of the downstream or upstream supply chain is of limited concern to the project owner who is contractually insulated from both sets of risks. It remains to be seen if this model can stand the test of time if the capacity holders find themselves in distress due to inability to secure feed gas on a price competitive basis or secure access to downstream markets.

Alternative methods of monetizing natural gas resources

Options for monetizing natural gas that are available to resource holders and developers include the following:

Local market. Depending on the location of the gas field and its proximity to populous areas, gas can be monetized by simply selling it into a local market. National governments generally prefer such a scheme, which allows the development and internal use of an indigenous resource. A steady supply of local gas can encourage industries that use natural gas as a feedstock (e.g., methanol and fertilizer production) or whose operations require large quantities of cheap power (e.g., aluminum smelters). It also promotes investment in infrastructure and creates jobs in the host country. Sometimes host nations will require project developers of remote natural gas resources to build infrastructure to supply internal markets with a share of the natural gas, as a prerequisite for the right to build LNG export facilities in parallel. This is another area where North American LNG export projects stand in contrast to the conventional integrated LNG project.

LNG or pipeline? In its early evolution, LNG was seen as a high-cost option and the last choice for countries with potential access to pipeline gas. LNG can generally compete with pipeline deliveries over longer distances (generally greater than 2,000 kilometers), and even over shorter distances where there are major impediments to pipeline construction. Today in Japan, LNG remains the preferred import approach because of the disconnected nature of the Japanese regional markets, which would make it difficult to bring a large pipeline to the country and arrange for the onward transmission and distribution. Japan's markets are presently viewed as prohibitively expensive and legally complex to integrate into a single grid, foreclosing the pipeline option.

Security advantages. LNG has security advantages that can override economics. LNG is generally transported across international waters, so that only the seller and buyer governments are involved. Pipelines often have to cross several international boundaries, which means that several governments may be involved, increasing the relative complexity of development. This may also raise the potential of supply interruptions through diversion of natural gas volumes, attack on the infrastructure, or simply having the intermediary country close the valve for political reasons. Gas buyers and sellers in a pipeline trade are essentially hostage to events in the countries crossed by their pipeline. A good example is the potential gas trade between Iran and India. The most obvious and economical method for bringing Iranian gas to Indian markets would be via an onshore pipeline across southern Pakistan, a scenario fraught with obvious political difficulties.

Another example of geopolitical risk is the delivery of Russian pipeline gas to Europe through Ukraine. It has resulted in temporary disruptions in the past because of payment disputes between the two countries. In 2014, these risks have been heightened by the overthrow of the pro-Russian leader in Ukraine, the subsequent annexation of Crimea by Russia, and the emergence of pro-Russian separatist movements in eastern Ukraine. On the other hand, unlike ships, pipelines are not vulnerable to the vagaries of weather and are easily maintained. They can provide a secure and stable method of delivering set volumes of gas on a year-round basis.

Some countries choose to import LNG in order to expand their supply portfolios. They may be diversifying away from another energy commodity, as in the case of Japan, which initially imported LNG to lessen its reliance on oil. Alternatively, they may be reducing their reliance on any single gas supplier, as in the cases of Thailand, Singapore, Poland, and Lithuania. Exporters may favor the LNG option because, unlike a pipeline project, they can diversify their revenue sources by selling to multiple buyers and markets. Exporters reliant on a pipeline are at the mercy of a single market that has to be fully developed at some cost prior to construction. If demand fails to materialize in that market, the asset may be underutilized, whereas LNG may be diverted to other markets when demand patterns change. More flexible LNG schemes can help secure smaller commitments from a number of buyers to reach the necessary threshold for project investment. Security of supply and markets is a central theme in the 21st-century energy market.

Gas-to-liquids. The term GTL refers to the reprocessing of methane into longer chain hydrocarbons that are liquid at atmospheric temperatures and pressures. The GTL process entails the generation of syngas, a mixture of CO and H_2 derived by combining methane with H_2O and/or O_2 at high temperatures. This syngas is then subject to the Fischer-Tropsch (FT) or another similar process in which the syngas is reacted with a catalyst to produce longer chain hydrocarbons that are liquid at normal pressures and temperatures. These hydrocarbons can then be processed by using standard refining techniques. The primary product, a little over 50% yield, is a form of very clean diesel fuel, which can be handled and used in the same way as diesel refined from crude oil. However, the capital cost of GTL is higher than that of LNG for the same energy output.

GTL competes primarily against more traditional refined oil products rather than LNG (although GTL and LNG do compete in terms of the gas monetization choice). An advantage of GTL is that it is a stable liquid under normal conditions (and therefore easily transportable) and very clean. Like crude oil and unlike LNG, GTL is a fungible commodity. These characteristics make GTL especially valuable in developed markets that are highly sensitive to environmental concerns. GTL competes against refined petroleum products for market share, making it highly sensitive to the oil price. Five GTL plants are in operation across the world: two in Qatar, one in

Malaysia, one in South Africa, and another in Nigeria. Shell and Sasol, prominent players in this industry, have proposed projects in North America, while Sasol and Malaysia's Petronas are promoting another one in Uzbekistan.

Compressed natural gas. For shorter-distance trades or for relatively small reserves (<2 Tcf, or 56.7 Bcm), another monetization option in the future may be compressed natural gas (CNG). In a CNG scenario, natural gas is compressed to between 2,000 to 4,000 pounds per square inch (psi) and transported aboard specially outfitted ships to market. These ships are essentially floating pipelines. A number of different containment systems for the compressed gas have been proposed, each consisting of a series of pipes into which gas is pumped at high pressure.

An advantage of CNG is that it does not require expensive infrastructure in the host countries. The facilities for a CNG project in the host nation downstream of the pipeline consist solely of compression and (if necessary) gas-processing facilities, so that substantially less capital is at risk than for an LNG or GTL project. However, a CNG tanker would cost roughly the same as an LNG ship, but could deliver only about a quarter of the volume. If CNG ships are built, the cost is expected to go down, but the shipping cost will still be at a premium to LNG. The nature of the CNG trade, which requires a constant rotation of vessel deliveries to maintain a steady gas flow, is also somewhat of a drawback (although this could be mitigated by gas storage on the receiving end). The weight and complexity of CNG vessels make maintenance difficult.

No CNG project has been developed to date, and most of those being considered are being pursued by independent companies, not oil and gas majors. CNG technology for long-term service is still unproven, which also raises financing issues and market acceptance. Another factor yet to be fully addressed is the acceptability of the high-pressure tankers in importing countries and the siting of the facilities to unload and connect them to the gas grid.

Notes

1. IEA (International Energy Agency), "Golden Rules for a Golden Age of Gas," *World Energy Outlook: Special Report on Unconventional Gas* (Paris: IEA, 2012), 9.
2. IEA, *World Energy Outlook: Are We Entering the Golden Age of Gas* (Paris: IEA, 2011).
3. Ibid.
4. BP, *Statistical Review of World Energy June 2014* (London: Pureprint Group, 2014).
5. IEA, *World Energy Outlook* (2011).
6. CEDIGAZ, *Natural Gas in the World* (Rueil Malmaison, France: n.p., 2013).
7. International Energy Agency (IEA), *World Energy Outlook 2014* (France: Organization for Economic Cooperation & Development, 2014).
8. Poten & Partners, *Global LNG Outlook* (2014), Q1.
9. Ibid.
10. BP, *Statistical Review.*
11. IEA, *World Energy Outlook* (2011).
12. EIA (US Energy Information Administration), *International Energy Outlook 2014* (Washington, DC: EIA, 2014).
13. Ibid.
14. IEA, *World Energy Outlook* (2011).
15. EIA, *International Energy Outlook.*
16. BP, *Statistical Review.*
17. Ibid.
18. Malcolm W. H. Peebles, *Natural Gas Fundamentals* (London: Shell International Gas Limited, 1992).
19. US EIA, "About U.S. Natural Gas Pipelines," https://www.eia.gov/pub/oil_gas/natural_gas/analysis_publications/ngpipeline/index.html.
20. Michael Tusiani, *The Petroleum Shipping Industry* (Tulsa, OK: PennWell Publishing, 1996).
21. Poten & Partners, LNGas Portal.
22. Ibid.

2

The LNG Chain: The Project Nature of the LNG Business

Introduction

The LNG business evolved as a collection of independent, discon-nected projects. It originated in an environment with few potential customers or suppliers, where gas was seen as a nuisance rather than as a valuable commodity. When trying to bring remote gas reserves to market, it was logical to develop each business opportunity as a stand-alone project, essentially simulating a long-distance pipeline. Each project was designed to bring dedicated reserves to specific markets through a chain of separate but closely linked stages: upstream gas production and gathering, liquefaction, shipping, and regasification. Each stage was connected to the other by long-term contractual relationships (fig. 2–1). Projects were developed with different technical and commercial considerations, government policies, financing, and fiscal terms so that project developers often had to create unique solutions for the specific problems and oppor-tunities presented by each environment.

Gas Production
Liquefaction Plant
Shipping
Regasification Terminal
Pipeline Delivery

Fig. 2–1. The LNG chain. (*Source:* Sempra LNG.)

The industry has changed significantly. Natural gas has become a valuable and widely available commodity. Markets have proliferated, with numerous buyers around the globe. Project sponsors are integrating downstream even as buyers are integrating upstream. Both project sponsors and buyers are purchasing LNG increasingly on a FOB basis without destination restrictions, in order to supply downstream market commitments and to seize trading opportunities. Spot and short-term trades have grown to about 30% of the global LNG market, introducing new flexibility into the industry. While traditionally the liquefaction venture marketed the LNG and usually arranged shipping to sell on a delivered basis, tolling arrangements at LNG export ventures are becoming more commonplace, particularly at prospective US projects. While upstream resource developers were the tolling capacity holders in Trinidad and Egypt, where this model was first established, buyers are now choosing to take tolling capacity in the US projects, and buying feed gas from the domestic market.

Because of the enormous capital investment at each stage of a project, the traditional structure with modest modifications still predominates, as project sponsors seek to maintain control to manage technical and commercial risks as much as is practicable. Technical risks can be lowered by adoption of proven designs, construction by experienced contractors, and operations that follow best practices and internationally accepted safety guidelines. All the energy companies involved, whether private or state owned, international or domestic, as well as the international construction and engineering firms that can undertake these demanding projects, have built up over time their preferred ways of tackling the manifold commercial, technical, safety, and organizational issues inherent in these projects. At the same time, they are adjusting to the new commercial realities.

The value of the gas and the overall economic viability of the project are only realized after regasification and sale in the final market. The final buyer of the LNG has traditionally been a major natural gas or power utility with a local or national franchise assuring it can place the gas in its market with a high degree of certainty and pass through the supply costs to its customers. The LNG SPA with this utility provides the credit support for the project's long-term cash flow. For the US liquefaction projects, which represent a major deviation from the traditional model, the fixed tolling fee provides similar credit support.

Establishing durable, long-term contracts between the different entities at each stage of a project reduces commercial risks. Reduced to its simplest elements, the LNG buyer has traditionally taken the volume risk through take-or-pay provisions, obligating the buyer to pay for the gas even if it does not take delivery, and the upstream parties have taken the price risk. Contracts have typically had terms of 20 years or longer with limited volume or destination flexibility. With less demand certainty, increasing train sizes, and greater shipping flexibility available in the market, buyers are now keen to sign FOB purchase contracts that permit destination flexibility to underpin their offtake guarantee. They are willing to absorb the extra shipping costs that might be involved and that could be recouped by trading profits. In the United States, tolling capacity holders are now shouldering both volume and price risk, often without assured access to markets to absorb all of the volumes.

The oil industry does not display similarly integrated project models because the oil market is more fungible and does not require dedicated infrastructure. The cost of oil transportation is relatively low compared with the price of the product, whereas LNG shipping costs represent a significant component of the total price. A developer of an oil field can reasonably look at a multitude of potential outlets or customers around the globe. There is, therefore, no need to enter into contracts with customers before undertaking the construction of oil production facilities, as is the case with LNG.

The Physical Chain

Upstream exploration and production

Natural gas is found onshore or offshore as nonassociated gas or as associated gas where it forms part of an oil reservoir and is produced in conjunction with the oil. The nonassociated gas can be dry—that is, nearly pure methane—or wet, containing higher chain hydrocarbons such as propane, butane, and condensates. In many oil-producing countries, such as Nigeria, associated gas is flared because it is too expensive to reinject into reservoirs or to gather for sale for domestic or export markets. Other countries, such as Algeria, Australia, Qatar, and Trinidad, possess large reserves of nonassociated gas, which have been developed independently for monetization via an LNG project. In all cases, the gas needs to be produced and treated before it can be processed in the liquefaction plant. This treatment removes condensate and other heavier hydrocarbons, water, CO_2, and other contaminants.

The cost of developing the gas can vary substantially from location to location depending on whether the gas is offshore (in which case, water depth and distance to shore are critical factors) or onshore, deep or shallow, wet or dry, associated or nonassociated. Gas well productivity and distance to the proposed liquefaction plant are also relevant. Compared with the other stages of the LNG chain, where most of the cost is incurred up front, gas production usually requires substantial ongoing investment throughout the lifetime of the project to offset the natural decline of the initial reservoirs. This is particularly true for liquefaction projects exploiting coalbed methane in Australia's Queensland, where wells deplete rapidly. While shale gas requires a similar ongoing investment to maintain production levels, the upstream is not an integrated part of the LNG ventures in the United States, though it is in Canada's British Columbia province.

When the natural gas is wet—that is, when it contains liquefied petroleum gases (LPGs) and condensates—the effective cost of the natural gas can be reduced significantly. While LPGs are typically removed at the liquefaction plant, condensates may need to be stripped out of the gas before they are transported to the plant. Sales of condensates can yield a substantial addition to the project's revenue stream. In some fields, such as those in Qatar, Oman, Australia (outside Queensland), and Russia (Yamal), the condensate revenue covers most of the upstream investment.

Exploration and production are usually covered by agreements between the host government and the energy companies. Such agreements set out the terms under which producers can explore for, develop, and produce oil and natural gas. The contracts often include provisions such as minimum exploration programs, cost recovery, tax treatment, and royalty payments. In the case of concession or license agreements, the oil and gas become the property of the producers, while in the case of production-sharing contracts, the government (often in the form of the state oil company) retains an ownership and participation right in the reserves once discovered.

Traditionally, oil and gas companies have funded exploration and production expenditure out of corporate funds because of the high risk involved. The large international majors tend to continue to prefer this approach, but there are an increasing number of cases where large production installations, such as offshore production platforms, have been project-financed, with the debt secured by the revenues and assets of the project itself.

Liquefaction

The largest and most important single investment in the LNG chain is the liquefaction plant. Gas composition, quantity, and location have an important bearing on the design of the plant, but in reality, liquefaction plants are nothing more than giant refrigerators. Although each of the world's large baseload liquefaction plants is unique in design, they all perform two basic common tasks: first treating the gas to remove impurities, and then cooling it to a temperature of around −163°C, where the gas becomes a liquid at atmospheric pressure.

Liquefaction plants are usually set up as a number of parallel processing units, called trains, each of which treats the gas and then liquefies it. The design capacity of each train is determined by the size of the equipment such as heat exchangers and gas or steam turbines, which drive the compressors essential for the liquefaction process. Historically, parallel trains were developed to enable the plant to continue operating if one of the trains was shut down for maintenance or repair. Taking advantage of economies of scale and larger equipment sizes, unit train capacities increased to 7.8 MMt/y (10.6 Bcm/y) in Qatar, compared to 1 MMt/y (1.4 Bcm/y) capacity in the early plants in Libya, Algeria, Alaska, and Brunei. However, the latest liquefaction plants have adopted smaller train sizes, around 4–5 MMt/y (5.4–6.8 Bcm/y), since these are less expensive to construct, easier to maintain, and the smaller output makes matching market demands easier.

After the gas leaves the upstream production facilities, it is metered and transported by pipeline to the liquefaction plant. Before entering the liquefaction plant, the gas must be treated to remove impurities (carbon dioxide, sulfur, mercury, and water) which can cause corrosion or freeze inside the heat exchangers. Many heavier hydrocarbons are stripped out at this point, leaving mainly methane, ethane (if present), nitrogen, and small quantities of other light hydrocarbons. Essentially, the gas at this stage needs to be of a quality that can be readily liquefied while being acceptable in the final market. Any heavier components removed in the plant (such as condensate and LPG) are shipped and sold separately, creating additional revenue for the project.

The liquefaction process equipment consists of compressors driven by steam or gas turbines or large electric motors, and heat exchangers, where heat from the incoming gas is transferred to refrigerant gases (such as propane or ethylene, or mixtures of both), which in turn transfer heat to an outside coolant (air or water or a combination). There are a number of proprietary processes, but the overall process concepts are similar. In earlier plants, the compressors were largely driven by steam turbines, but today gas turbines are the standard compressor drivers.

The storage and loading facilities are an important part of a liquefaction plant. After the gas is cooled to a liquid state, it must be stored in double-walled, insulated tanks. The tanks are designed to keep the LNG cold until it can be loaded onboard an LNG tanker. LNG tanks are usually built with an inner tank of cryogenic nickel/steel, surrounded by several feet of insulation and an outer tank constructed from prestressed concrete or mild steel. In the case of mild steel tanks, the tank is further surrounded by a berm to contain the LNG in the unlikely event of a failure of the inner tank and to provide a measure of protection to the remainder of the plant.

Such a berm is not required for a reinforced concrete tank, because the concrete can contain the LNG if the inner tank fails. Earlier built tanks had a relatively small capacity, about 50,000 m³ of liquid, but tank sizes have grown in scale to 200,000 m³, with 240,000 m³ tanks now under consideration. The number of tanks and the design chosen is a function of the site size, planned production rates, safety regulations, LNG tanker capacities, frequency of tankers calling at the plant, and the construction cost.

The plant also will include a jetty and loading facilities that allow safe access to LNG carriers. Often breakwaters or harbors need to be constructed to protect the marine berths as well as the LNG carriers. Jetties have articulated loading arms, which can handle the cryogenic temperatures. They are connected to the ship's manifold with special quick-disconnect couplings that can be released in the event of an emergency.

As LNG plants are often located in remote areas, significant infrastructure (roads, airport, staff accommodation with schools, hospitals, laboratories, maintenance facilities, etc.) might be required during the construction and the operations phase of the project. All the existing liquefaction plants have, until now, been built onshore at coastal locations. However, project developers are increasingly adopting modular construction for high labor cost environments such as Australia, where the labor force is a significant constraint, and in geographically hostile locations like northern Norway. In this construction approach, the modules are constructed elsewhere and then shipped to the site for assembly into the LNG plant. Designs have also been prepared for offshore floating plants integrated with offshore production facilities. Currently, floating LNG production units are being built for projects for Australia, Malaysia, and elsewhere.

Safety is paramount during design, construction, and operation. Liquefaction plants have proven to be very safe and reliable, and most have produced above their initial design capacity and for well beyond their initial design life. For example, the Camel plant in Algeria, completed in 1964, was shut down only in 2010, after 46 years of operation.

Shipping

After liquefaction and storage, the LNG is loaded in specially designed, insulated ships. LNG is sold either FOB, in which case the buyer is responsible for arranging the shipping and cargo title transfers on loading, or delivered ex-ship (now known as DAP, or "Delivered at Place")[1] or CIF (cost, insurance, freight), where the seller arranges the shipping, and title is transferred at the destination or after loading. LNG ships historically were custom built for, and dedicated to, specific projects, sailing in regular service between the LNG supplier and one or more customers. The size of LNG ships has steadily increased since the mid-1990s. Currently, ships with cargo capacities of up to 266,000 m³ are serving Qatari megatrain projects, while the 155,000–170,000 m³ size has become the current de facto standard.

LNG ships are specially designed with double hulls and other safety features. Unlike the double hulls of oil tankers, the double hulls on an LNG tanker provide structural rigidity and allow the ship to carry large quantities of ballast water because the cargo is so light (LNG has half the density of water). There are two types of cargo systems employed in the LNG fleet: spherical tanks, which are essentially self-supporting, and membrane tanks, which rely on the ship and their own insulation for support. These systems are described in more detail in chapter 6.

Traditionally, LNG ships used steam turbines for propulsion, but as the relative prices of oil and gas have moved and engine technology has improved, newer ships are more likely to employ slow-speed diesel engines, driving the propeller(s) either directly or indirectly through electrical generators. These diesel engines can be tri-fueled, dual-fueled, or may only run on marine diesel fuel. While they cost more up front, they are less expensive to operate and maintain. They are also more compact, allowing for more storage within the hull.

Regardless of the cargo system, some boil-off occurs during the voyage. On older ships, up to 0.15% of the cargo boils off every day; however, modern tankers have significantly lower boil-off rates. For LNG tankers without onboard reliquefaction facilities, the gas that boils off during a voyage is used as fuel for propulsion. (Ships with onboard reliquefaction facilities powered by diesel-only engines were built in South Korea to service the Qatari megatrains. These vessels have the capability to deliver their entire loaded cargo.) After unloading, ships usually retain a small amount of LNG, called the heel, on board. This LNG keeps the cargo tanks cool for the next loading. The colder the tanks are prior to loading, the less boil-off occurs during loading and transport, which in turn affects loading time and the amount of LNG delivered. Warmer tanks may require cooldown—a time-consuming task.

Ship design, construction, and operation are subject to very high standards. As a result, the safety record of LNG ships is exceptionally good. There has never been an LNG cargo lost or spilled. Very few LNG carriers have been scrapped, and there is general acceptance that, with appropriate maintenance, an LNG vessel can have a working life of 40 years or more. Older vessels are also finding new life as FSRUs, and at least one is being converted for FLNG service.

Regasification terminals

LNG cargoes are discharged at regasification terminals (also called receiving or import terminals) that are located in the customer's country. They are usually owned by the customer and operated on a proprietary basis. In some nations, particularly in Europe, third-party access is required unless the regulator grants an exemption at the request of the developer, and then under limited circumstances.

A regasification terminal consists of one or more marine berths, each with a set of unloading arms, LNG storage tanks, and vaporization equipment to move the regasified LNG into the pipeline system. It may also include facilities for loading LNG directly onto tanker trucks for road delivery. Many terminals have now added LNG reload capability for delivery to other import facilities, which is proving an attractive option since it makes use of underutilized terminal capacity and facilitates spot cargo trading. Traditionally, terminals were funded by utilities, just as other gas distribution infrastructure assets were, and were included in the utilities' rate bases, where they earned a regulated rate of return. With the advent of third-party or open-access terminals, project financing is increasingly used to fund the construction of these import facilities.

More recently, terminals have been developed using FSRUs deployed offshore or berthed in nearshore arrangements. Typically, FSRUs are deployed in emerging markets, such as Argentina, Brazil,

Dubai, Indonesia, and Kuwait, or where there have been obstacles to onshore development (northeast United States, Israel, and Puerto Rico). The variety of technologies and commercial approaches associated with the FSRU segment will be covered in detail later.

Markets

LNG sales from a project have traditionally been on a long-term basis (20 years or more) to a large utility, either a natural gas company or a power generator. These companies are often state owned or have a monopoly franchise area. They provide the long-term financial underpinnings that enable shareholders and lenders to commit very large amounts of capital with confidence.

Pricing of LNG has until recently depended on the specific conditions of the customers' markets. In markets such as Japan and Korea, with utility buyers and no imported pipeline supplies and little or no domestic production, prices have been mainly set at a discount to crude oil as a proxy price for the LNG. (Many formulas include an S-curve, where the price formula varies above and below a certain oil price range, to dampen the impact of high oil prices on the buyer, and low oil prices on the seller.) In turn, supply arrangements and prices that enabled the utilities to recover their gas supply costs from their captive customers were subject to approval by government authorities. The relative linkage to oil prices in long-term contracts tends to weaken or strengthen, depending on the tightness of the LNG market. Spot and short-term market prices tend to be far more volatile, often commanding premiums to long-term LNG purchases in the winter and summer months and discounts during the shoulder months. While there are ongoing discussions around the development of gas trading hubs in Asia, it remains unclear if the fundamental structure of the Asian markets (at least today) could permit such a development in any meaningful way.

In Europe, LNG competes with pipeline gas; therefore, LNG pricing tends to follow pipeline pricing, both being linked to oil or oil products when they are supplied under long-term contracts. However, European buyers, particularly in northwest European nations, are demanding and winning more market-responsive prices tied to European natural gas trading hubs, which are growing in liquidity. In the United Kingdom, LNG is priced against the National Balancing Point (NBP), a virtual trading hub. Hubs at Zeebrugge in

Belgium and Title Transfer Facility (TTF) in the Netherlands are both increasing in importance in setting European pricing. The shift of both pipeline and LNG pricing to gas indices gives the European LNG market some similarities to North America, especially in providing LNG suppliers with more flexibility and more options. In the United States, LNG pricing for imports is driven by the local natural gas market and is generally priced against the Henry Hub index, often with adjustments for basis (or location) differentials. Prospective US LNG exports will also be linked to Henry Hub with adjustments to reflect basis and pipeline transportation costs to the LNG export point.

Whether LNG will ultimately become a fungible commodity like oil, or whether long-term contract pricing will become driven by local gas market prices, is a matter of debate. With little or no gas-on-gas competition in Asian LNG markets—and arguably, limited competition in some European markets—most of the world's long-term market for LNG is insulated from competitive forces. But post-2020, US LNG exports at Henry hub–linked prices will contribute to commoditizing the business and, at a minimum, lower the overall portion of the LNG market indexed to crude oil prices.

The Ownership Chain

Every project tends to have its own organizational structure, determined by unique circumstances such as legal and fiscal regimes, ownership of the gas reserves, national ambitions, and the way the value chain is divided among the different project components. In view of the technical complexity of the business and its need for large up-front investments, gas producers, often including the national oil company as sole or part owner, tend to be involved in the production and liquefaction stages, frequently taking control of the LNG shipping and selling on a DAP basis. This structure evolved from the producers' desire to ensure control over the project's implementation and protect their upstream interests. It also helped guarantee high utilization of infrastructure while minimizing surplus facilities, including shipping. Projects were designed conservatively to assure that the venture could meet its contractual commitments, which in turn allows projects to produce above their nameplate capacity.

There is now a trend toward upstream participation by the customers as a way of increasing their security of supply and gaining the right to invest in the more profitable components of the LNG chain—which they perceive as being financed using their credit and their markets. Increasingly, LNG buyers are gaining control of shipping, and are moving towards purchasing on an FOB basis (attempting to capture upside from trading activity) as well as taking equity participation in the venture in proportion to their LNG purchases. At the same time, LNG producers have moved downstream to secure market outlets through ownership or the leasing of LNG regasification capacity. The Qatar projects have followed this pattern, developing their own terminals in Europe (Rovigo in Italy, South Hook in the United Kingdom) and the United States (Golden Pass in Texas).

The entity owning the liquefaction plant might be controlled by just the natural gas producers, and might sell the LNG (and other products) on an FOB or delivered basis, taking price risk. Alternatively, the liquefaction plant can be established as a tolling facility, in which case the gas producers pay a fee to convert their gas into LNG, at the same time keeping control of the product until its ultimate sale. The liquefaction plant might be controlled in whole or in part by the gas producers. US liquefaction ventures are taking this concept one step further. The US tolling plants are for the most part owned by independent companies who are not taking upstream, downstream, or price risk, but simply leasing capacity to third parties. The capacity holders will process feed gas they purchase on the domestic market at the plant, for their own marketing programs. They are shouldering both price and volume risks, betting that they will be amply compensated for their risk exposure.

While early projects were financed largely by shareholder equity, it has by now become normal practice for liquefaction plants to be project-financed by international banks and export credit agencies, typically with debt on the order of 60% to 70% of capital costs. Because of the long-term contract structure and the participation by experienced and highly regarded international companies, the credit rating of an LNG project can be higher than that of the host country, since the project's credit rating is often governed by the quality of the LNG buyers' contracts and credit ratings, and the reputation and credit quality of the sponsors/owners.

In view of the high up-front capital costs, and because LNG is rarely, if ever, an energy price setter in any market, let alone one over which the project participants have much control other than through the pricing formulas embedded in the LNG sales contracts, optimal project economics tend to depend on maximizing sales volumes throughout the project's lifetime. Variable operating costs for the project chain are essentially zero, since the primary variable cost, the gas used for fuel along the chain, remains in the ground indefinitely with minimal value if not produced and sold. Labor, shipping, and other cash operating expenses are not truly variables unless plants or ships are mothballed for extended periods, which rarely happens. As a result, project participants put a high premium on maintaining volumes even during periods of low prices. The opportunities created by the spot and short-term LNG market and by the increasing possibilities of cargo diversions and swaps are being used more aggressively by LNG sellers to maximize their sales volumes and by LNG buyers to manage market fluctuations. Shipping to markets that have temporary gas shortages and are prepared to pay relatively high prices can generate substantial additional revenues, even with increased shipping distances, and those revenues can ultimately be shared by chain participants.

As a result, the LNG market is becoming more complex, with many new types of commercial transactions, often borrowed from other segments of the energy industry. Both buyers and sellers are trying to carve out trading roles, moving away from the traditional point-to-point arrangements of the past. These changing conditions have encouraged producing and purchasing companies to extend their reach up and down the LNG chain.

Oil and gas majors, for example, have taken positions in various downstream markets worldwide (fig. 2–2) through equity participation and/or leasing capacity at regasification terminals. Open access to import terminal and pipeline capacity allows companies on both sides of the traditional buyer/seller relationship to buy LNG for their own account, acquire LNG terminal capacity, and compete for end-use sales. This trend has evolved quickly in Atlantic Basin gas markets. National oil companies (NOCs) such as Sonatrach, Qatar Petroleum, and Petronas have gained access to UK markets through capacity agreements and/or direct investment in receiving terminals. International oil companies (IOCs) such as the super-majors

ExxonMobil, Shell, Total, ConocoPhillips, and others are integrating forward as well and securing access to and/or ownership in various North American and European LNG import terminals. In Asia-Pacific, there have been comparatively limited opportunities for international companies to participate in the downstream sector, as markets in Japan, South Korea, Taiwan, China, and India have been much slower than their European and US counterparts in implementing energy market liberalization measures.

LNG purchasers have taken similar positions in liquefaction projects throughout Southeast Asia and now in proposed projects in the United States and British Columbia (fig. 2–3). They are also leasing tolling capacities at liquefaction projects in the United States with the intention of arranging their own feed gas supply, and are taking the LNG produced into their supply portfolios both for their home markets and for trading and marketing in general.

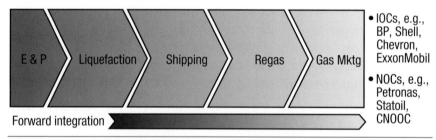

Fig. 2–2. Forward integration. (*Source:* Poten & Partners.)

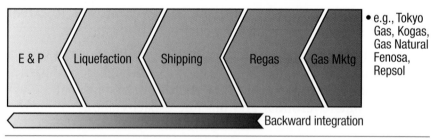

Fig. 2–3. Backward integration. (*Source:* Poten & Partners.)

Main ownership structures

There are four primary models of project structure:

A. The same participants throughout the chain, with the point of sale either FOB or delivered, and with the liquefaction profits integrated with the upstream profits.

B. The liquefaction plant as separate profit entity (company), owned mainly by the resource owners, buying gas and selling LNG, with margins reflecting both volumes and prices of the LNG sales. This company can sell LNG either on an FOB or on a delivered basis.

C. The liquefaction plant as a cost center owned either by the national oil company or by producers and customers. The liquefaction plant company can buy gas and sell LNG or operate on a tolling basis, earning a utility margin in either case.

D. Capacity holders reserve liquefaction capacity for a fee (tolling, as in the US liquefaction projects). These companies may also be equity investors in the project. They typically arrange their own feed gas supply and market the LNG.

Model A has the advantage of a seamless structure and avoids the need to negotiate a transfer price and conditions for the gas supplied to the plant. However, the unincorporated joint venture variations, as used in the Australian NWS project for example, can lead to management by committee and can also complicate project funding as each participant needs to raise its own financing. Since the NWS model was driven by Australian legal and fiscal considerations, it may not be optimal in other settings.

Model B uses established companies and management structures based on well-proven fiscal and legal principles. While this structure requires the negotiation of the price of gas supplied to the liquefaction plant, it is flexible in accommodating shareholding interests that differ from the resource ownership interests. The liquefaction company can also be used as a financing vehicle. The revenue of the company is generated by its sale of LNG (and often condensate and LPG). Its main cost is the purchase of gas from the producers. The gas price is often set as a percentage of the LNG price.

Model C is normally based on a fixed or semifixed tolling fee structure that covers the plant's operating costs, which assures a profit margin. This margin can be independent of fluctuations in the volume or price of the LNG sold, or have some exposure to both, to align ownership interests along the chain. Deducting the tolling fees from the FOB LNG price permits a clear transfer price to be established for the feed gas. Model C has the same advantages as model B in that it uses established corporate forms and existing fiscal regimes and permits flexibility in ownership. Properly structured and given the secure and relatively stable revenue stream, it can be easier and cheaper to raise project financing for this model. Often the tolling fees are subject to "hell or high water" provisions, payable independent of any upstream or downstream problems. It can also facilitate the inclusion of new feed gas reserves, owned by third parties, in expansion projects.

Model D is quite similar to model C, but with little or no connection between the ownership of the liquefaction plant and the ownership of the feed gas supply or the LNG. There is no exposure to volume or price risk on the part of the owners. There is no need to set transfer pricing for the feed gas, as this is determined on the open gas market. This model has so far shown itself to be amenable to project financing.

Except for model A, ownership interests can differ between the stages of a project. Major differences in shareholding between production, liquefaction, and shipping can create interface problems, such as transfer pricing, risk sharing, and project management issues, which have to be resolved through the project's commercial contract structure. Because of the different kinds and levels of risk in each stage, reaching agreement on the allocation of volume, price, and technical and political risks can be time-consuming. The shareholding in the liquefaction plant and shipping is often based on the sharing of the resource base; in practice, this may be modified by ambitions on the part of the host government or by LNG buyers seeking a stake in these components. Multiple participants and varying interests throughout the chain may create conflicts of interest and delays in agreeing to terms, but adding participants to the project can enhance the project's financial attractiveness, and might be a condition the project has to accept in order to achieve government support and market access.

Commercial and funding pressures can cause the need for more complicated structures. For example, in Egypt, Oman, Trinidad, and Qatar, there are different shareholdings in each train in the plant, and, in Trinidad, individual trains have different commercial structures.

LNG shipping

Sellers or buyers can own ships, either directly or through a special purpose company, or the ships can be owned by third parties who charter out the ship to the sellers or buyers. Ships are increasingly chartered from independent owners following competitive tenders, and are often financed on a nonrecourse basis by international banks. In many ways, LNG shipping has come to look more like a utility function (once a long-term charter is signed), with the shipowners assured of a modest but stable long-term return—truly a gas pipeline equivalent. Independent owners are also ordering LNG carriers on a speculative basis, or without long-term employment contracts. They prefer to find long-term charters for their ships, but in the absence of securing these commitments on attractive terms, they employ them in spot and short-term trading opportunities.

Examples of delivered (DAP or CIF) sales are Australia NWS, where the ships are owned mainly by the participants and chartered by the joint-venture; Brunei, where BLNG charters ships long-term from Brunei Tankers, which owns the ships; Nigeria LNG, where most of the ships are owned by an associated company, Bonny Gas Transport, and chartered in by NLNG; and Qatar, where the ships are owned by Nakilat (an independent Qatari company) and chartered to the RasGas and Qatargas ventures.

Examples of FOB sales are Oman LNG to South Korea, where the ships are owned by Korean interests and chartered to Kogas; Qatar's RasGas LNG to Korea, where the ships are also owned by Korean companies and chartered to Kogas; and Trinidad's Atlantic LNG, where ships are owned by a variety of owners, including the buyers, and chartered back. Operation and management of the ships is often delegated to shipping companies.

The Commercial and Financial Chain

Traditional LNG buyers generate cash flow from the sale of regasified LNG to their customers, such as industrial and commercial gas users, power generators, and residential consumers. From this cash flow, they cover regasification terminal costs and earnings, shipping (in case of an FOB purchase), and LNG purchases. The revenue from the LNG sales covers charter hire payments for the ships (if the sale is on a delivered basis), the operating costs of the LNG plant, the feed gas, the servicing of any debt used to finance the project, taxes and royalties to the host government, and hopefully the profits to the project shareholders.

The project shareholders in an integrated project are usually most interested in their netback—that is, the value of the gas at the wellhead—as that value will determine their profitability. The netback generates tax and royalties for the host government. Other profit factors are the costs to bring the gas from the wellhead to the liquefaction plant and shipping if the sale is on a delivered basis.

Costs along the LNG chain

Over the history of the LNG business, increasing skills and experience in the technical and commercial aspects of the industry have resulted in more cost-competitive and cost-effective infrastructure and services. The LNG industry has shown itself to be subject to the experience curve and economies of scale (to a point). LNG facility costs dropped significantly all along the LNG chain from the early 1980s into the early years of the 21st century. All stages contributed to this—exploration and production, liquefaction, shipping, and regasification.

However, this trend has since reversed. Escalating raw material costs—combined with intensified worldwide competition for management, labor, key equipment items, and engineering resources—have outweighed continued gains from improved technologies and economies of scale. Although all projects benefit from technological progress, these benefits have not entirely offset the problems faced by projects disadvantaged by long shipping distances to markets, high location cost factors, or expensive production costs because of hostile, sensitive, or remote environments. Recent projects in Australia, a high construction cost country with strong labor unions, are costing between $3,000/ton and $4,000/ton of annual production capacity, including upstream and liquefaction. For example, the three-train, 15.6 MMt/y Gorgon LNG project being built by Chevron, ExxonMobil, and Shell is reported to be costing a jaw-dropping $54 billion to build.

Another factor increasing the cost of these facilities is the inability of owners in many environments to secure lump sum turnkey contracts for the construction of liquefaction plants. The financial scale of the projects exceeds the ability of the contractors to underwrite the types of risks (such as currency and labor costs) associated with lump sum contracts, and the result has been a shift in risks (and costs) away from contractors to owners.

Exploration and production

Because of major differences between gas reserves (in size, depth, complexity, and location), and varying legal, fiscal, and royalty regimes, the economics of upstream gas production varies substantially from project to project. This variation can be further affected by the benefits of major condensate and LPG streams or the costs of dealing with undesirable constituents such as CO_2. Upstream exploration and production costs have benefitted from increasingly widespread use of three-dimensional seismic, horizontal drilling, and subsea completion technology. Development can be successfully completed at far greater reservoir depths—and in the case of offshore development, far greater water depths—than was the case 30 years ago, opening up new deepwater basins to cost-effective development. Horizontal drilling and multistage hydraulic fracturing ("fracing" or "fracking") have also opened up vast unconventional natural gas resources that were previously uneconomic.

Gas fields that are developed in LNG projects typically result in feed gas costs in the range of $3.00 to $4.00/MMBtu, depending on local geologic considerations and tax regimes. As noted above, associated liquids revenue streams can offset upstream costs, sometimes providing the liquefaction plant with very low-cost feed gas. For example, the upstream costs at ConocoPhillips's Darwin LNG project in Australia are substantially covered by a first-phase liquids recovery and gas reinjection project at the field. The LNG partners did, however, have to pay for a long and expensive underwater pipeline to deliver the feed gas from the Bayu-Udan field to the liquefaction plant.

At US liquefaction projects, where capacity holders buy gas from the domestic market at prices linked to Henry Hub, the feed gas cost will be more volatile than in an integrated project, since it depends on market conditions and not on the upstream investment. Here, though, the LNG project incurs no upstream investment costs. However, the liquefaction capacity holders have exposure to pipeline transportation costs and basis risk.

In Canada's British Columbia, feed gas is essentially stranded gas and will have its own economic drivers, including up-front upstream investment costs, and a substantial fee for pipeline delivery to the liquefaction plant, which can add more than $1/MMBtu to the cost of the feed gas at the plant gate.

Liquefaction

During the last three decades of the 20th century, the unit cost of liquefaction plants decreased by about a third because of improved scale, design efficiencies, better project management, and increasing competition among engineering, procurement, and construction (EPC) contractors. Plant costs bottomed out around $200 per ton per year (t/y) of capacity in some cases. Hence, at the time, a large LNG plant of around 8 million tons per year cost between $1.5 and $2 billion. However, this downward cost trend has reversed, particularly in Australia, where EPC contract awards have increased to multiples of the low benchmark at approximately $2,000/t/y of liquefaction capacity. At this unit cost, the same sized Australian plant would cost $16 billion to build.

Construction costs for LNG plants are lower elsewhere but are still formidable. Early indications suggest that unit liquefaction costs in Canada's British Columbia will be approximately $1,300 to $1,500/t/y, with East Africa coming in at approximately $1,250/t/y. By contrast, US liquefaction projects, benefiting from existing storage and marine infrastructure as well as higher labor productivity, are approximately $650/t/y to $750/t/y. While many projects are struggling to secure lump sum EPC bids, this has not been the case in the United States, where construction risks are more manageable, the up-front engineering and permitting is more intense, and the locations are near major industrial centers with deep pools of skilled construction labor.

The availability of EPC project management, engineering, and construction resources has not entirely kept pace with the LNG industry's rapid growth, and this has played a part in driving up construction costs. Up until the early 2000s, construction on most LNG liquefaction projects was led by the LNG industry's "club" of EPC contractors—Chiyoda, KBR, JGC, and Bechtel, with one to two liquefaction trains completed per year on average. Now, in early 2015, more than a dozen trains are being constructed concurrently. Yet there has not been a corresponding fourfold increase in available construction resources. The last decade has seen Technip grow to prominence in EPC consortia, while tank specialist CB&I has also emerged as a new club member, completing the Peru LNG project in 2010 and currently (early 2015) constructing two further projects at Cameron and Freeport on the US Gulf Coast. Additional contractors such as Foster Wheeler and Worley Parsons have assumed more important roles on some projects, but without taking lump sum risk. Major construction contractors such as Fluor have also stepped up their presence in construction support roles, but leadership on EPC projects has largely remained within the slightly expanded club.

Shipping

LNG ship costs have shown a similar cycle; costs declined into the early 2000s and have since increased. During the 1980s, prices were well over $200 million for 125,000 m^3 to 135,000 m^3 tankers. Then the entry of South Korean shipyards into the business helped to drive down costs. In March 2000, Korea's Daewoo Marine Engineering and Shipbuilding shipyard agreed to build a 138,000 m^3 carrier for Belgian shipping company, Exmar, at a price of under $145 million. Admittedly, this price represented an all-time record low for a large LNG new build. Ship prices have rebounded since then. In 2015, a new 174,000 m^3 M-type electronically controlled gas injection engine (MEGI) LNG carrier, which promises to become the industry workhorse, costs slightly more than $205 million.

As noted earlier, shipping is a significant cost component in the delivery chain. Shipping costs depend directly on distance to the market, and will vary between $0.30/MMBtu for shipping within the Mediterranean or from Russia's Sakhalin Island to Japan, to over \$2.40/MMBtu from the United States to the Far East and \$1.00/MMBtu from the United States to Europe. It costs around \$0.85/MMBtu to ship Australian LNG to Far East outlets, with shipping from British Columbia to the Far East slightly more.

Perhaps the most challenging LNG project from a shipping perspective is Yamal LNG, which will require a fleet of ice-breaking LNG tankers and an elaborate logistical challenge that will see LNG moving on two entirely different routes depending on the time of the year. Shipping costs for Yamal LNG to Asia could approach \$3.00/MMBtu.

Regasification terminals

The costs of regasification terminals fluctuate considerably depending on local construction costs, the cost of land, regasification technology, the capacity of the terminal, and the total amount of storage installed. The terminal owner decides on the required storage, baseload, and peak vaporization capacity, according to the seasonality of the market and the degree of security of supply required. The cost of a conventional onshore terminal with 200,000 to 300,000 m^3 of storage is typically a minimum of \$750 million. This does not include land and downstream pipeline costs. Larger terminals are now costing more than \$1 billion to build. An alternative that has proven attractive is FSRUs that can be implemented on a fast-track basis with smaller up-front costs. Onshore infrastructure costs around \$100–\$300 million, excluding primary storage and piping, but the FSRU charter can cost approximately \$175,000 per day.

As a rule of thumb, regasification terminals add about \$0.60/MMBtu to delivery costs. The figure is slightly higher at Asian terminals, given the large storage facilities required, and somewhat lower at European terminals where there is typically less storage. Given the current low utilization rates in European terminals, the effective terminalling cost might be even lower: capacity holders will take almost any contribution towards their fixed costs, even if it means discounting the terminalling rate from the actual tariff. On paper, regasification costs are lowest at US terminals, but many of these terminals are no longer actively importing LNG or are significantly underutilized. Floating facilities, fully utilized, could be about the same cost, or approximately \$0.60/MMBtu, which reflects the lower investment costs, but also the higher operating expenses of the FSRU charter. Frequently, FSRU operating costs are a bit higher due to a lower throughput than that of an onshore facility.

Value generation along the delivery chain

As discussed above, the cost of service of the delivery chain or the financial breakeven at a modest discount rate varies significantly for each project because of the nature of the gas reserves, liquefaction costs, and local taxes and duties. The cost of shipping will depend directly on distance to the market. This yields a wide range of delivery break-even prices (including returns on the investments) for new LNG export projects, from a low around $10.00–$11.00/MMBtu from wellhead to ex-terminal in Europe for US and East African projects to as much as $15.00 or even $16.00/MMBtu for high-cost Australian projects for sales into the Far East.

These Australian projects reached final investment decision (FID) after signing long-term contracts with buyers in the Far East, and most have their financial performance linked to crude oil prices. For example, during 2012 and 2013, when crude oil traded at over $100/bbl, the LNG import price for deliveries into Japan, linked to crude oil prices, averaged $16.30/MMBtu—enough to justify investment in these projects. By comparison, Canadian projects in British Columbia, with break-even prices of $13–$14/MMBtu, could prove to be more competitive in the Far East, and US projects at $10.50–$11.50/MMBtu look to be even stronger competitors. East African projects, depending on still-to-be-determined fiscal regimes, could save another $0.50/MMBtu in delivery costs into Asia. Steep drops in oil prices in the second half of 2014 into early 2015 have sent investors back to the drawing board to recalculate the economic viabilities of their projects.

Sales into Atlantic Basin LNG outlets, which will in all likelihood be required to absorb some of the new LNG production, are more problematic. With European buyers reluctant to sign term contracts at oil-linked prices, hub prices—with the National Balancing Point (NBP) as the deepest and most liquid—are closely watched. From 2011 to 2013, the NBP price averaged $9.70/MMBtu, compared to $3.50/MMBtu at Henry Hub. The $6.20/MMBtu premium would have justified LNG exports from the US Gulf Coast, but Poten & Partners estimates that the premium in 2014 dipped to about $4.50/MMBtu, with NBP declining to $8.50/MMBtu and Henry

Hub steadying at $4.00/MMBtu. This is significantly below a trading break-even margin of $5.25–$5.50/MMBtu, which would be required to cover liquefaction, shipping, regasification, and a trading margin. Other northwest European hubs (Zeebrugge and TTF) trade at levels very close to NBP. In early 2015, in the new low oil price environment, the arbitrage spread between NBP and oil-linked sales to Asian buyers has almost disappeared.

Long-term SPAs for pipeline gas and for LNG in Europe continue to have significant linkage to crude oil and product prices, but these prices are under pressure. During the 2011 to 2013 period, the German border price, a benchmark for term gas sales on the Continent, averaged $10.75/MMBtu. Given that spot gas was available at approximately $1.00/MMBtu less than this, end users are shopping for spot gas and forcing the buyers of term gas to discount their prices (and incur large financial losses). Even at this price level—if a seller could achieve it—Europe would be the market of last resort. In a lower oil price environment, the European market could become even less attractive as an LNG outlet.

As both the EU economy and gas demand recover, Europeans may have to pay more to assure a secure supply of LNG. The deteriorating EU–Russian relationship due to the 2014 Russian incursion into Ukraine could force the EU to rethink its gas/LNG supply strategy, leading to higher LNG import prices to attract supply and back out Russian pipeline gas.

Buyers in South America and the Middle East are often willing to pay a premium equal to that paid by Asian buyers, but this is often for spot and short-term supplies. These markets are still relatively modest in size, and can vary dramatically in demand from year to year and season to season.

Note

1. In the previous edition, we discussed three standard forms of LNG SPAs, namely FOB, CIF, and delivered ex-ship (Ex ship or DES). These terms were all developed by the International Chamber of Commerce and included in the Incoterms publications. In 2010, the ICC issued a new Incoterms guide, which eliminated the DES term (among others) and replaced it with the term DAP (Delivered at Place). In this edition, we have adopted the modern Incoterm of DAP in place of ex-ship, although we recognize that many legacy agreements in the industry still use the ex-ship terminology.

3

Global Gas and LNG Markets

Introduction

In most downstream markets, the natural gas business developed within a regulated environment that reflected the principle that the transmission and distribution segments of the business are natural monopolies; as such, they were generally franchised and regulated by governmental authorities. Transmission lines and gas distribution systems were built under cost-of-service tariff structures, which guaranteed that the owners would recover their costs, their capital, and a fair or just and reasonable return on their investments. Often, the components of the supply chain—production, gathering, transmission, and distribution—were functionally and legally separated from one another. In countries with indigenous production, regulators often set the price of gas at the wellhead. In the case of imported gas, regulators often had the right to approve supply contracts before they could come into effect.

Natural gas producers and suppliers sold to transmission companies under long-term purchase contracts with high take-or-pay obligations in order to justify the exploration risks and high upstream capital investments. In turn, the transmission companies (unless they were integrated with the distribution companies, as was usually the case in Europe) resold the gas under similar arrangements to the distributors. The regulatory bargain or compact was the guiding concept. The monopoly operator was permitted to recover its costs and earn a reasonable and assured return on its investments (provided these were deemed prudent) but in turn was obligated to provide a public service—the assured supply of natural gas to the consumers (the obligation to serve). The actual cost of the natural gas purchases themselves, whether from producers (in the case of the transmission companies) or from transmission companies (in the case of distributors), was treated as a pass-through, fully recovered from the next buyer along the chain all the way to the burner tip. The purchase agreements themselves were often subject to regulatory scrutiny and approval, and pass-through costs were also subject to prudency tests, all designed to protect the end consumers from excessive pricing.

Liberalization of Downstream Natural Gas Markets

Although the traditional cost-plus structure provided security of supply, it is widely believed that it contributed to higher prices than would have been the case under competition, especially in those markets where the business was mature. Opening up gas and power markets to competition came to be seen as a way of lowering energy costs to industrial, commercial, and residential consumers. Regulators have adopted policies separating supply and transportation functions, and providing open access to the pipeline infrastructure (transmission and distribution) to all prospective suppliers and customers—thus bringing competition to supply and marketing. In turn, these policies are contributing to a more competitive marketplace for gas and have led to the emergence of gas trading hubs—prerequisites for the development of a commodity market with transparent pricing. The hubs usually develop where several pipelines meet, often near large storage facilities. Examples include Henry Hub, Louisiana, in the United States, Zeebrugge in Belgium (where there is an LNG terminal, the terminus of the Bacton-Zeebrugge Interconnector, and the landfall of a large pipeline from the North Sea), and the Title Transfer Facility (TTF) on the Dutch-German border. In the case of the United Kingdom, the hub (the National Balancing Point or NBP) is a notional rather than a physical one.

While real price reductions are often realized in a competitive environment, they can be accompanied by greater price volatility and higher prices in a tight natural gas market. Liberalization also complicates funding the capital-intensive investments that ensure natural gas deliveries to consumers. While large consumers with sophisticated purchasing departments have benefited from competition, the record for smaller consumers, especially residential ones, is less clear.

The move toward liberalized gas and electricity markets appears irreversible in the Western world. Market liberalization is now advanced in the United States, Canada, and the United Kingdom. In western Europe, it is being promoted by the European Union (EU) and at the national level, though the pace of implementation varies widely by country. Natural gas trading hubs have developed in US regional markets and across Europe. In Asia, on the other hand, the traditional natural gas market structure predominates. Progress towards market liberalization is moving slowly if at all. Trading hubs have not yet developed there, even though the concept is being promoted in China and Singapore. Internationally, there are widely disparate regional natural gas price trends (see table 3–1), reflecting the state of liberalization as well as the diversity of supply sources and local market conditions.

Table 3–1. Global comparative gas prices ($/MMBtu)

	LNG		Natural Gas			Crude Oil
	Japan cif	Avg. German Import Price	UK (Heren NBP Index)	US Henry Hub	Canada (Alberta)	OECD countries cif
2005	6.05	5.88	7.38	8.79	7.25	8.74
2006	7.14	7.85	7.87	6.76	5.83	10.66
2007	7.73	8.03	6.01	6.95	6.17	11.95
2008	12.55	11.56	10.79	8.85	7.99	16.76
2009	9.06	8.52	4.85	3.89	3.38	10.41
2010	10.91	8.01	6.56	4.39	3.69	13.47
2011	14.73	10.48	9.04	4.01	3.47	18.55
2012	16.75	11.03	9.46	2.76	2.27	18.82
2013	16.17	10.72	10.63	3.71	2.93	18.25

(*Source:* BP Energy Statistics.)

Whatever the pace of natural gas and power sector liberalization in global markets, natural gas demand will in all likelihood continue to grow for the foreseeable future. However, where natural gas and

LNG prices are linked to oil prices, natural gas demand growth has been constrained in the face of persistent high prices linked to oil, particularly in Europe. This has led to natural gas demand destruction or migration (especially in the power generation and industrial sectors of developed economies) and conservation measures. Still, the future of natural gas appears favorable, with the International Energy Agency (IEA) projecting that the share of natural gas in the global energy mix will rise from 21% today to 25% in 2035.[1] That percentage translates into a 50% increase in global natural gas demand between 2010 and 2035.

Some analysts have questioned whether the energy industry can supply such large and growing volumes of natural gas on a reliable basis and at acceptable prices over long periods of time, even though worldwide reserves are adequate to support this development (table 3–2). Unconventional natural gas development supplementing conventional resources gives confidence in the promise of ample natural gas reserves to cover demand for the foreseeable future, as discussed in chapters 1 and 5.

On balance, these resources promise more competitive natural gas prices that will stimulate demand growth. However, reserves are not evenly distributed around the world and certainly not where demand is the greatest. For example, North America holds approximately 6% of proven global reserves, but accounts for 28% of world consumption according to BP energy statistics for 2013.[2] In contrast, Middle Eastern nations hold 43% of global gas reserves but consume only 13% of the world's supply. While in the very early stages of development outside of North America, unconventional natural gas resources are more geographically diverse, with implications for international natural gas trades.

Table 3–2. Worldwide natural gas reserves vs. production

	2000	2005	2010	2011	2012	2013
Gas Reserves (Tcm)	139.2	156.9	176.2	185.6	185.3	185.7
Gas Production (Bcm/y)	2,410.3	2,778.6	3,190.8	3,287.7	3,343.3	3,369.9
Gas Reserves/Production Ratio (years)	57.7	56.5	55.2	56.5	55.4	55.1

(*Source:* BP Energy Statistics.)

LNG import market prospects

The outlook for global LNG consumption remains robust, but the nature of the market is evolving rapidly as newer and more challenging markets emerge among the developing economies and the old economies undergo their own transformations. Increasing flexibility is the new world LNG order; flexibility coupled with more challenging buyer and supplier arrangements, supplementing and even beginning to supplant the traditional "point-to-point" relationships that characterized the industry during its first four decades. The emergence of South America and the traditional exporting regions of the Middle East, Indonesia, and Malaysia as LNG importers represent the new dynamics of the LNG industry, presenting exciting opportunities while posing challenges to financing and market security.

Asia

In Asia, which accounts for three-quarters of global LNG demand, most nations are taking a cautious approach to gas market liberalization. The region's traditional LNG importers (Japan, South Korea, and Taiwan) have very little access to pipeline gas, and measures to introduce competition into their gas markets tend to take a backseat to security concerns. Liberalization efforts also face infrastructural obstacles not encountered in Western markets (for example, the lack of an integrated pipeline grid in Japan). In Japan, the late 1990s saw the introduction of laws that cautiously opened gas and electricity markets to competition. However, incursions by gas utilities into the franchise areas of other gas marketers as well as into the power sector, and vice versa, have been relatively modest. Moreover, since the various companies typically import LNG under similar price terms, competition is curtailed by limited price differentiation, and Japan has made no moves to open LNG terminals to third-party access. In South Korea and Taiwan, after initial enthusiasm for privatizing and restructuring the gas and electricity businesses, the governments are adopting go-slow restructuring programs. As yet there is no natural gas trading hub in the Far East, although Singapore intends to establish a trading hub at its new LNG import terminal. Price linkage to oil remains prevalent even as Asian buyers begin to acquire US LNG at Henry Hub-linked prices.

Asia remains the backbone of the global LNG business with consumption of 177 MMt (241 Bcm) in 2013.[3] Demand growth remains robust, nearly doubling from 91.5 MMt (124 Bcm) in 2005 (fig. 3–1). New markets in Southeast Asia are emerging, including at several LNG exporting countries. Regasification capacity increased from 252.0 MMt/y (343 Bcm/y) in 2005 to 334.4 MMt/y (455 Bcm/y) in 2013. In spite of this rapid expansion, annual capacity utilization in 2013 was only 74%—though this compares well to 23% in Europe and much less than that in North America.

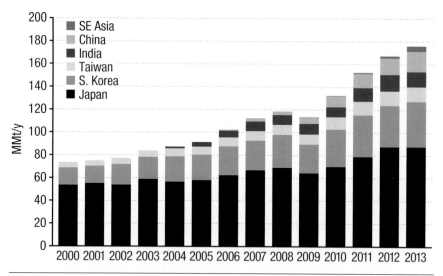

Fig. 3–1. Asian LNG imports by country. (*Source:* Poten & Partners.)

Asia—Traditional LNG markets

LNG demand by the three traditional importers, Japan, South Korea, and Taiwan, climbed from 87.4 MMt (118.9 Bcm) in 2005 to 139.9 MMt (190.3 Bcm) in 2013.[4] Still, their share of the global LNG market shrank by 13% during this time frame to 58% of the total as new markets opened up. Nevertheless, these three countries remain the most desirable markets for LNG exporters and traders, who prize their LNG appetite and their buyers' credit quality.

Japan. In 2013, Japan imported 87.5 MMt (119 Bcm) of LNG, giving it 36% of global trade and making it by far the world's largest importer.[5] The country's preeminent role as the global leader in LNG purchases emerged after the oil crisis of 1973, which caused a revolution in Japanese energy planning. Japan found itself overexposed to oil both in terms of price and security. Policy makers decided that this was unacceptable and implemented plans to increase gas supplies in the form of LNG and expand nuclear power. LNG was seen as a fuel that offered more diversified supply sources than oil, especially outside the Persian Gulf, and whose suppliers would be more closely bound to their Japanese buyers than was the case with oil suppliers. In 1970, Japan was dependent on oil for over 70% of its primary energy supply, and oil represented 60% of the fuel used in power generation. By 2011, oil's share of primary energy had been reduced to 43% according to the IEA, and its share of power generation to only 9%. The big winners in this transition were nuclear and LNG.

Japanese companies—utilities and trading houses—established a coordinated effort colloquially known as "Japan Inc." as the means to develop the Far East LNG industry. Although the nation's first LNG cargo was imported from Kenai, Alaska, in 1969, the rapid buildup did not begin until 1972, with a new liquefaction plant at Lumut, Brunei. By 1977, this plant was supplying over 7 MMt/y (9.8 Bcm/y) to Japanese buyers. In the same year, Abu Dhabi and Indonesia joined the ranks of Japanese LNG suppliers. In 1983, Malaysia entered the trade, followed by Australia in 1989, then Qatar in 1997, and Oman in 2000.

In the LNG trades to Japan, LNG export project sponsors often enlisted Japanese trading houses as shareholders in the project company to facilitate sales. For example, Mitsubishi partnered in projects in Brunei and Malaysia, as did Mitsui in Abu Dhabi. Both firms participated in Australia's North West Shelf (NWS) venture and in Russia's Sakhalin II project. Nissho Iwai assisted in Indonesia, and Mitsubishi, Mitsui, and Itochu joined the Oman project. In Qatar, Mitsui and Marubeni hold shares in Qatargas, and Itochu and Nissho Iwai are participants in Ras Laffan LNG (RasGas). In the earlier ventures, extremely creditworthy Japanese buying consortia frequently lined up to purchase output from new production trains at oil-linked prices favored by both sellers and buyers, and Japanese shipyards built the ships to deliver the cargoes to Japan. Japanese contractors and suppliers participated in the construction of the liquefaction plants. Meanwhile, Japanese banks financed the entire supply chain.

However, this comfortable structure has eroded, for a number of reasons. The end of Japan's postwar economic "miracle" moderated growth in Japanese energy demand. At the same time, the potential for gas to displace other fuels diminished. By the turn of the century, the requirement for large new LNG purchases appeared more limited, especially when compared to the number of export projects in development. Japan's LNG imports plateaued at 65 MMt/y to 70 MMt/y (88.4 Bcm/y to 95.2 Bcm/y) during the 2007 to 2010 period. With limited demand growth, Japanese buyers resorted to longer buildup periods in their purchase contracts to ease new volumes into their supply portfolios. They also sought more FOB purchases coupled with destination flexibility. Japanese buyers could no longer simply band together and buy out new liquefaction projects or trains, particularly as the size of the trains increased to 5 MMt/y (7.8 Bcm/y) and then to 7.8 MMt/y (10.6 Bcm/y).

The Japanese LNG demand outlook, according to most projections, would be limited. Then, the Fukushima Daiichi nuclear disaster in March 2011 prompted the closing of the nation's nuclear plants, which propelled LNG consumption as the nation's thermal power plants were called upon to replace lost nuclear power. Utilities scoured the market for additional LNG, which tightened the Pacific LNG supply situation dramatically. Japan, in 2013, imported LNG from 19 countries, led by Australia (17.9 MMt [24.3 Bcm]), Qatar (16.1 MMt [21.9 Bcm]), and Malaysia (14.9 MMt [20.3 Bcm]). The resulting tight Pacific Basin market saw large volumes flow from LNG exporters in the Atlantic Basin, totaling 10.8 MMt (14.7 Bcm) in 2013. Deliveries from the Atlantic Basin should increase even further as Japanese buyers tie up US LNG supplies at Henry Hub–linked prices. In all, the loss of the nuclear reactors resulted in a 17.5 MMt/y increase in Japan's LNG consumption over a three-year period (table 3–3).

To facilitate these purchases, Japanese companies have developed impressive import infrastructure, which includes 40 LNG terminals, many among the largest in the world (fig. 3–2). They have an aggregate 192 MMt/y (262 Bcm/y) of regasification capacity, mainly owned and operated by regional gas and power utility companies.[6]

Table 3-3. Japanese LNG imports by source

	2005	2006	2007	2008	2009	2010	2011	2012	2013
Australia	10.2	12.2	12.1	12.0	11.9	13.3	14.0	15.9	17.9
Brunei	6.3	6.5	6.4	6.1	6.1	5.8	6.3	5.9	5.1
Indonesia	14.3	14.4	13.6	14.1	13.0	12.8	9.3	6.2	6.3
Malaysia	13.6	12.0	13.3	13.1	12.6	14.0	15.0	14.6	14.9
Oman	1.0	2.4	3.6	3.2	2.6	2.9	3.8	4.0	4.0
Qatar	6.3	7.5	8.2	8.2	7.7	7.6	11.9	15.7	16.1
Russia	–	–	–	–	2.8	6.0	7.1	8.3	8.6
UAE	5.1	5.2	5.6	5.6	5.1	5.2	5.5	5.5	5.4
Other	1.3	2.4	4.1	6.9	2.7	2.5	5.8	11.5	9.3
Total Imports	58.0	62.5	66.8	69.2	64.5	70.1	78.8	87.5	87.6

(*Source:* Poten & Partners.)

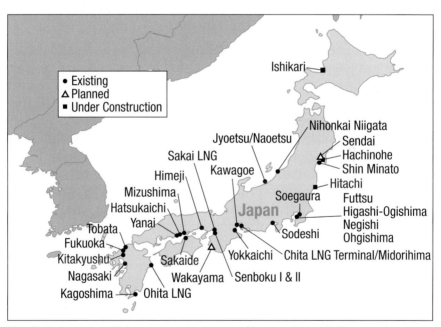

Fig. 3-2. Japan's LNG import terminals map. (*Source:* Poten & Partners.)

Before the Fukushima accident, Japan ranked as the third-largest nuclear power generator in the world, behind the United States and France. Japan had 50 nuclear reactors at 17 power plants with a total installed generating capacity of 46 GW. Over 10 GW of nuclear capacity at the Fukushima, Onagawa, and Tokai facilities ceased operations immediately following the earthquake and tsunami; Fukushima will never return to service. Maintenance standards in Japan require facilities to come offline every 13 months for inspections. After the Fukushima disaster, the Japanese government required facilities to pass stress tests as well as gain local government approval. As reactors were removed from operation, they remained offline. By May 2012, Japan had no nuclear generation for the first time in over 40 years. The government returned two of the reactors, Kansai Electric's Ohi #3 and Ohi #4, to operation in July 2012, leaving Japan with only 2.4 GW of capacity for slightly more than a year. These two reactors were again removed from service in September 2013, leaving Japan with no functioning nuclear capacity for a second time.

The 2010 Energy Plan, in light of the country's concern to reduce greenhouse gas emissions, had called for at least 12 new nuclear reactors to be constructed by 2020 and the nuclear share of electric generation to increase to 50% by 2030. However, in 2012, after Fukushima, the government, led by Prime Minister Noda, pledged a no-nuclear policy. This initiative encountered opposition from the business sector. The current administration, headed by Prime Minister Abe, supports a plan for nuclear power to provide at least 15% of power generation. Prime Minister Abe and industrial interests favor recommissioning nuclear power to lower energy costs. However, they must balance safety concerns and resistance from antinuclear government factions and the public.

In September 2012, Japan established the Nuclear Regulatory Authority (NRA) to replace two other nuclear agencies, the Nuclear Safety Commission and the Ministry of Economy, Trade, and Industry's (METI's) Nuclear and Industrial Safety Agency. The NRA, which is intended to provide a more independent assessment of nuclear safety, adopted more stringent nuclear safety guidelines and procedures in July 2013, and is now in charge of their enforcement. Any application to restart operations at a nuclear facility must be submitted to the NRA, which takes up to six months to review an application. The NRA safety guidelines are meant to ensure that facilities can withstand natural disasters by being located away from active earthquake fault lines. The guidelines demand installation of larger seawalls, air vents, and safety control rooms. Also, the new standards include decommissioning any reactor older than 40 years. Ultimately, this standard will contribute to a long-term decline in nuclear capacity unless new reactors are constructed or the lifetimes of the existing reactors are extended.

According to METI, natural gas contributed over 27% of the electric generation in Japan in 2010, playing a significant role in the power sector. Post-Fukushima, a majority of the lost nuclear generation was replaced by natural gas–fired power plants, and this share has increased significantly since then. To meet power demand, at least 7 GW of capacity at three new gas-fired, combined cycle units are expected to come on line by 2016. Capacity utilization in gas-fired power facilities in Japan's top 10 regional utilities was 66% in June 2013 according to the government. LNG use will, in the long run, be determined by the speed of building new gas-fired capacity—the lead-time on greenfield plants being 7 to 10 years, mainly due to environmental permitting—and by how much nuclear capacity is restored. Small shifts in nuclear power generation can have magnified impacts on the demand for LNG.

The expansion of LNG imports has come at a major price to Japan's balance of payments. Japan has therefore announced that it will move to expand renewable power generation—until now, not a significant factor given restrictions on land use, which makes it difficult to site major onshore infrastructure such as large-scale wind and solar generating facilities.

South Korea. The earliest challenge to Japanese dominance in the Far East LNG business came from South Korea (table 3–4). With state-owned Korea Gas Corporation's (Kogas) first purchase of LNG, Far East LNG export ventures were no longer totally dependent on Japan for sales. Kogas signed its first contract with Indonesia for LNG from the Arun complex in 1983, with deliveries starting in 1986. The Korean firm subsequently signed two more contracts for Indonesian LNG before entering into a short-term contract with Malaysia that was subsequently converted to a long-term arrangement. In the late 1990s, Kogas entered into several contracts for Middle Eastern LNG supplies. LNG supplies from Qatar's RasGas project began in 1999, followed by Oman in 2000 and then Australia.

Table 3–4. South Korea's LNG imports by source (MMt)

	2005	2006	2007	2008	2009	2010	2011	2012	2013
Indonesia	5.5	5.0	3.8	3.0	3.0	5.5	7.9	7.4	5.6
Malaysia	4.7	5.7	6.1	6.3	5.8	4.8	4.1	4.1	4.3
Nigeria	–	0.1	0.2	0.3	0.2	0.9	1.2	1.8	2.8
Oman	4.2	5.2	5.1	4.8	4.1	4.6	4.2	4.1	4.3
Qatar	6.2	6.6	8.1	9.0	6.9	7.5	8.2	10.3	13.5
Russia	–	–	–	–	1.0	2.9	2.9	2.2	1.8
Yemen	–	–	–	–	0.2	1.7	2.7	2.6	3.6
Other	1.7	2.7	2.5	5.2	4.0	4.9	5.5	3.8	4.0
Total Imports	**22.3**	**25.4**	**25.9**	**28.4**	**25.2**	**32.6**	**36.7**	**36.3**	**39.9**

(*Source:* Poten & Partners.)

South Korea quickly became the world's second largest LNG importer, and Kogas is now the largest buyer in the world. LNG imports into South Korea surged 75%, from 22.3 MMt (30.3 Bcm) in 2005 to 39.9 MMt (54.4 Bcm) in 2013, when the country imported LNG from 18 countries.[7] Qatar supplied 13.5 MMt (18.4 Bcm), followed by Indonesia with 4.3 MMt (5.8 Bcm), and Malaysia and Oman, each with 3.6 MMt (4.9 Bcm). Another 4.5 MMt (6.1 Bcm) was imported from the Atlantic Basin led by Nigeria (2.8 MMt [3.8 Bcm]). South Korean gas demand has large seasonal swings, which presents a significant logistical supply challenge and requires major investments in storage at the LNG terminals. Kogas is a very active buyer of spot cargoes during the winter months, and the firm has also entered into several medium-term contracts with deliveries heavily weighted to the winter months as well. South Korean regasification capacity reached 63.6 MMt/y (86.5 Bcm/y) in 2013 and is projected to reach 79 MMt/y (107.4 Bcm/y) by 2020.[8]

Kogas terminals. Kogas dominates the nation's import infrastructure with three large import terminals. South Korea's first LNG import terminal at Pyeongtaek was commissioned by Kogas in 1986 and can now process 28.24 MMt/y (38.4 Bcm/y) of LNG. The nation's largest terminal, at Incheon, was commissioned by Kogas in 1996 and has expanded to a current capacity of 31.32 MMt/y (42.6 Bcm/y). The last of South Korea's big three terminals was completed in Tongyeong in 2002 and has expanded to a current capacity of 15.0 MMt/y (20.4 Bcm/y). With further expansion limited at these import facilities, Kogas completed a new import terminal at Samcheok in 2014 with initial capacity of 9.45 MMt/y (12.9 Bcm/y), and expansions planned for 2016 and 2017. Kogas is also planning to build a small (1.5 MMt/y, 2.0 Bcm/y) LNG import terminal on Jeju Island.

Privately owned terminals. The Korean gas market is gradually being opened to private firms (fig. 3–3). Steel giant POSCO was the first firm to exploit gas market liberalization measures by building the Kwangyeong terminal at its massive steel complex in the south in 2005. Capacity at the 3.0 MMt/y (4.1 Bcm/y) import facility is shared with SK E&S to facilitate LNG imports from Indonesia. SK E&S has linked up with GS Energy to build the nation's second privately owned LNG import terminal at Boryeong. The two firms will share its 1.5 MMt/y (2.0 Bcm/y) import capacity.

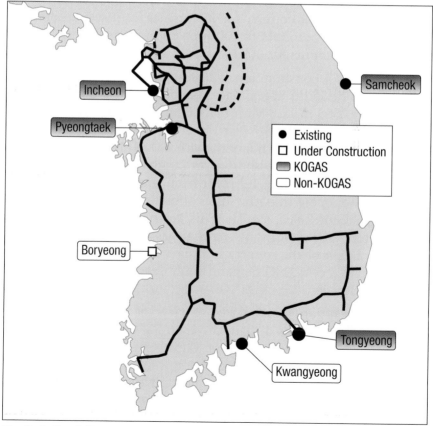

Fig. 3–3. South Korea's LNG import terminals map. (*Source:* Poten & Partners.)

Taiwan. In Taiwan, Chinese Petroleum Corp. (CPC) remains the country's sole LNG importer. The firm owns and operates Taiwan's two LNG import terminals: the 8 MMt/y (10.9 Bcm/y) Yung An facility, located in the island's southwest, and the 3.2 MMt/y (4.4 Bcm/y) Taichung terminal, located mid-island (fig. 3–4). Yung An was commissioned in 1990 and Taichung in 2009. CPC wants to expand the country's receiving capacity by nearly 60% to 20 MMt/y (27.2 Bcm/y) by 2024. In addition to the expansion of its Taichung terminal to 5 MMt/y (6.8 Bcm/y) in 2016, CPC is considering a long-delayed third import facility in the north, capable of handling up to 6 MMt/y (8.2 Bcm/y). It would likely be built near Taipei, adjacent to Taiwan Power (Taipower) Company's 4,348 MW TaTan combined-cycle gas turbine (CCGT) power plant in northern Taoyuan County. The third terminal could be jointly developed with the state-owned power generator, potentially breaking CPC's monopoly on LNG imports for the first time. Taipower consumes more than half of the country's LNG imports for power generation. Power generation accounts for 70% of natural gas use in Taiwan, while city gas accounts for 30%.

Taiwan's LNG consumption, albeit small in scale compared to that of Japan or South Korea, has expanded by 80% from 7.1 MMt (9.7 Bcm) in 2005 to 12.8 MMt (17.4 Bcm) in 2013.[9] Qatar was the top supplier at 6.4 MMt (8.7 Bcm) in 2013, followed by Malaysia at 2.9 MMt (3.9 Bcm) and Indonesia with 2.0 MMt (2.7 Bcm). CPC relies heavily on spot and short-term purchases—which hit nearly 3 MMt in 2013, including 1.3 MMt from the Atlantic Basin (table 3–5). The monopoly importer has contracted future LNG supply from the US Gulf Coast, similar to its Japanese counterparts. LNG accounts for 98% of Taiwan's natural gas supply, which in turn represents 11.6% of the country's primary energy supply. If nuclear power is phased out, LNG demand could reach 20 MMt/y as early as the next decade.

Power generation, particularly in view of the uncertainty of nuclear development, is a main driver in Taiwan of LNG demand growth for the future. The fate of Taipower's controversial 2,700 MW Longmen nuclear plant is uncertain. Taipower hopes to bring it on line after it meets all the requirements mandated by the Atomic Energy Council. But antinuclear sentiment in Taiwan is strong, and the legislature's Economic Committee has passed a nonbinding resolution calling for the complex to be converted to gas-fired generation. Taipower's existing nuclear plants may be shut down when their existing operating licenses expire. Nuclear power accounted for about 18% of Taiwan's electricity mix in 2012, while gas-fired units generated 30% and coal 41%.

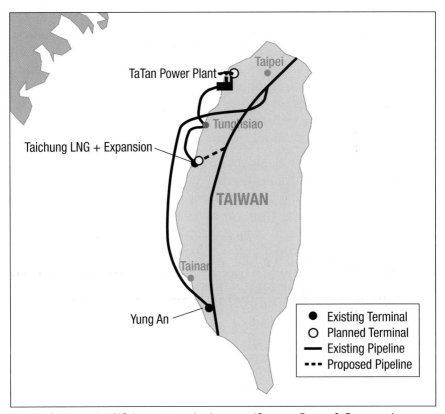

Fig. 3–4. Taiwan's LNG import terminals map. (*Source:* Poten & Partners.)

Table 3–5. Taiwan's LNG imports by supplier (MMt)

	2005	2006	2007	2008	2009	2010	2011	2012	2013
Indonesia	3.6	3.2	3.5	3.1	2.8	2.0	2.0	1.8	2.0
Malaysia	3.0	3.3	3.0	2.7	2.6	2.9	3.3	2.8	2.9
Qatar	–	0.4	0.4	0.8	1.2	2.8	3.9	6.0	6.4
Other	0.5	0.9	1.6	2.5	2.2	3.6	2.8	2.1	1.6
Total Imports	7.1	7.8	8.5	9.1	8.8	11.3	12.0	12.7	12.8

(*Source:* Poten & Partners.)

Asia—New LNG markets

The International Energy Agency (IEA) forecasts that China and India will register the biggest gains in global natural gas use by 2035, with Chinese demand growing by 7% per year to 545 Bcm (52.7 Bcf/d) and the Indian market expanding by nearly 5% per year to 180 Bcm (17.4 Bcf/d).[10] This outlook reflects the large populations of the two countries, their relatively low per capita energy consumption, and their strongly growing economies. In both countries, the heavy dependence on coal, a significant source of energy and pollution in major cities, is a growing problem, particularly in China. So another motivation for increasing gas use is improving air quality, and with demand outstripping indigenous production, both governments see a role for LNG in their supply portfolios.

India imported its first LNG in 2004, and China followed in 2006. In 2013, the two countries imported 31.4 MMt (42.7 Bcm) between them.[11] Their share of global LNG trade climbed to 13%, and should become much larger, as they are investing heavily in LNG import facilities. Their combined imports could reach 80 MMt/y (108.8 Bcm/y) by 2020 and could account for 22% of global LNG trade.[12] Authorities in both countries are struggling with price reform (a prerequisite for a healthy gas business), promoting domestic production (including from shale gas), and developing pipeline imports from neighboring countries.

China. Domestic production has failed to keep up with demand in China, and the country is feeling the pressure to expand available natural gas supplies. Even though China's natural gas production has increased an impressive three and a half times from 32.7 Bcm (3.2 Bcf/d) in 2002 to 116.8 Bcm (11.3 Bcf/d) in 2013,[13] it has failed to keep up with consumption, which surged five and a half times from 29.2 Bcm (2.8 Bcf/d) in 2002 to 162.9 Bcm (15.8 Bcf/d) in 2013. China has made up the difference with imports of pipeline gas and LNG reaching 46.1 Bcm (4.5 Bcf/d) in 2013, a new record. Pipeline gas imports in the same year from central Asia and Myanmar contributed approximately 21.3 Bcm (2.1 Bcf/d), while LNG imports of 24.8 Bcm (18.2 MMt) rounded out the import picture.

With dependence on foreign imports growing, China is pushing forward with increases in city-gate and retail prices to promote indigenous production. Under the National Development and Reform Commission's price reform policy, local governments started raising prices in 2014 to reflect rising costs to local gas companies. Gas to meet incremental demand will be priced at a higher rate than current supplies, which fill existing consumption. At the same time, air quality concerns are prompting the government to adopt tough plans to reduce pollution, with gas emerging as a major part of the solution. According to the IEA in its 2014 mid-term outlook, the power, industrial, and transport sectors will drive overall Chinese gas demand to 315 Bcm (30.5 Bcf/d) in 2019, an increase of 90% over 2013 consumption.[14] The IEA outlook sees half of new gas demand in China being met by domestic resources, most of which are unconventional; Chinese production is set to grow by 65%, from 117 Bcm (11.3 Bcf/d) in 2013 to 193 Bcm (18.7 Bcf/d) in 2019. This growth still yields a supply gap of 122 Bcm/y (11.8 Bcf/d) by 2019 to be met by imports.

China has taken measures to meet its future import requirements (table 3-6). Contracts have been signed with central Asian producers for the supply of 80 Bcm/y (7.7 Bcf/d) of pipeline gas.[15] These include three contracts for Turkmen gas totaling 65 Bcm/y (6.3 Bcf/d) starting in 2010, another for 10 Bcm/y (1.0 Bcf/d) of Uzbek gas from 2012, and one for 5 Bcm/y (0.5 Bcf/d) of Kazak gas from 2013. Another 12 Bcm/y (11.6 Bcf/d) will come from Myanmar. China has signed one deal with Russia to import 38 Bcm/y (3.7 Bcf/d) of gas via the "eastern" route, and another for similar volumes to be supplied via the "western" route. But major production centers still have to be developed and an expensive pipeline built before the Russian gas can be delivered, and it is unlikely that significant Russian volumes will be delivered before 2020.

Table 3-6. China's LNG imports by source (MMt)

	2005	2006	2007	2008	2009	2010	2011	2012	2013
Australia	—	0.7	2.5	2.7	3.5	3.9	3.6	3.6	3.6
Indonesia	—	—	—	—	0.5	1.7	2.0	2.4	2.4
Malaysia	—	—	—	0.0	0.6	1.2	1.6	1.9	2.7
Qatar	—	—	—	—	0.4	1.2	2.3	5.0	6.8
Other	—	—	0.4	0.6	0.4	1.3	2.7	1.9	2.6
Total Imports	—	0.7	2.9	3.4	5.4	9.4	12.2	14.7	18.0

(*Source:* Poten & Partners.)

Since the country built its first regasification terminal, Dapeng LNG in Guangdong province, in 2006, imports have risen dramatically, making China the third largest LNG consumer in the world. More than half of China's total natural gas imports, 18.2 MMt, were in the form of LNG (24.8 Bcm) in 2013. At the end of that year, China had 10 operating terminals with 37.7 MMt/y (51 Bcm/y) of regasification capacity.[16] Another 7 terminals are under construction, which will add 20 MMt/y (27 Bcm/y) of capacity by 2017; and another 16 terminals are either planned or proposed, which may see capacity reach 85 MMt/y (115 Bcm/y) by the beginning of the next decade (fig. 3–5). Expansions of existing terminals could also contribute even more volume.

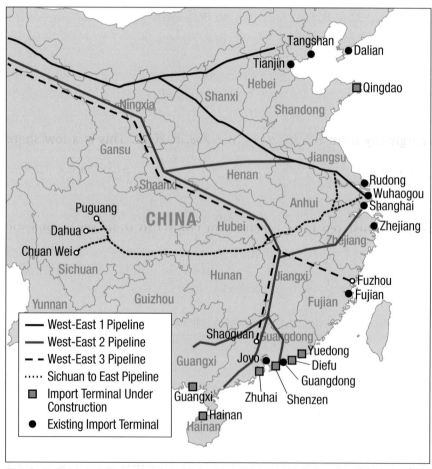

Fig. 3–5. China's LNG import terminals map. (*Source:* Poten & Partners.)

Chinese state oil and gas companies (NOCs) have been required to secure supply prior to gaining government approval to build a regasification terminal. In part to fulfill this requirement, Chinese firms, as of January 2014, had signed purchase agreements for 37.8 MMt/y (51.4 Bcm/y) of LNG from Australia, Malaysia, Indonesia, Papua New Guinea (PNG), Qatar, and aggregators' portfolios. These agreements virtually assure a doubling of LNG imports by 2020. Some contracts are tied to new liquefaction projects primarily located in Australia and Papua New Guinea, and slated to come on line after 2014. In addition to purchasing supply, Chinese companies are investing in significant equity stakes in Australia's liquefaction projects, particularly those involving coalbed methane. In North America, too, China is actively seeking potential LNG opportunities from shale gas by investing in upstream developments and LNG projects in western Canada.

India. India's expanding economy and growing population have led to a 6% per year growth in primary energy consumption, reaching 563.5 MMt of oil equivalent (MTOE) in 2012 compared to 420.1 MTOE in 2007. The share of natural gas in this mix increased marginally from 8% in 2008 to 8.7% in 2012. This is a low share compared to the global average of 24%, primarily because of supply constraints. It also reflects a complex pricing regime, which renders LNG uncompetitive in some prominent demand sectors and hinders investment in domestic upstream gas resources. As of early 2014, the Administered Price Mechanism (APM) in India set a domestic wellhead price of $4.20/MMBtu against term LNG prices of $8.50 MMBtu, and average spot LNG prices of $15 MMBtu. At the end of 2014, the recommendations of the Rangarajan Committee to raise the price of APM gas to $8.40/MMBtu were pending approval by the government. The new policy would promote domestic production even as it helps rationalize pricing in the domestic market. But Indian consumers in the chemical sector have much to lose from rising domestic prices and may resist the change, as might private consumers facing potentially higher electricity prices.

India's gas consumption more than doubled from 27.6 Bcm (2.7 Bcf/d) in 2002 to 63.0 Bcm (6.1 Bcf/d) in 2010, buoyed by Krishna-Godovari D6 block (KG-D6) production in offshore East Indian waters. Natural gas consumption subsequently declined to 51.4 Bcm (5.0 Bcf/d) in 2013, largely owing to a sharp decline in KG-D6 production. KG-D6 production, which started in 2009, peaked at 69 MMm³/d (25 Bcm/y at an annual rate) in March 2010, slid to just 11.7 MMm³/d (4.3 Bcm/y) in 2013, and was at 13.6 MMm³/d (5.0 Bcm/y) in early 2014. This reduction contributed to a sharp decline in Indian natural gas production from a peak of 50.8 Bcm (139 MMm³/d) in 2010 to just 33.7 Bcm (92 MMm³/d) in 2013.[17]

Priority allocation of KG-D6 natural gas was given to the fertilizer and power sectors at the APM domestic price of $4.20/MMBtu. These two end-use sectors account for more than 60% of Indian gas demand. When KG-D6 gas production collapsed, the allocation to the power sector ended. In some instances, power generators switched to coal-fired generation. In others, power shortages resulted—generators could not competitively produce electricity using higher-priced LNG imports, since power prices are also regulated. Approximately 7,000 MW of gas-fired power generation capacity was off-line in 2013, and the country suffered from widespread power blackouts. The 12.5 Bcm/y drop in natural gas use from 2010 to 2013 largely reflects lower use in power generation. The new pricing regime promises to increase domestic gas supply, which can be used to bring these power plants on stream, but this will take some time. Still, demand for natural gas for gas-based power generation is highly price-sensitive, and will remain so until power price reforms are also implemented.

Unlike China, India has been unable to develop any pipeline imports. Without pipeline gas import capacity, any shortage of gas supply can only be filled by LNG imports, which tripled from 4.1 MMt (5.6 Bcm) in 2005 to 13.9 MMt (18.9 Bcm) in 2012. The following year, imports declined to 13.2 MMt (18 Bcm) as power generators no longer had inexpensive domestic gas supply to blend with more expensive LNG.[18] Of the LNG imported in 2013, 11.1 MMt (15.1Bcm) were imported from Qatar at an average price of $12.50/MMBtu. Most of the Qatari volumes were delivered through the 9.9 MMt/y (15.5 Bcm/y) Dahej terminal in the state of Gujarat (table 3–7).

By the end of 2013, India had four LNG import terminals with 22.1 MMt/y (30.1 Bcm/y) of regasification capacity on the country's west coast.[19] By the end of this decade, the country's import capacity promises to more than double to nearly 50 MMt/y (68 Bcm/y) with facilities serving both coasts (fig. 3–6). Numerous import facilities are now planned for the east coast, which is suffering from natural gas shortages after the collapse of KG-D6 production. Poten & Partners expects LNG imports to double to 28 MMt/y by 2020. Indian companies are diversifying their LNG dependence from Qatar by lining up term supplies from Australia (Gorgon LNG), LNG aggregators (BP and Gazprom), and two US liquefaction projects (Sabine Pass and Cove Point) at Henry Hub–linked prices.

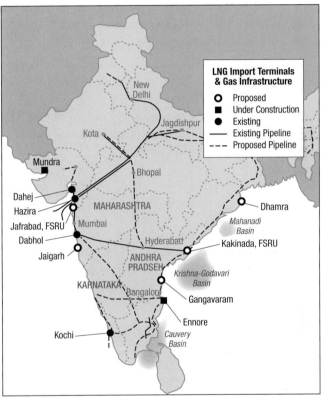

Fig. 3–6. India's LNG import terminals map. (*Source:* Poten & Partners.)

Table 3–7. India's LNG imports by source (MMt)

	2005	2006	2007	2008	2009	2010	2011	2012	2013
Egypt	–	0.4	–	0.2	0.1	0.1	0.7	0.6	0.4
Nigeria	–	–	0.5	0.3	0.2	0.2	0.6	1.2	0.9
Qatar	3.9	5.2	6.9	6.3	6.6	7.7	9.8	11.8	11.1
Other	0.2	0.5	1.1	1.8	2.5	0.8	1.0	0.7	0.9
Total Imports	4.1	6.0	8.4	8.6	9.4	8.8	12.1	14.3	13.2

(*Source:* Poten & Partners.)

Southeast Asia

When considered individually, emerging southeast Asian LNG markets (Indonesia, Malaysia, the Philippines, Singapore, Thailand, and Vietnam) are modest in size. But in aggregate they represent a significant new market. By the end of 2013, the region had 10.8 MMt/y (14.7 Bcm/y) of LNG terminal capacity, which could expand to 36 MMt/y (50 Bcm/y) by 2020 based on projects under construction and planned, as seen in figure 3–7.[20]

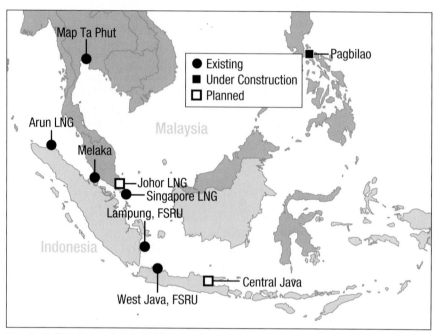

Fig. 3–7. Southeast Asia's LNG import terminals map. (*Source:* Poten & Partners.)

The region, long a source of LNG exports from Indonesia and Malaysia, is now emerging as a major LNG market, with Singapore declaring ambitions to develop an LNG trading hub (fig. 3–8). LNG imports were first introduced into the region in 2011, reached 4.2 MMt (5.7 Bcm) in 2013, and could triple by 2020.[21] Intraregional pipelines have been the main form of gas transport, but the archipelagic nature of the region makes it suitable for LNG trades, and for floating storage and regasification units (FSRUs) in particular.

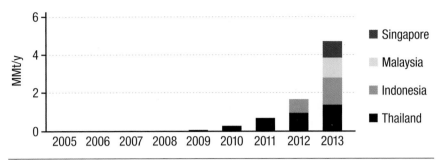

Fig. 3–8. Southeast Asia's LNG imports by country. (*Source:* Poten & Partners.)

Thailand

Indigenous production supplemented by pipeline imports from Myanmar supplied all of Thailand's natural gas needs until 2011. In that year, the country became the first Southeast Asian nation to import LNG. There have been no major gas resource discoveries in recent years, and the remaining reserve life at current production rates is just 15 years, even as gas demand grows rapidly. Indigenous natural gas production has doubled from 20.5 Bcm (2.0 Bcf/d) in 2002 to 41.8 Bcm (4.0 Bcf/d) in 2013, but has failed to keep pace with consumption, which increased from 26.9 Bcm/y (2.6 Bcf/d) to 52.2 Bcm/y (5.1 Bcf/d) during this time frame.[22] The balance was made up by pipeline imports from Myanmar, which totaled 8.5 Bcm (0.8 Bcf/d) in 2013, and LNG imports of 1.9 Bcm (1.4 MMt).

With dwindling prospects for increasing domestic natural gas production, and competition from China for Myanmar natural gas, state-owned PTT, formerly known as the Petroleum Authority of Thailand, decided to diversify its gas supply by building an LNG import terminal at Map Ta Phut. The 5 MMt/y (6.8 Bcm/y) onshore regasification plant received its first cargo in May 2011, and deliveries reached 1.4 MMt (1.9 Bcm) in 2013—including 1.0 MMt (1.3 Bcm) from Qatar.[23] PTT is advancing plans to double the size of Map Ta Phut by 2017, while studying the possibility of establishing a second terminal.

Indonesia

For Indonesia, a top LNG exporter, plans include at least four regasification plants. The country's 3.0 MMt/y (4.1 Bcm/y) West

Java FSRU received its commissioning cargo in May 2012. The terminal has been allocated supplies from the country's liquefaction complexes. In 2013, the facility received 1.4 MMt (1.9 Bcm) of LNG.[24] The country's second regasification project is the Arun receiving terminal in Aceh. Pertamina has converted the existing LNG export complex into a 3.0 MMt/y (4.1 Bcm/y) import facility with completion at the end of 2014. Next in line is PGN's 2 MMt/y (2.7 Bcm/y) Lampung FSRU, which is due to start up in 2015, followed by Pertamina's 3.0 MMt/y (4.1 Bcm/y) Central Java FSRU in 2016. The country has ample gas resources, but they are not located near demand centers, which are scattered around the archipelago. In 2013, the country produced 63.4 Bcm (6.1 Bcf/d) of gas against consumption of 34.6 Bcm (3.3 Bcf/d).[25] LNG exports in 2013 totaled 18 MMt (24.4 Bcm).[26]

Singapore

Singapore had relied exclusively on imports of natural gas by pipeline from Indonesia and Malaysia, but in the face of potential shortfalls in these deliveries, has diversified its supply sources to include LNG. The market, primarily for power generation, has increased two and a half times from 3.6 Bcm (0.3 Bcf/d) in 2002 to 9.5 Bcm (0.9 Bcf/d) in 2013.[27]

The 6.0 MMt/y (8.1 Bcm/y) LNG import terminal on Jurong Island received its first cargo in March 2013. BG has exclusive supply rights at the terminal, providing cargoes from Trinidad and Equatorial Guinea. Hoping to capitalize on the growing LNG trading business, Singapore LNG was designed for both unloading and reloading LNG cargoes. The government is supporting the LNG trading initiative hoping to replicate Singapore's role as a regional oil product center. Plans are being advanced to more than double the size of the Jurong terminal by 2018, and the government is considering developing a second terminal. Many LNG firms have established trading operations in Singapore.

Malaysia

Like Indonesia, Malaysia is a major LNG exporter, with sizable natural gas reserves. In 2013, indigenous gas production totaled 62.1 Bcm (6.0 Bcf/d), while the country consumed 30.6 Bcm (0.3 Bcf/d).[28]

Also similar to Indonesia, Malaysia's demand centers are far from its gas resources on Kalmantan (the Indonesian section of Borneo) and the country has a large LNG export business that it wishes to maintain. LNG exports totaled 34.0 Bcm (24.9 MMt) in 2013.[29]

Malaysia's first LNG import facility, comprising a floating storage unit and a jetty regasification facility, received its first LNG cargo in June 2013. The 3.8 MMt/y (5.1 Bcm/y) Malacca LNG terminal received 1.0 MMt (1.3 Bcm) of LNG in 2013.[30] Petronas is developing two more LNG import facilities: the 0.7 MMt/y (1.0 Bcm/y) Sabah LNG project scheduled for commissioning in 2015 and the 3.8 MMt/y (5.1 Bcm/y) Johor LNG plant to be commissioned by 2018.

Philippines

There are several LNG import projects under evaluation in the Philippines. Shell started engineering work on a 4 MMt/y (5.4 Bcm/y) FSRU to be located adjacent to its Tabangao refinery in Batangas in 2014 and hopes to start operations by 2017. Other regasification projects are being proposed by utility First Generation Corporation, among others, but so far have failed to gain traction. The Philippines gas market is rather small at 3.5 Bcm/y (0.3 Bcf/d) and is scattered throughout the archipelago.

Vietnam

Vietnam envisions having two small LNG import facilities up and running by 2020. The country's gas demand has increased from just 2.4 Bcm (0.2 Bcf/d) in 2002 to 8.8 Bcm (0.85 Bcf/d) in 2013, all supplied from indigenous production. The first LNG import facility, the 1 MMt/y (1.3 Bcm/y) Thi Vai terminal in Ba Ria-Vung Tau Province is supposed to be on line by late 2015 and the second, the 1.8 MMt/y (2.4 Bcm/y) Son My terminal in Binh Tuan, by 2019, with expansions depending on demand. Problems with arranging financing could introduce significant delays in this schedule.

Arabian Gulf and Pakistan

The countries surrounding the Arabian Gulf are rich in hydro-carbons, which makes them appear to be an unlikely LNG market. According to BP, their proven natural gas reserves totaled 80.5 Tcm (2,835 Tcf) at year-end 2013, accounting for 43% of the global total.[31] But these reserves are not evenly distributed across the region and there is a very limited intraregional gas pipeline infrastructure. Nearly three-quarters of these reserves are based on the same giant nonassociated gas field in Iranian waters (South Pars) and Qatar (North Field), leaving some regional nations short of gas supply. The region, led by Qatar, has emerged as a premier LNG export center as well as a very large gas market (fig. 3–9).

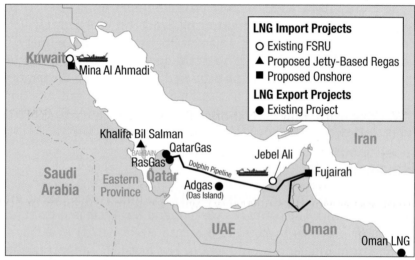

Fig. 3–9. Arabian Gulf LNG import terminals map. (*Source:* Poten & Partners.)

Regional consumption doubled from 218 Bcm (21 Bcf/d) in 2002 to 428 Bcm (41 Bcf/d) in 2013, with the consumers largely being gas-based industry, desalination, and power-generating plants.[32] Natural gas is typically provided to domestic markets at extremely low prices, which has encouraged consumption but discouraged investment in natural gas resource development. Because of the uneven distribution of natural gas resources, and the limited pipeline infrastructure, significant LNG import opportunities have emerged. Import terminals utilizing FSRUs are now operating in Dubai and

Kuwait, and an onshore facility is being advanced in Fujairah by Abu Dhabi sponsors (fig. 3–10). Bahrain is considering importing LNG, while Pakistan, just outside of the Gulf region, has already implemented its own import terminal. LNG imports into regional outlets were 3.3 MMt (4.5 Bcm) in 2013, and Poten & Partners projects imports could triple by the end of the decade.[33]

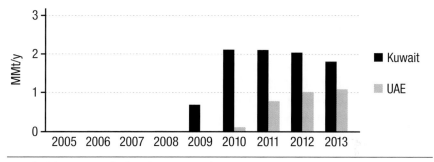

Fig. 3–10. Arabian Gulf LNG imports. (*Source:* Poten & Partners.)

Kuwait

Working with Kuwait National Petroleum Company (KNPC), Excelerate Energy designed and constructed the region's first LNG import terminal, the Mina Al-Ahmadi GasPort, approximately 20 miles south of Kuwait City. The terminal was commissioned in August 2009. Mina Al-Ahmadi delivers regasified LNG at a baseload rate of 500 MMcf/d (3.6 MMt/y), with a peaking capability of 600 MMcf/d (4.3 MMt/y). It was developed in just 18 months. The capital cost for the project, even with extensive refurbishments and enhancements made to existing jetty facilities, was significantly less than that of a land-based facility of similar capacity. It also had the added benefit of seasonal flexibility. It was intended to supply gas during the summer months to meet peak power demand. In 2013, the import season extended from April through October, during which time 1.8 MMt (2.5 Bcm) were delivered—including 1.5 MMt (2.0 Bcm) from Qatar.[34]

Mina Al-Ahmadi was originally intended to be an interim solution to bridge the gap between Kuwait's existing natural gas needs and the development of domestic gas, but LNG seems to have become a permanent part of the country's natural gas supply mix.

In March 2014, KNPC awarded a pre-front-end engineering design and a front-end engineering design contract to Foster Wheeler for a new onshore LNG import terminal with a design sendout capacity of around 1.5 Bcf/d (11.4 MMt/y). The design will allow for expansion(s) up to 3.0 Bcf/d (22.8 MMt/y). KNPC plans to start commercial operations in 2020.

Dubai

Dubai Supply Authority (DUSUP) completed the construction of its LNG terminal in Jebel Ali port at the end of 2010, with a capacity of nearly 500 MMcf/d (3.8 MMt/y) of LNG via a FSRU, the converted *Golar Freeze*. DUSUP buys LNG both via a long-term agreement and by purchasing spot cargoes to support seasonal swings in demand. The combination of DUSUP's FSRU terminal and the Margham gas storage facility provides import flexibility. In 2013, DUSUP imported 1.1 MMt (1.5 Bcm) of LNG between May and October to meet peak power generation demand. Of this, 0.9 MMt (1.2 Bcm) was delivered from Qatar.[35]

Fujairah

A gas supply shortfall in Abu Dhabi has spurred the development of an LNG import project in Fujairah. State enterprises Mubadala and International Petroleum Investment Corporation had been fast-tracking an FSRU in Fujairah. In late 2013, the FSRU scheme was overruled by the Emirate's Executive Council, and Emirates LNG is now pursuing a larger 1.2 Bcf/d (9.2 MMt/y) onshore terminal rather than a phased development. This will likely delay first imports beyond 2017, pending Emirates LNG lining up supplies at price and delivery terms satisfactory to state-owned power generator ADWEC. Also, ADWEC only wants supplies to meet peak power demand, necessitating large storage capacity which, in turn, would boost the cost of the terminal.

Western Europe

Although western European nations individually represent modest-sized gas markets, in aggregate, the 28 countries comprising the European Union (EU) are approximately two-thirds the size of the US market. Six countries account for about three-quarters of this total. In 2013, Germany, the largest, consumed 83.6 Bcm (3.0 Tcf), followed by the United Kingdom at 73.1 Bcm (2.6 Tcf), Italy at 64.8 Bcm (2.3 Tcf), France at 42.8 Bcm (1.5 Tcf), the Netherlands at 37.1 Bcm (1.3 Tcf), and Spain at 29.0 Bcm (1.0 Tcf).[36] According to the European natural gas association, total natural gas consumption in the EU fell 1.4% in 2013, following 10% and 2% declines in 2011 and 2012 respectively.[37] In 2013, gas demand remained under pressure as subsidized renewables and coal (including cheap American coal displaced by shale gas) replaced gas in the power sector. At the same time, continuing economic recession—evidenced by a 0.5% decrease in EU28 average industrial production in 2013 compared with 2012—again affected gas demand from industry.

On the gas supply side, indigenous production registered a slight decrease (1%) from 2012 to 156 Bcm (10.1 Tcf) in 2013, or 33% of the total net supplies.[38] Import dependence remains high as UK North Sea natural gas production declines. Russia met 27% of EU28 natural gas demand in 2013, followed by Norway at 23% and Algeria with 8%. The rest, about 9%, was supplied as LNG.

Historical perspective

The discovery of the vast Groningen gas field in the Netherlands in 1959 provided the foundation for international gas trade in continental Europe. The field was the basis for the early development of continental Europe's gas pipeline grid and gas market. Supplemented by smaller offshore finds, annual exports from the Netherlands grew from only 10 Bcm/y (0.4 Tcf/y) in the early 1970s to 68.7 Bcm (2.4 Tcf) in 2013, according to BP.[39]

The oil crisis of 1973 provided a large boost to the European gas business. Indigenous and regional gas was viewed as a more secure energy source than Middle Eastern oil. By the late 1970s, the development of markets, infrastructure, and higher prices encouraged new supplies from the Union of Soviet Socialist Republics (USSR), Algeria, and Norway. Gas-buying consortia comprising national monopolies typically contracted for gas supplies at oil-linked prices. Security and supply diversity were also key considerations. These powerful consortia provided critical credit support and volume assurances that underpinned the construction of international pipelines to deliver gas into Europe from the USSR, Algeria, and Norway.

In the late 1990s, the EU enacted legislation to end gas and power monopolies and to bring competition to end-user gas and power markets. This liberalization was already well advanced in the United Kingdom, which had developed separately from the continental pipeline grid. It further evolved after the privatization and subsequent breakup of the old British Gas. Liberalization of gas and power businesses is now being implemented in European markets as well, even though the entrenched monopolies (state and private) have battled to retain their market control as long as possible.

The EU is intent on establishing a unified energy market. National gas and power monopolies have been privatized and split up. At the same time, national regulatory authorities have implemented third-party access to supply and transmission infrastructure. Incumbent electric and gas utility monopolies, already competing in other national markets, are now facing competition from new entrants on their home turf. Market structures vary greatly by state, depending on the maturity and physical characteristics of each market, the level of competition, the number of established players and new entrants, and the way nations implement EU directives. The European gas market has undergone major changes, gas trading hubs have been established, and oil linkages in term contracts are being challenged by buyers who value the apparent competitiveness of hub-based prices.

In early 2006, concerns about the reliability of Russian supplies became an issue when Gazprom temporarily cut off supplies to Ukraine over a pricing dispute. The cutoff in turn affected other European buyers who depended on transit routes through Ukraine to deliver their Russian supplies. In 2014, security of supply and geopolitical considerations over EU dependence on Russian gas have again moved to the fore with the annexation of Crimea by Russia and Kremlin support for pro-Russian separatists in eastern Ukraine.

The ongoing crisis in Ukraine could prove to be a catalyst for European leaders to reinforce energy security and to decrease Europe's dependence on Russian gas. Not only did Russia provide 162.4 Bcm (5.7 Tcf) of pipeline gas to western, central, and eastern Europe, and Turkey in 2013, Russia supplied another 48.9 Bcm (1.7 Tcf) to former Soviet states including Ukraine. Ukraine is also the major conduit for natural gas to western Europe. Approximately

40% of Russia's natural gas exports to western Europe are delivered via pipelines that traverse Ukraine.

The Russian-sponsored crisis in eastern Ukraine, and the annexation of Crimea has encouraged politicians and analysts to point to LNG as an alternative to Russian gas in the EU energy mix. LNG had already made significant inroads before the current crisis. LNG imports into the EU trebled from 21.5 MMt (29.3 Bcm) in 2000 to a peak of 60.1 MMt (81.7 Bcm) in 2011. But LNG imports into Europe collapsed by half to 33.9 MMt (46.1Bcm) in 2013 because of the decline in European gas demand and the strength of demand in Asia from buyers willing to pay premium prices (fig. 3–11).

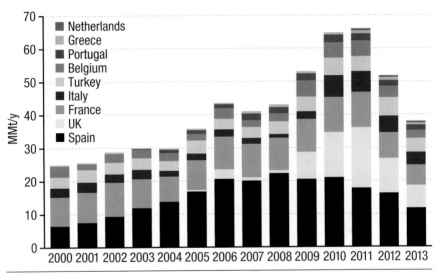

Fig. 3–11. European LNG imports by country. (*Source:* Poten & Partners.)

There is ample terminal capacity to expand LNG imports into the EU. There is currently 185 Bcm/y (169 MMt/y) of regasification capacity. Regasification plants either under construction or sanctioned by project developers will add another 46 Bcm/y (34 MMt/y) to import capacity. Terminal utilization averaged only 25% in 2013, compared to a high of 60% in 2006 (fig. 3–12).

The challenge is inelastic supply and the length of time needed to build a new LNG export plant. The high capital intensity and near-zero marginal operating cost of liquefaction plants lend themselves to maximum output at all times. Most of this LNG is already selling at higher than European prices, primarily in Asia. This situation has led to the curtailment of deliveries into Europe, and it is unclear whether the Europeans are ready to pay the price premium needed to increase LNG imports and replace Russian gas.

Fig. 3–12. European LNG terminal map. (*Source:* Poten & Partners.)

Some relief may be forthcoming, as new Asian supplies, primarily from Papua New Guinea and Australia, appear to be running into flattening demand in the Asian markets in 2015. Either some of this LNG or other supplies displaced by it may make their way to Europe. Politically, US LNG exports to Europe would appear to offer an attractive alternative, but this is unrealistic in the near term.[40] First, there is no US LNG export capability in place today, and there will not be any until late in 2015, regardless of US government policy. The first US Gulf Coast export project is under construction, and more will follow, but they are unlikely to produce significant export volumes before 2020, and much of this has already been earmarked for eastern buyers.

Second, suppliers are not in the business of facilitating government policy. They are commercial entities that sell their LNG to the highest bidder. Even if more of the proposed projects are approved, they will not be built without firm and bankable commitments to purchase their output over a long period. European buyers appear less ready to make these commitments than Asian ones are. While US-produced LNG appears to be a winner in Asia, its competitiveness in Europe is less clear. The cost of liquefying, transporting, and regasifying US natural gas for injection into the European grid would add about $5/MMBtu to the US benchmark price at Henry Hub in Louisiana. At best, US LNG can barely compete with Russian gas in Europe, whose price will eventually reflect lower crude oil prices being experienced in early 2015. Furthermore, replacing Russian gas with US supply may prove to be complicated. While there have been vast improvements in European pipeline grid interconnectivity, it is unclear how quickly that grid would be able to accommodate the new flow patterns that greatly expanded LNG deliveries would require. The problem would be particularly acute in eastern Europe, where dependence on Russian gas is the highest.

Offsetting this potential problem, however, are indications that some of the potential US exports might initially find their way to Europe, as their intended markets might not be able to absorb them in the initial delivery period. Some prospective exporters from Sabine Pass, Cove Point, and Cameron have indicated that they are considering European sales, and others may follow.

United Kingdom

The United Kingdom initiated the world's first commercial LNG imports when a cargo of Algerian LNG was delivered into Canvey Island in 1964. By the 1970s, the discovery of large gas resources in the Southern Basin of the North Sea ended this trade and the United Kingdom's only terminal was mothballed and eventually dismantled. In recent decades, UK policy has strongly encouraged the development of North Sea oil and gas to stimulate the nation's economy and to reap tax revenues. Over time, UK supply sources expanded to include pipeline gas from Norway. With ample indigenous and Norwegian gas supplies, prices fell, and government policy encouraged the use of gas in power generation. The original "dash-for-gas" by power generators was first witnessed in the United Kingdom. The United Kingdom was also one of the first nations to open up its gas and power businesses to competition, and it arguably has seen a successful implementation of competition in end-use markets. According to IEA estimates, by 2010 natural gas fueled approximately 56% of the country's electricity generation—up from 39% in 2000 and only 2% a decade earlier.[41]

The United Kingdom's isolation from the continental grid ended in 1998, when the Interconnector pipeline linking the United Kingdom's Bacton terminal to Zeebrugge, Belgium, was completed. The Interconnector was initially conceived to flow surplus North Sea gas to the Continent. It was designed to transport 20 Bcm/y (1.9 Bcf/d) in forward flow from the United Kingdom to Belgium, and 8.5 Bcm/y (0.82 Bcf/d) in reverse flow. As UK North Sea production fell, compression was added to increase the reverse flow on the Interconnector to 16.5 Bcm (1.59 Bcf/d). By 2006, a further expansion raised UK import capacity to 23.5 Bcm/y (2.3 Bcf/d). Another interconnector pipeline, connecting Balgzand in the Netherlands to Bacton, with an import capacity of 16 Bcm/y (1.5 Bcf/d), started up at the end of 2006.

After decades of self-sufficiency, UK energy authorities realized that falling North Sea production and steadily increasing gas demand would transform the country into a net importer. North Sea production had declined from 103.6 Bcm (10 Bcf/d) in 2002 to 80 Bcm (7.7 Bcf/d) by 2006. It has since fallen precipitously to just 39 Bcm (3.8 Bcf/d) in 2013.[42] While the interconnector pipelines can be used to supply Dutch and Norwegian gas to the United Kingdom, they could provide a link for Russian gas as well. Other import routes have also been developed—notably the Ormen Lange field, which commenced deliveries in 2007 of 20 Bcm/y (1.93 Bcf/d) of gas to British consumers via the 1,166-kilometer Langeled pipeline to the Easington terminal in eastern England. The giant Ormen Lange field has gas reserves of approximately 400 Bcm (14 Tcf).[43]

LNG, reintroduced into the UK energy mix in 2005, provided an additional way of bridging the United Kingdom's growing supply gap. Imports surged to 18.2 MMt (25 Bcm) in 2011, largely supplied from Qatar, before dropping to just 6.9 MMt (9 Bcm) in 2013 as suppliers diverted cargoes to higher value markets.[44] At the same time, UK demand was curtailed by a bad economy (table 3–8). After peaking at 99.2 Bcm (9.6 Bcf/d) in 2010, the gas market declined by 22% to 77.7 Bcm (7.5 Bcf/d) in 2013,[45] but it appears to have flattened as the economy has started to recover. The downturn has resulted in underutilization of regasification facilities, operating at only 18.5% of annual capacity in 2013. There are currently four

LNG import terminals in the United Kingdom, with a regasification capacity of 37.3 MMt/y (51 Bcm/y).[46]

Table 3–8. UK LNG imports by source (MMt)

	2005	2006	2007	2008	2009	2010	2011	2012	2013
Qatar	–	0.1	0.2	0.1	4.5	10.3	15.7	10.3	6.2
Other	0.4	2.7	0.8	0.8	3.6	3.4	2.7	0.3	0.6
Total Imports	0.4	2.9	1.0	0.9	8.1	13.7	18.3	10.6	6.9

(*Source:* Poten & Partners.)

Grain LNG, Isle of Grain. National Grid Transco, the pipeline grid's owner and operator, finished converting its peak shaving plant on the Isle of Grain in Kent into an LNG import terminal, called Grain LNG, in 2005. Initial capacity was 4.3 Bcm/y (3.2 MMt/y), but the terminal was expanded to 20.4 Bcm /y (15.0 MMt/y) in two phases: the first phase, completed in 2008, added 9.0 Bcm/y (6.6 MMt/y) capacity; the second expansion, completed in 2010, added another 7.1 Bcm/y (5.3 MMt/y). A subsequent expansion proposal was shelved due to lack of interest.

Dragon LNG and South Hook LNG, Milford Haven, Wales. Two separate groups built LNG terminals a few miles apart in Milford Haven, Wales: Dragon LNG (BG [50%] and Petronas [50%]), and South Hook LNG (a joint venture between Qatar Petroleum [67.50%], ExxonMobil [24.15%], and Total [8.35%]). Both terminals were commissioned in 2009 with initial annual capacities of 5.7 Bcm/y (4.2 MMt/y) at Dragon and 10.4 Bcm/y (7.65 MMt/y) at South Hook. In 2010, the capacity at South Hook was doubled to 20.8 Bcm/y (15.3 MMt/y), adequate to accommodate deliveries from two Qatari megatrains.

There were a number of other proposals to build LNG import facilities in the United Kingdom Of these only one has material-ized—Excelerate Energy's 4.4 Bcm/y (3.2 MMt/y) Teesside GasPort project, a dockside regasification facility, which has remained virtu-ally unused.

Iberia

During the 1990s, Spain was western Europe's fastest growing market for natural gas, virtually all of it imported. Consumption increased nearly threefold in that decade, following a similar rise in the 1980s. The increase was driven by industrial and power-generation demand and facilitated by a major expansion of the national pipeline system and import infrastructure. But gas consumption peaked at 38.6 Bcm (3.7 Bcf/d) in 2008, when the Spanish market was ranked third in size in western Europe, behind Germany and France and just ahead of the United Kingdom. In 2012, consumption fell by 18.9% to 31.3 Bcm (3.0 Bcf/d), and it fell another 7.2% to 29.0 Bcm (2.8 Bcf/d) in 2013[47] according to BP, as the economy faced a persistent recession.

Spain has only limited connectivity with the western European gas grid, although this is being expanded. With tiny domestic gas resources, the country is heavily dependent on imports from Algeria by pipeline, and LNG from a number of producers around the globe. In 2013, LNG accounted for approximately half of Spain's overall imports (30.2 Bcm).[48]

Algeria was the main supplier of natural gas into Spain, contributing 50% of imported natural gas via a mix of pipeline natural gas and LNG. Spain is connected to Algeria via two pipelines: the Maghreb-Europe (GME) pipeline (capacity of nearly 13 Bcm/y [1.2 Bcf/d], most of it targeted for Spain with small volumes delivered to Portugal) and the Medgaz pipeline (capacity of 8 Bcm/y [0.8 Bcf/d]). In addition to Algeria, Nigeria and Qatar are major LNG suppliers.

Originally, Gas Natural (GN) dominated the Spanish gas market, not only as a supplier but also via its 91% ownership of the nation's gas infrastructure owner and operator, Enagás. However, GN was forced in 2002 to reduce its ownership in Enagás in keeping with European liberalization laws, and it now owns only 5% of the company. Taking into account GN's loosening grip over gas sales and the growing influence of newcomers such as power generation utility companies Union Fenosa (which merged with GN in 2008), Endesa, and Iberdrola (not to mention international firms such as BP, ENI, Shell, and Total), the Spanish gas market has seen dramatic changes since 2000.

Enagás provides nondiscriminatory third-party access to its infrastructure, including the nation's pipeline grid and Enagás's four LNG import terminals.[49] Two more terminals are fully owned by third parties, bringing total capacity to 49.2 MMt/y (66.9 Bcm/y). Meanwhile, LNG imports, which peaked at 22.2 MMt (30.1 Bcm) in 2008, collapsed to just 11.6 MMt (15.8 Bcm) in 2013, or by nearly half. Qatar (2.7 MMt or 3.7 Bcm) was the top supplier in 2013, followed by Algeria and Nigeria (2.5 MMt or 3.4 Bcm each) as seen in table 3–9. Terminal utilization dropped to just 23% in 2013. To benefit from ample LNG supplies under contract and unused terminal capacity, an active reexport trade has been developed. In 2013, 2 MMt (2.7 Bcm) of LNG were reexported, yielding a net import figure of just 9.6 MMt (13.1 Bcm).

Table 3–9. Spain LNG imports by source (MMt)

	2005	2006	2007	2008	2009	2010	2011	2012	2013
Algeria	4.1	2.6	3.5	3.8	4.0	3.7	2.9	2.9	2.5
Nigeria	3.7	5.8	6.8	5.8	3.8	6.0	5.0	3.9	2.5
Qatar	3.5	4.7	3.9	3.7	3.4	4.0	3.5	2.9	2.7
Trinidad	0.4	2.8	1.7	3.4	2.9	2.3	1.8	1.8	1.5
Other	5.0	4.8	4.1	5.5	6.4	4.9	4.4	4.5	2.4
Total Imports	16.8	20.6	20.0	22.2	20.4	20.9	17.7	16.0	11.6

(*Source:* Poten & Partners.)

Enagás terminals. Spain's older LNG facilities at Barcelona (12.45 MMt/y or 16.9 Bcm/y), Huelva (8.55 MMt/y or 11.6 Bcm/y), and Cartegena (8.55 MMt/y or 11.6 Bcm/y) are all owned by Enagás. The firm mothballed a fourth terminal at El Musel in Gijon (5.85 MMt/y or 8.0 Bcm/y) in 2012 prior to commissioning because of lack of demand.

Third-party terminals. The Bahía de Bizkaia Gas consortium commissioned its 5.1 MMt/y (6.9 Bcm/y) terminal near Bilbao in northern Spain in 2003, while the Regasificación de Sagunto (SAGGAS) joint venture commissioned its 6.4 MMt/y (8.7 Bcm/y) facility near Valencia in southern Spain in March 2006. Enagás has taken a significant shareholding in Bahía de Bizkaia, buying out some of the original partners. A third greenfield terminal, known as Regasificadora del Noroeste (REGANOSA), started up in 2007. This 2.6 MMt/y (3.6 Bcm/y) facility is located at Murgados, in northwest Spain.

Sines, Portugal. Elsewhere on the Iberian Peninsula, Portugal's only receiving terminal began operating at Sines in 2004 with a capacity of 3.75 MMt/y (5.1 Bcm/y). It was expanded to 4.77 MMt/y (6.5 Bcm/y) in 2012. LNG imports, primarily from Nigeria, peaked at 2.3 MMt (3.1 Bcm) in 2010, but declined to 1.7 MMt (2.3 Bcm) in 2013, of which 250,000 tons were re-exported.

France

With limited use of natural gas in the power sector, dominated by nuclear generation, natural gas consumption has not experienced the large declines experienced elsewhere in Europe. Still, natural gas consumption has flattened over the past decade owing to improvements in energy efficiency. In 2002, consumption reached 40.7 Bcm (3.9 Bcf/d), and has been within 5% of this level ever since. In 2013, consumption totaled 42.5 Bcm (4.1 Bcf/d).[50] With limited domestic production and a ban on fracking, France will remain heavily dependent on imports by pipeline as well as LNG for its gas supply. In 2013, France imported 30.5 Bcm (3.0 Bcf/d) in pipeline gas, including 16.5 Bcm (1.6 Bcf/d) from Norway, 6.5 Bcm (0.6 Bcf/d) from the Netherlands, and 8.1 Bcm (0.8 Bcf/d) from Russia. LNG imports contributed 8.7 Bcm (6.1 MMt) to French gas supplies that year, including 5.3 Bcm (3.9 MMt) from Algeria (table 3–10).

Table 3–10. France LNG imports by source (MMt)

	2005	2006	2007	2008	2009	2010	2011	2012	2013
Algeria	5.8	6.3	6.0	5.8	6.0	4.9	4.5	3.5	3.9
Nigeria	2.9	3.0	3.0	2.7	1.8	2.8	2.8	2.2	0.8
Qatar	0.0	0.3	0.0	0.0	0.2	1.7	2.3	1.3	1.3
Other	0.4	0.3	1.1	1.3	2.0	1.2	1.2	0.8	0.1
Total Imports	9.1	9.9	10.1	9.8	10.0	10.6	10.7	7.8	6.1

(*Source:* Poten & Partners.)

Even as French natural gas demand has held reasonably steady, LNG imports have declined in recent years. After peaking at 10.7

MMt (14.6 Bcm) in 2011, they declined by 31% in 2012 to 7.4 MMt (10.6 Bcm), and in 2013, they went down another 16% to 6.4 MMt (8.4 Bcm).[51] Meanwhile, regasification capacity has increased from 12.3 MMt/y (16.7 Bcm/y) in 2005 to 18.0 MMt (24.5 Bcm/y) in 2013, with average capacity utilization at the nation's import terminals dropping to 34%. In 2013, 364,000 tons of LNG were reexported from French terminals. There are two LNG import terminals on the Mediterranean coast and a third on the Atlantic. While these three terminals, owned and operated by GDF Suez, provide ample regasification capacity to facilitate LNG imports into France, a fourth one, sponsored by EDF, Fluxys, and Total, is under construction on the country's northern Atlantic coast at Dunkirk.

GDF Suez terminals (Fos Tonkin, Montoire de Bretagne, Fos Max). The dominant French gas company owns and operates the country's three existing LNG import terminals with a combined capacity of 18.0 MMt/y (24.5 Bcm/y). The oldest, Fos Tonkin, was commissioned at Fos, near Marseilles, in 1972, to facilitate small ship cargoes from Algeria. The 5.1 MMt/y (6.9 Bcm/y) import facility can only accommodate 75,000 m³ LNG ships, which are considered an anachronism in today's LNG industry. The terminal at Montoir de Bretagne on the Atlantic coast, with a capacity of 7.3 MMt/y (9.9 Bcm/y), opened up the French market to larger LNG carriers, facilitating large cargoes from Algeria as well as longer-haul deliveries from Nigeria. In 2010 the 6 MMt/y (8.2 Bcm/y) Fos Max terminal on the island of Cavaou, not far from Fos Tonkin, was built to facilitate larger cargo deliveries from Egypt. Both Montoir and Fos Max can receive cargoes on Q-Max LNG vessels, which are currently the largest LNG tankers on the waters.

EDF terminal. Even as other LNG import projects in France were abandoned owing to inadequate demand, EDF decided to construct a new regasification terminal at Dunkirk in the far northeast of the country. This was partly a political decision to replace jobs lost by closure of Total's refinery nearby. Dunkirk, scheduled for completion in 2015, will be the largest French terminal and the largest terminal on the Continent, at 9.5 MMt/y (13 Bcm/y) of capacity. It will be able to berth Q-Max LNG tankers and will have direct pipeline linkage to the Zeebrugge trading hub in Belgium.

Italy

Italy is the largest gas market in the Mediterranean and is supplied by pipelines from Algeria, Russia, Libya, and the Netherlands, plus indigenous production. The Italian gas market peaked at 71.2 Bcm (6.9 Bcf/d) in 2005, but has since declined by 19% to 57.8 Bcm (5.6 Bcf/d) in 2013.[52] Indigenous production fell from 13.4 Bcm (1.3 Bcf/d) in 2002 to 7.1 Bcm (0.7 Bcf/d) in 2013, contributing to a growing reliance on gas imports. During 2013, pipeline gas comprised the majority of supply (51.6 Bcm or 5.0 Bcf/d). Russia supplied 24.9 Bcm (2.4 Bcf/d) through European pipeline connections, while Algeria delivered 11.4 Bcm (1.1 Bcf/d) through the 30 Bcm/y TransMed Pipeline (running from Algeria through Tunisia into southern Italy). Libya contributed another 5.2 Bcm/y (0.5 Bcf/d) over the 8 Bcm/y Green Stream line. Italy also received 8.6 Bcm (0.8 Bcf/d) from the Netherlands. An additional Algerian-Italian link through the Algeria-Sardinia Galsi Pipeline could provide a further 8 Bcm/y to Italy, but this project is stalled at present.

LNG imports make a relatively modest contribution to the nation's gas supply. LNG imports peaked at 6.5 MMt (8.8 Bcm) in 2010 and have since fallen by 38% to 3.9 MMt (5.4 Bcm) in 2013—including 3.7 MMt (5.0 Bcm) from Qatar, shown in table 3-11.[53] LNG's modest contribution to Italy's gas supply mix is in large part due to the country's notoriously difficult regulatory approval process. In spite of numerous proposals, only three terminals have been built; the most recent two offshore. The nation's LNG import capacity totals 14.5 Bcm/y (10.7 MMt/y).

Table 3-11. Italy LNG imports by source (MMt)

	2005	2006	2007	2008	2009	2010	2011	2012	2013
Algeria	1.9	2.2	1.7	1.1	0.9	1.3	1.3	0.7	0.0
Qatar	–	0.0	–	–	1.3	4.8	4.5	4.0	3.7
Other	0.2	0.1	–	–	0.1	0.5	0.5	0.2	0.2
Total Imports	2.0	2.3	1.7	1.1	2.3	6.5	6.3	5.0	3.9

(*Source:* Poten & Partners.)

Panigaglia LNG, Portovenere. Snam's 1971 vintage Panigaglia LNG terminal will require a significant upgrade if it is to remain in operation. The 2.5 MMt/y (3.4 Bcm/y) terminal, built to facilitate small ship deliveries from Algeria, received just one cargo from the North African producer in 2013. The occasional Spanish reload is still being discharged at Panigaglia. The timetable for a decision on a terminal modernization program has been extended to 2015.

Adriatic LNG, Offshore Rovigo, Adriatic Sea. Qatar Petroleum and ExxonMobil linked up with Italy's Edison Gas to construct a gravity-based LNG terminal offshore near Rovigo. The 5.5 MMt/y (7.5 Bcm/y) structure, which can accommodate LNG carriers as large as 152,000 m³, was commissioned in 2009 to facilitate deliveries from Qatar's RasGas II venture. In 2013, virtually all of the nation's LNG imports were delivered through Adriatic LNG.

OLT Offshore LNG Toscana, Offshore Livorno. The OLT (Offshore LNG Terminal) Offshore LNG Toscana, located 22 miles off the Tuscan coast near Livorno, was commissioned in late 2013 after years of delay. Neither E.ON nor its OLT partners have lined up firm LNG supplies for the 2.7 MMt/y (3.75 Bcm/y) FSRU, suggesting low utilization for at least its first year of operation. LNG carriers up to 137,000 m³ can berth at OLT. The *Golar Frost* was converted into an FSRU for service at OLT, and the overall cost of the project, including the pipeline connection to shore, is reported to have exceeded $1 billion.

Germany

Germany, the industrial heart of Europe, is also the largest natural gas market on that continent, consuming 83.6 Bcm (8.1 Bcf/d) in 2013 according to BP.[54] German gas demand has fluctuated between 75 Bcm/y (7.2 Bcf/d) and 87 Bcm/y (8.4 Bcf/d) over the past decade. Chancellor Angela Merkel's decision in 2011 to shut all 17 of Germany's nuclear power stations by 2022 promises to be a driver of future German gas demand growth, though overall demand growth tends to be muted by energy efficiency gains and by renewable energy development.

Approximately 90% of Germany's natural gas demand is met with imports; less than 10% is produced domestically. Domestic production has declined by nearly half from 15.3 Bcm (1.5 Bcf/d) in 2002 to 8.2 Bcm (0.8 Bcf/d) in 2013. German natural gas imports are geographically diversified. In 2013, they were broadly split between Norway (33.5 Bcm or 3.2 Bcf/d), Russia (39.8 Bcm or 3.9 Bcf/d), and the Netherlands (22.4 Bcm or 2.2 Bcf/d). Because of its comprehensive crossborder pipeline infrastructure and its central location within Europe, Germany is an important natural gas transit hub, with significant amounts of natural gas from Russia and Norway transiting the country for delivery to other markets. The TTF on the German-Dutch border is the largest trading hub on the continent.

Germany's natural gas imports are supplied via crossborder pipelines. Some German companies have booked capacities in overseas LNG terminals, and there is a permitted site for an LNG terminal in Germany, which spells a future possibility.

The Netherlands

The Netherlands is central to European gas trade. In 2013, Dutch pipeline exports were 53.2 Bcm (5.1 Bcf/d), with pipeline imports of 21.5 Bcm (2.1 Bcf/d).[55] There could be some troubling developments, as future production at the giant Groningen field may be reduced due the occurrence of earthquakes linked to gas production. Production from Groningen and smaller offshore fields has leveled out after peaking at 70.5 Bcm (6.8 Bcf/d) in 2010. After declining to 63.9 Bcm (6.2 Bcf/d) in 2012, production recovered to 68.7 Bcm (6.6 Bcf/d) in 2013. During this period, domestic gas demand fell from 43.6 Bcm (4.2 Bcf/d) to 37.1 Bcm (3.6 Bcf/d), helping to maintain export volumes. In 2011, the Netherlands added an LNG import terminal to its gas infrastructure, which could help sustain the nation as a gas exporter and trading hub.

GATE LNG, Rotterdam. The 9 MMt/y (12 Bcm/y) Gateway to Europe (GATE) LNG terminal received its first cargo in June 2011. Without firm supply commitments, GATE has never received more than three cargoes in a month, and in many months, there is no activity. Operators Gasunie and Vopak are exploring LNG bunkering opportunities at GATE and started reexporting LNG during the fourth quarter of 2013. The terminal is struggling to establish commercial viability, with imports of just 574,000 tons and reexport of 218,000 tons in 2013.

Belgium

With a small domestic market (16.8 Bcm or 1.6 Bcf/d in 2013) and without indigenous gas resources, Belgium is a major natural gas trading nation with large gas pipeline connections to the North Sea and to neighboring markets. Fluxys Belgium, the independent operator of both the natural gas transmission grid and storage infrastructure in Belgium, has developed the port of Zeebrugge as the heart of its own natural gas transmission grid and of the northwestern European natural gas system. Zeebrugge serves as a crossroads of two major axes in European natural gas flows: the east/west axis from Russia to the United Kingdom (via the Interconnector) and the north/south axis from Norway to southern Europe (via Zeepipe). It also hosts the Zeebrugge Hub, one of Europe's leading natural gas trading hubs, and the Zeebrugge LNG terminal, which provides access to global LNG supplies.

In 2013, Belgium imported 26.2 Bcm (2.5 Bcf/d) in pipeline gas.[56] Russia was the top supplier with 12.3 Bcm (1.2 Bcf/d), followed by Norway (9.4 Bcm or 0.9 Bcf/d), the Netherlands (5.4 Bcm or 0.5 Bcf/d), and the United Kingdom (2.5 Bcm or 0.2 Bcf/d). LNG, all from Qatar, contributed 3.3 Bcm (2.4 MMt), as shown in table 3–12. The Zeebrugge LNG terminal has established a very active reexport trade.

Table 3–12. Belgium LNG imports by source (MMt)

	2005	2006	2007	2008	2009	2010	2011	2012	2013
Qatar	–	0.3	1.7	2.2	4.7	4.3	4.5	3.3	2.4
Other	2.0	2.9	0.4	0.2	0.2	0.4	0.3	0.0	0.0
Total Imports	2.0	3.2	2.1	2.4	5.0	4.7	4.8	3.3	2.4

(*Source:* Poten & Partners.)

Zeebrugge LNG. The Zeebrugge LNG import terminal was commissioned in 1987 with an initial capacity of 3.3 MMt/y (4.5 Bcm/y). An expansion project doubled capacity to 6.5 MMt/y (9 Bcm/y) in 2008. The terminal has been modified to accommodate Q-Max tankers and to reload cargoes for export. LNG imports, primarily from Qatar, peaked in the range of 4.7 to 5.0 MMt/y (6.4 Bcm/y to 6.8 Bcm/y) during the 2009 to 2011 time frame, but declined to 2.4 MMt (3.3 Bcm) in 2013.[57] Meanwhile, the reexport trade has flourished. Approximately 1.1 MMt (1.5 Bcm) of LNG were reexported in 2013.

Eastern Mediterranean

Natural gas markets in the eastern Mediterranean are attracting growing interest. Turkey has by far the largest regional gas market, and LNG is a growing component of the country's gas supply portfolio, with two LNG import terminals in operation. LNG is also an important component of gas supplies into Greece, and Israel began importing LNG in 2013. Lebanon and Cyprus are exploring LNG imports, even as Egypt has initiated LNG imports (figs. 3–13 and 3–14).

Fig. 3–13. Eastern Mediterranean LNG import terminal map. (*Source:* Poten & Partners.)

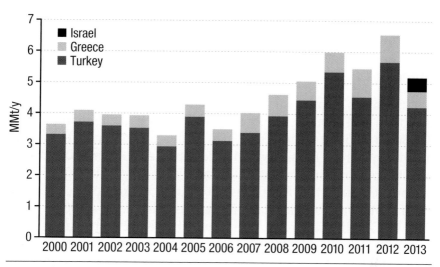

Fig. 3–14. Eastern Mediterranean LNG imports by country. (*Source:* Poten & Partners.)

Turkey

Consumption of natural gas in Turkey more than doubled from 17.4 Bcm (1.7 Bcf/d) in 2002 to 45.6 Bcm (4.4 Bcf/d) in 2013, when the country imported 38.2 Bcm (3.7 Bcf/d) of pipeline gas.[58] Russia was by far the largest supplier at 26.5 Bcm (2.6 Bcf/d), followed by Iran (8.7 Bcm or 0.8 Bcf/d). LNG contributed another 6.1 Bcm (4.5 MMt) in 2013.[59] There is significant import of spot cargoes to meet seasonal demand, which is particularly strong during the winter months when natural gas shortages are often heightened by disruptions to pipeline supplies from Iran.

Turkey's two LNG terminals have an aggregate annual capacity of around 8.9 MMt (12 Bcm). State-owned BOTAŞ owns the Marmara Ereglisi LNG Terminal, and privately owned EgeGaz owns the Aliaga Terminal. Term supplies from Algeria and Nigeria are primarily delivered into the 4.5 MMt/y (6 Bcm/y) Marmara Ereglisi facility, which was commissioned in 1994. Algeria supplied Turkey with 2.8 MMt (3.8 Bcm) and Nigeria with 0.9 MMt (1.2 Bcm) in 2013 under term arrangements. BOTAŞ and EgeGaz share the capacity at Aliaga for the delivery of spot and short-term LNG, which come primarily from Qatar (0.3 MMT in 2013). Aliaga was completed in

2002, but lay idle until 2006 in the face of resistance from BOTAŞ to opening its grid to third-party access.

Greece

Greece has a small gas market. Demand peaked at 4.4 Bcm (0.4 Bcf/d) in 2011, but severe economic recession cut demand to just 3.6 Bcm (0.35 Bcf/d) in 2013.[60] The market is primarily supplied by pipeline gas from Russia, which accounted for 2.4 Bcm (0.2 Bcf/d) of total pipeline imports of 3.0 Bcm (0.29 Bcf/d) in 2013. LNG imports through the Revithoussa terminal make up the balance. The facility was completed on Revithoussa Island in 2000 to facilitate LNG deliveries from Algeria, which provides term supplies of approximately 0.5 MMt/y (0.7 Bcm/y). Spot cargo imports totaled approximately 0.3 MMt (0.4 Bcm) in 2012. Severe economic contraction eliminated the need for extra cargoes in 2013.

Israel

Israel, which has emerged as a potential LNG exporter, installed the first FSRU in the eastern Mediterranean under contract with Excelerate. The 3.1 MMt/y (4.2 Bcm/y) Hadera LNG facility, commissioned in January 2013, received 0.4 MMt (0.5 Bcm) during its first year of operation.[61] The Excelerate regasification vessel picks up cargoes via ship-to-ship transfer in sheltered water and returns to the buoy to discharge regasified LNG into the Israeli pipeline grid. Hadera is an interim measure to supply the Israeli market in the face of disruptions to Egyptian supplies, and until production at the Tamar field starts.

Egypt

In early 2015, Egypt initiated LNG imports when the *Höegh Gallant* was stationed at Ain Sukhina in the Gulf of Suez. Supplies for 2015 and 2016 have been arranged with Trafigura (33 cargoes), Vitol (9 cargoes), Noble (7 cargoes), and Sonatrach (6 cargoes). Another 35 cargoes will come from Russia's Gazprom, all to be delivered over a five-year period.

Other eastern Mediterranean

Elsewhere in the eastern Mediterranean, Lebanon is exploring plans to install a 3.5 MMt/y (5 Bcm/y) FSRU to supply power plants in the Baddawi region with a targeted 2016 start-up.

Eastern and Central Europe

With an extreme reliance on Russian gas supplies, eastern and central European countries are seeking to implement natural gas supply diversification measures, including building LNG import terminals. The most advanced projects are in Poland and Lithuania. Other Baltic states are exploring their LNG import options as well.

Poland

The Polskie LNG Terminal, under construction at Swinoujscie near Szczecin in the western part of Poland's Baltic coast (West Pomeranian region) is expected to be completed by the end of 2015. It has an initial capacity of 3.7 MMt/y (5 Bcm/y) but can be expanded to 5.5 MMt/y (7.5 Bcm/y) if needed. In 2013, Poland imported 11.4 Bcm (1.1 Bcf/d) of pipeline gas, including 9.4 Bcm (0.9 Bcf/d) from Russia.[62] The first tranche of LNG is expected to be delivered into the terminal from Qatar under a term purchase agreement. Plans to expand the terminal are being explored to further reduce reliance on Russian gas. Initial optimism regarding shale gas potential in Poland has faltered as drilling targets fail to be realized.

Lithuania

A Höegh FSRU docked in Lithuania in late autumn 2014. The 2.2 MMt/y (3 Bcm/y) import facility is expected to start at about 0.7 MMt/y (1 Bcm/y), ramping up to capacity over time. Lithuania consumes approximately 3 Bcm/y (0.29 Bcf/d) of natural gas, which is solely supplied via pipeline from Russia.

North America

North America, which a decade ago was viewed as a promising LNG market, has experienced a reversal of fortune, moving rapidly from LNG importer to LNG exporter. In 2005, LNG deliveries into this region peaked at 18 MMt, and many analysts were projecting this market to rival the size of Japan's in a couple of decades. Instead, the surge in unconventional gas production has reversed the decline in domestic supply, turning the United States from the next big LNG importer into the next big LNG exporter (fig. 3–15). This leaves Mexico as the most promising LNG importer in North America.

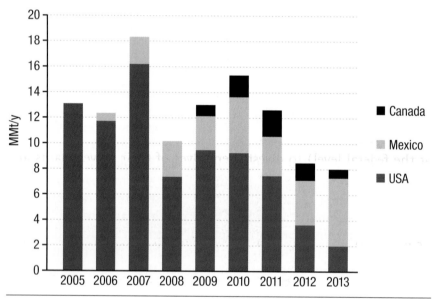

Fig. 3–15. North American LNG imports. (*Source:* Poten & Partners.)

United States

Deregulation of the US natural gas industry. The process of deregulation in the United States began with the Natural Gas Policy Act (NGPA) of 1978, which provided for the phased removal of price controls from wellhead natural gas that had been in place since the 1950s. Wellhead deregulation was completed with the Wellhead Decontrol Act of 1989. During the same period, the Federal Energy Regulatory Commission (FERC) developed new regulations for interstate pipelines, which subsequently changed their role in the delivery of natural gas. In 1985, faced with rising prices and falling demand, FERC issued Order 380 allowing customers (local distribution companies) to break their take-or-pay obligations with the transmission companies, while leaving the transmission companies committed to their suppliers for the corresponding contractual supply obligations they had made. Order 380 also implemented transportation-only services on the transmission lines. The transmission industry was severely impacted by Order 380, with several companies going bankrupt and almost all suffering serious financial losses as they could not find markets willing to accept the gas supplies they had contracted at the prices they had agreed to pay. The regulatory compact, in many people's minds, was broken. In 1992, FERC Order 636 formally separated the merchant (gas supply) and transportation functions of interstate pipeline companies and essentially took the transmission companies out of the gas supply and marketing business.

Simultaneously, many states began forcing electric power companies (which are largely regulated at the state level, in contrast to interstate natural gas transmission companies, which are regulated at the federal level) to divest themselves of their power plants and create a competitive market in power generation. A new group of merchant energy firms emerged to take advantage of investment and trading opportunities crossing both business sectors. These firms were initially the favorites of Wall Street, but following the collapse of Enron, they experienced a loss of investor confidence, and the value of their shares sharply declined. Abuses of trading positions and market power (most notoriously during the California energy crisis of 2000–2001) led FERC to become more vigilant about monitoring gas and power transactions, and in June 2004, FERC established the Market Oversight and Investigations office. Its purpose is to protect customers by assuring compliance with FERC rules and regulations.

Natural gas prices. Following the deregulation of wellhead natural gas prices and the unbundling of natural gas transportation and supply functions, the Henry Hub natural gas trading hub developed in Louisiana. Henry Hub is the most closely watched benchmark price for natural gas in the United States and is a reference price for US trades. The New York Mercantile Exchange (NYMEX) prices for futures contracts are based on Henry Hub prices. Besides Henry

Hub, there are growing liquid trading hubs at other locations reflecting the expansion of unconventional natural gas supplies. North American wholesale natural gas prices are generally set by market conditions, and supply contracts are, by and large, not contractually linked to competitive oil product prices or other indices, as is still the case in many other nations. Transportation and storage rights are also actively traded separately from the physical commodity.

During the latter part of the 1980s and most of the 1990s, natural gas prices were low relative to other fossil fuels. During this decade, the Henry Hub benchmark price moved in a range of about $1.50 to $2.50/MMBtu. This led to a strong increase in natural gas demand, particularly in the industrial and power sectors. The year 2000, however, signaled the end of low, stable natural gas prices in the United States. Early cold in the 2000–2001 winter sent gas prices soaring, and Henry Hub prices reached a then-extraordinary peak of around $10/MMBtu in January 2001. High prices caused considerable demand loss and stimulated a major drilling effort, which helped restore a market balance. By early 2002, gas prices fell back to the $2.00 to $2.50/MMBtu level. Then a cold 2002–2003 winter again sent Henry Hub prices to over $7.50/MMBtu. While prices decreased in the spring of 2003, gas demand by the power sector kept prices above $4.50/MMBtu that summer.

It appeared that higher prices were here to stay. Henry Hub remained above $5.00/MMBtu during 2004, and crept inexorably past $6.00/MMBtu as the winter of 2004–2005 approached. In the summer and fall of 2005, the loss of production in the Gulf of Mexico caused by Hurricane Katrina drove gas prices to all-time highs, above $14.00/MMBtu. By February 2006, aided by warmer-than-normal winter weather in most of the country, they had settled at $7.00/MMBtu and remained at or near that level for the next three years. Prices began falling in 2009 and averaged less than $4.00/MMBtu through 2013. But strong gas demand boosted by switching from coal to gas and then a very cold 2013–2014 winter caused natural gas prices to rebound to approximately $4.50 per MMBtu on average during the first half of 2014 before falling below $3.00/MMBtu in early 2015 (fig. 3–16). At this price level, US gas prices are a bargain by global standards, at half the prices in the United Kingdom and in continental Europe, and about a third of LNG prices delivered to Asia. The growing discrepancy between North American and global prices arose on a single phenomenon, the emergence of shale gas.

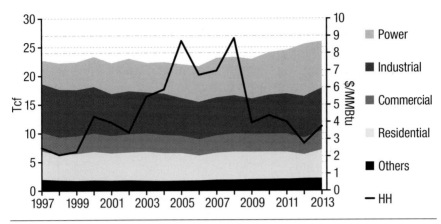

Fig. 3–16. Henry Hub natural gas price vs. US demand. (*Source:* Poten & Partners, EIA.)

Shale gas revolution. The consensus view in the 1980s and 1990s was that the country's natural gas resource base was waning and that production was in secular decline. However, technological innovations in the form of horizontal drilling coupled with multistage hydraulic fracturing, or fracking, reversed this seemingly inevitable trend. These innovations have turned the country from a potentially large LNG importer to an even larger prospective exporter.

Proven natural gas reserves, which had been in decline, increased from 4.7 Tcm (166 Tcf) at year-end 1992 to 10.00 Tcm (354 Tcf) by the end of 2013, according to EIA statistics.[63] However, this is a very conservative estimate that understates the impact of unconventional gas. The Potential Gas Committee estimated the total potential natural gas resource including probable (2P) and possible (3P) to be far larger, at 75 Tcm (2,650 Tcf) at year-end 2012, including 30 Tcm (1,059 Tcf) of shale gas.[64]

The unconventional gas phenomenon has reversed the slide in production of US natural gas (fig. 3–17). Production had declined to around 480 Bcm/y (16.9 Tcf/y) in the mid-1980s, when most forecasters agreed it was in long-term decline. In 2013, however, it reached a record high of 687.6 Bcm (24.3 Tcf), making the United States the world's top natural gas producer and consumer.[65] Much of this reversal is attributed to shale gas production, which surged to 275 Bcm (9.7 Tcf) in 2012 from just 29 Bcm (1.0 Tcf) in 2006 while conventional natural gas production declined. In 2014, the Marcellus gas field alone produced more natural gas than the entire country of Iran, sitting on the world's second largest gas reserves. The United States, on the verge of becoming a major LNG importer in 2005, has buyers lining up to purchase its potential LNG exports.

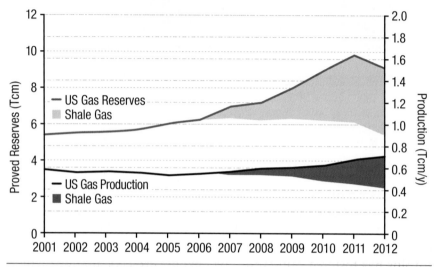

Fig. 3–17. Shale gas reverses US gas decline. (*Source:* Poten & Partners, EIA.)

The US gas market accounts for 22% of the global consumption of natural gas, and is the world's largest and probably most competitive wholesale gas market. Natural gas represents about 28% of the country's primary energy, second only to oil's 38%.[66] While oil use is concentrated in transportation, gas is used for generation of electricity, as plant process fuel, and as an industrial feedstock (table 3–13). Natural gas also has the largest share of residential heating and cooking. Natural gas has had the fastest growth of all fuels in the power generation sector since the mid-1990s, reflecting the impact of wholesale electric market competition on fuel preferences of power generators. By 2040, the US EIA projects, the natural gas share of primary energy will increase to 30%—nearly as much as oil's 33%.

Table 3-13. US natural gas consumption by end use (Tcf)

	2005	2006	2007	2008	2009	2010	2011	2012	2013
Residential	4.8	4.4	4.7	4.9	4.8	4.8	4.7	4.1	4.9
Commercial	3.0	2.8	3.0	3.2	3.1	3.1	3.2	2.9	3.3
Industrial	6.6	6.5	6.7	6.7	6.2	6.8	7.0	7.2	7.5
Vehicle Fuel	0.0	0.0	0.0	0.0	0.0	0.0	0.0	0.0	0.0
Electric Power	5.9	6.2	6.8	6.7	6.9	7.4	7.6	9.1	8.2
Pipeline & Distribution Use	0.6	0.6	0.6	0.6	0.7	0.7	0.7	0.7	0.7
Lease and Plant Fuel	1.1	1.1	1.2	1.2	1.3	1.3	1.3	1.4	1.4
Total Consumption	**22.0**	**21.7**	**23.1**	**23.3**	**22.9**	**24.1**	**24.5**	**25.5**	**26.0**

(*Source:* EIA.)

In its outlook published in 2014, the EIA projects that dry natural gas production will reach 31.9 Tcf (905 Bcm) in 2025 and 37.5 Tcf (1,062 Bcm) per year by 2040.[67] These figures surpass its natural gas demand outlook of 28.4 Tcf (1,003 Bcm) in 2025 and 31.6 Tcf (1,117 Bcm) per year by 2040, and will facilitate a steady growth in exports—reaching 5.3 Tcf (150 Bcm) in 2025 and 7.9 Tcf (225 Bcm) per year in 2040 in the EIA's view. The expectation of this surplus has fostered numerous LNG export projects, many of which are utilizing LNG import facilities that are no longer needed (see chapter 5).

The EIA outlook and prospects for LNG export depend on the successful exploitation of the nation's vast shale gas resource, which is encountering some opposition. Some call it the "Gasland" effect, after a 2010 film about America's shale gas industry in which a Pennsylvania man puts a match to his water tap and then reels back from the dramatic gas flame. The film blames fracking for such incidents, in spite of a lack of evidence and a host of regulations designed to prevent methane from getting into the groundwater from natural gas wells. In reality, methane can find its way into water wells, irrespective of drilling activity. While the film's premise has been shown to be exaggerated or downright erroneous, its impact lingers and grows. As of 2014, however, shale gas is being developed largely unimpeded, and shale gas and oil are seen as major drivers of employment and investment in the United States.

Shale gas has a good environmental record in America, according to a report from MIT.[68] With more than 20,000 wells drilled in the past decade, there have been only a few instances of groundwater contamination, most of them due to breaches of existing regulations, associated with poor completion practices or water storage practices at or near the surface of the ground. There does not appear to be systemic risk. Fracking takes place thousands of feet below the water table, and fracking zones are typically separated from groundwater by fairly impermeable rocks, making fracking an unlikely direct source of contamination.

There is no question that fracking involves a high degree of public controversy. Limited science, coupled with high public emotion and mistrust of oil and gas companies, has resulted in a noisy debate, with negative connotations of shale gas overwhelming what could be perceived as the positives, such as the development of cleaner energy supplies. During the period between 2005 and 2013, the United States experienced a decline in CO_2 emissions of about 11% in spite of continuing economic and energy consumption growth. Nearly all of this drop is attributed to the displacement of coal by natural gas in US generating plants.

Outlook for supply and demand. In the early part of the 21st century, the US Department of Energy (DOE) and other organizations had anticipated a 30 Tcf/y (850 Bcm/y) US gas market by 2010, but an era of higher gas prices moderated these expectations. Consumption of 26 Tcf (737 Bcm) in 2013 lags well behind earlier projections. While it may take longer to get there, the current lower price environment has the DOE again projecting that the US gas market will reach 30 Tcf/y, albeit far later than anticipated. In its 2014 reference case forecast, the DOE projects a 28 Tcf (990 Bcm) gas market in 2025, up from 25.6 Tcf (905 Bcm) in 2012, and 32 Tcf (1,130 Bcm) in 2040.[69]

In the 1980s and 1990s, robust growth in US gas consumption increasingly outpaced domestic supply, creating a growing dependence on natural gas imports. During this period, the supply gap was largely filled by imports from Canada. But the prolific Western Canada Sedimentary Basin was also maturing and facing the same depletion trends as the US lower-48 states. More and more wells had to be drilled just to offset declines in producing fields. Thus, at the beginning of the 21st century, North American self-sufficiency came to an end and the industry turned to more distant gas resources. Market expectations opened up opportunities for LNG imports and the development of frontier natural gas resources, such as Alaska's North Slope and Canada's Arctic resources.

These trends led to the view that LNG, benefiting at the time from impressive cost reductions in production and delivery infrastructure, was competitive with these frontier gas supplies and would be called upon to fill the supply gap. The Trinidad LNG project was built largely to supply the nearby US market, and Qatar established mega-sized liquefaction chains benefiting from economies of scale to deliver large quantities of gas to the US market at costs that were at the time competitive with indigenous North American gas supplies.

Mothballed LNG import terminals were reopened and new ones built. Of course, this assumed that Henry Hub prices would remain high for a prolonged period of time, a risky presumption, as those who survived the price collapse of the late 1980s could attest. Not foreseen was the surge in unconventional natural gas production that would reverse the decline in domestic natural gas production, bring lower Henry Hub natural gas prices, and curtail the need for the import of pipeline gas and LNG. Imports of LNG peaked at 16.2 MMt in 2007 and then collapsed to just 2.0 MMt in 2013 (table 3–14).

Table 3–14. US LNG imports by source (MMt)

	2005	2006	2007	2008	2009	2010	2011	2012	2013
Egypt	1.5	2.1	2.3	1.2	3.4	1.5	0.7	0.1	–
Nigeria	0.2	1.2	2.0	0.3	0.3	0.8	0.1	–	0.1
Qatar	0.1	–	0.4	0.1	0.3	0.9	2.0	0.7	0.2
Trinidad	9.1	8.1	9.7	5.5	5.0	4.2	2.7	2.4	1.5
Other	2.2	0.4	1.8	0.4	0.6	1.8	2.0	0.5	0.3
Total Imports	13.1	11.7	16.2	7.4	9.5	9.3	7.5	3.7	2.0

(*Source:* Poten & Partners.)

LNG: The experience of the 1970s. Even as LNG import projects proliferated, skeptics were quick to point to the failed Algeria-to-US LNG trades of the 1970s. At that time, four LNG import terminals and a number of tankers were built to deliver LNG from Algeria under long-term take-or-pay contracts. These investments were made under a regulatory policy that "rolled in" high-priced new gas with low-priced domestic gas already under contract to interstate pipelines. This policy effectively subsidized new natural gas, primarily LNG and imports from Canada, at the expense of domestic production. The 1978 NGPA signaled the beginning of the end of this approach. New domestic natural gas prices initially rose, and domestic production increased in response. But higher prices led to reduced demand, and wellhead prices then dropped sharply in the face of a large surplus of domestic supply. The collapse of US wellhead prices coupled with price decontrol in the 1980s resulted in Algerian LNG being priced out of the market, the collapse of the LNG supply contracts (with extensive accompanying litigation), and the mothballing of terminals (except for the Everett Terminal in Boston) and LNG ships for significant periods of time. The regulators had approved the contract conditions supporting the earlier LNG trade and then, with Order 380, removed these underpinnings. It was ultimately the market, and not regulation, that provided the foundation for the revived LNG trade into the United States in the late 1990s.

The four import and regasification plants built to serve the earlier trade provided the foundation infrastructure for the revival of LNG imports. The cost of reactivating a mothballed terminal was less than building a new terminal and could be done more quickly. By mid-2003, all four terminals were receiving LNG cargoes, and plans were in place to add capacity to some of these facilities at a relatively low cost, even as new terminals were being planned.

The tumultuous histories of these facilities have continued, with only Everett still operating as a viable LNG import facility (fig. 3–18). The others are all severely underutilized and plan to add liquefaction capacity (see chapter 5).

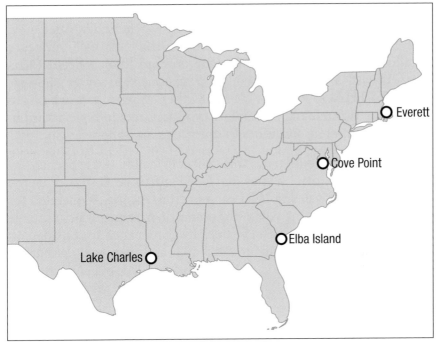

Fig. 3–18. First generation US LNG import terminals. (*Source:* Poten & Partners.)

First generation LNG import terminals in the United States.

Everett, Massachusetts. This is the sole onshore LNG import terminal in the US Northeast and has operated almost continuously since it was built in Boston Harbor in 1971. The terminal provides approximately 20% of New England's annual needs and an even larger share during the winter. The terminal was opened by Distrigas Corporation, a subsidiary of Cabot Corporation, to receive LNG from Sonatrach, Algeria's national oil company. It imported Algerian LNG until 1985, when FERC Order 380 allowed its customers to abandon their take-or-pay obligations. Unable to recover the costs of LNG imports priced at then above-market prices, the terminal was placed into bankruptcy reorganization. Following a settlement between Distrigas and Sonatrach in 1987 providing for new supply arrangements, the facility has remained in continuous operation. Sonatrach supplied the bulk of the LNG to the facility until Trinidad's Atlantic LNG was commissioned in 1999.

In September 2000, Cabot sold the terminal and related assets to Belgium's Tractebel, which was subsequently acquired by Suez, now GDF Suez, which still owns and operates the terminal.[70] Everett's installed nameplate sendout capacity is more than 1 Bcf/d (10.3 Bcm/y), with a sustainable daily throughput capacity of approximately 0.7 Bcf/d (7.4 Bcm/y). Unlike other US terminals, which are vastly underutilized and are exploring options to add liquefaction to export LNG, Everett continues to operate as an import facility.

Lake Charles, Louisiana. This import facility was opened by Trunkline LNG Corp., a subsidiary of Panhandle Eastern Pipeline Corp., in 1980. Receiving LNG cargoes from Algeria, it operated for only three years, closing when low domestic natural gas prices made this trade uneconomic.[71] The terminal was reopened for business in 1987. Duke Energy subsequently acquired and then sold Panhandle to CMS Energy in the spring of 1999. In early 2001, BG LNG Services Inc. won the rights to all available capacity at the terminal for 22 years from January 2002.

CMS Energy sold Panhandle to Southern Union in 2003, and the terminal's new owner expanded baseload sendout capacity in two phases from 0.63 Bcf/d (6.5 Bcm/y) to 1.5 Bcf/d (15.5 Bcm/y) by the second quarter of 2006.

LNG imports through Lake Charles came to a halt in 2010, spurring Energy Transfer Equity (ETE), which had by then acquired Trunkline, and BG to initiate plans to build a three-train liquefaction project capable of producing up to 16.2 MMt/y of LNG (see chapter 5). Lake Charles Exports, a jointly owned subsidiary of ETE and BG, has received 25-year authorizations to export up to 15 MMt/y of LNG to Free Trade Agreement (FTA) and non-FTA countries. Adding liquefaction at the Lake Charles facility would provide bidirectional capabilities, allowing the terminal to liquefy and export natural gas as well as import and regasify LNG. FTA and non-FTA export authorizations are conditional, pending environmental approval from the FERC. FERC approval is also a prerequisite for the FID on the liquefaction project.

Elba Island, Georgia. This terminal was opened by Southern LNG, a subsidiary of Southern Natural Gas (Sonat) Inc., in 1978, and received Algerian LNG for two years before a pricing dispute ended this trade.[72] After operating briefly in peaking service, the facility remained dormant until it was reopened for LNG imports in late 2001. El Paso merged with Sonat in 1999, and assumed control over Sonat's pipeline system as well as the Elba Island Terminal. An affiliate, El Paso Merchant Energy (EPME), won 100% of the capacity rights at the terminal in an open season. Enron subsequently challenged the award and negotiated supply rights for 43% of the capacity. These rights were later acquired by Marathon when Enron went bankrupt. El Paso also entered into a supply agreement with BG for LNG from Trinidad.

Like many US energy trading companies, El Paso experienced financial difficulties in the wake of the Enron debacle, and, in December 2003, the firm sold its entire capacity at Elba Island to BG, together with its related agreements for LNG supply from Trinidad. In June 2003, Southern LNG started an expansion of the terminal, adding 0.36 Bcf/d (3.7 Bcm/y) of baseload sendout capacity. This expansion, which was placed into service in early 2006, increased sendout capacity to 0.806 Bcf/d (8.3 Bcm/y). Shell acquired all the rights to this expansion capacity.

In December 2005, Southern LNG announced plans to expand Elba Island once again. This expansion would add 0.9 Bcf/d (9.3 Bcm/y) of sendout capacity, thereby doubling the sendout capabilities of the facility. The first phase of this expansion increased sendout capacity by 0.90 Bcf/d to a total of 1.7 Bcf/d. But deliveries into Elba Island slowed, and Phase II of this expansion was shelved. Just five cargoes were unloaded at the terminal in 2013. Kinder Morgan acquired El Paso and with Shell is now pursuing the addition of small liquefaction trains at Elba Island for a total capacity of 1.5 MMt/y.

Cove Point, Maryland. The LNG terminal on the Chesapeake Bay near Baltimore was originally built jointly by Columbia Gas and Consolidated Natural Gas to serve the mid-Atlantic market.[73] It began operations in 1979 but was mothballed in 1980. Consolidated sold its interest to Columbia in 1988, which partially reactivated the plant in 1995 for natural gas peaking services (following the installation of a small liquefaction plant).

In July 2000, Williams Corp. purchased the terminal, and then resold the facility in 2002 to Dominion Resources. It reopened as an LNG terminal in 2003. BP, El Paso, and Shell won equal capacity rights at the facility under an open season bidding process. El Paso subsequently sold its rights to Norway's Statoil.

By 2009, Dominion had boosted throughput capacity from 1 Bcf/d to 1.8 Bcf/d. All of Cove Point's expansion capacity was assigned to Statoil. By 2011, modifications had been made to allow 267,000 m³ LNG vessels to dock at the pier, enlarging from the previous maximum ship size of 148,000 m³. Deliveries into Cove Point have collapsed, with only one cargo unloaded in 2013.

Dominion Resources now plans to construct a 5.25 MMt/y liquefaction train at Cove Point, and construction started in 2014. This would put the LNG train in service in 2017. Dominion has received DOE permission to export 5.7 MMt/y of LNG to FTA and non-FTA countries. The capacity at the liquefaction facility has been fully subscribed by international buyers, who will arrange their own feed gas and pay a tolling fee for liquefaction and other services.

Second generation LNG terminals in the United States. As the industry gained confidence—a false confidence as it turned out—that the revived LNG trade into the United States was not simply a replay of the 1970s, companies floated numerous plans to build new LNG import terminals (fig. 3–19). These proposals were frequently frustrated by the difficult regulatory process. Still, eight were built, including one offshore facility on the US Gulf Coast and two offshore facilities in waters off Massachusetts.

Cameron, Louisiana. California-based Sempra Energy commenced commercial operations at its Cameron LNG facility in Hackberry, Louisiana in October 2009.[74] The plant has two marine berths that can accommodate Q-Flex LNG ships, and 1.5 Bcf/d (15.5 Bcm/y) of vaporization capacity. A project to expand the sendout capacity from 1.5 Bcf/d to 2.65 Bcf/d (27.4 Bcm/y) was shelved because of a lack of LNG import demand. Cameron has not received a cargo since November 2012.

Without LNG import demand, Sempra is adding three liquefaction trains, each with nameplate capacity of 4.5 MMt/y, to convert Cameron LNG into a bidirectional facility capable of importing and regasifying LNG and liquefying feed gas and exporting LNG. Sempra achieved the necessary regulatory approvals in 2014 with first LNG production planned in late 2017.

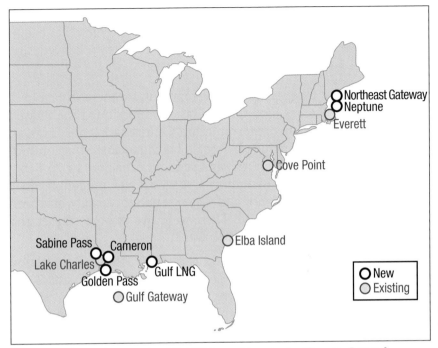

Fig. 3–19. Second generation US LNG import terminals. (*Source:* Poten & Partners.)

Sabine Pass, Louisiana. Cheniere Energy's fully owned subsidiary, Sabine Pass LNG, developed, owns, and operates the Sabine Pass LNG import terminal.[75] Sabine Pass operated as a tolling facility, processing LNG under terminal use agreements (TUAs) with Total and Chevron. According to the agreements, the two TUA customers each reserve LNG import and regasification capacity of approximately 1.0 Bcf/d. Cheniere Marketing keeps a capacity of approximately 2.0 Bcf/d.

The regasification terminal was built in two phases. Phase I, commissioned in the spring of 2008, had a peak capacity of approximately 2.6 Bcf/d. Phase II, commissioned in the summer of 2009, boosted peak capacity to approximately 4.0 Bcf/d. Sabine Pass has two marine berths capable of accommodating 266,000 m³ LNG carriers. Within a year of commissioning Phase II, Cheniere decided to add liquefaction capacity at Sabine Pass in response to the changing market conditions. Construction of the first two trains commenced in August 2012, with completion of the first train scheduled for late 2015, and the second train following in 2016. The second phase of construction, adding two more trains, began in June 2013, with the trains scheduled for start-up in 2016 and 2017. Cheniere has begun preliminary work on a fifth and sixth train. Each train has a design capacity of 4.5 MMt/y.

Freeport, Texas. Freeport LNG Development, now a partnership of Michael Smith, Zachry, and Osaka Gas, commissioned the Phase I development of the import terminal on Quintana Island, Texas, in April 2008.[76] Phase I included a single berth capable of accommodating 267,000 m^3 LNG tankers, and sendout capacity of 1.75 Bcf/d (18.1 Bcm/y). Phase II, including a second LNG unloading dock and 0.5 Bcf/d (5.2 Bcm/y) of additional sendout capacity, has been shelved due to lack of demand. There were two import cargoes and one reload for export at Freeport in 2013.

Freeport LNG now plans to add three 4.4 MMt/y liquefaction trains at its import terminal. Construction began in late 2014. Under a 48-month construction schedule, the first train will be commissioned in late 2018, with the second train following six to nine months later. Separate entities will own each train, with tolling partners taking ownership in the trains.

Golden Pass LNG, Texas. This 2 Bcf/d (20.7 Bcm/y) terminal is majority-owned by Qatar Petroleum (70%), with ExxonMobil and ConocoPhillips owning the remaining 30%.[77] It was intended as the prime destination for volumes from the integrated RasGas III project in Qatar, which is owned by Qatar Petroleum and ExxonMobil. However, the terminal was also designed to receive a portion of output from Qatar Petroleum and ConocoPhillips' Qatargas III project. The first cooldown cargo was received in October 2010, with commercial start-up in March 2011. Golden Pass has two berths capable of accommodating LNG tankers up to 267,000 m^3 in size. Like other import terminals on the US Gulf Coast, Golden Pass is vastly underutilized. It has only received 11 cargoes since its inaugural shipment in October 2010, the last of which was delivered in June 2011.

In August 2012, Golden Pass Products LLC, an entity set up by Qatar Petroleum and ExxonMobil to develop a liquefaction project at the terminal, unveiled plans to build a three-train liquefaction plant capable of producing 15.6 MMt/y of LNG at the site. Regulatory authorization was initiated in 2014.

Gulf LNG Clean Energy, Mississippi. This project on the Pascagoula Bayou Ship Channel in Pascagoula, Mississippi, was commissioned in June 2011. It is owned by El Paso Corporation (now a subsidiary of Kinder Morgan; 50%), Crest Group (30%), and the Angolan national oil company Sonangol (20%). The 1.3 Bcf/d (13.4 Bcm/y) terminal was built to receive cargoes from Angola LNG.[78] It has a single marine berth equipped to accommodate 170,000 m^3 LNG carriers. The terminal received two commissioning cargoes in June 2011 and has not received a cargo since.

Gulf LNG, now Gulf LNG Liquefaction Co., has initiated the development of 11.5 MMt/y of liquefaction. Gulf LNG Liquefaction is owned by El Paso (50%), GE Financial Services (who purchased their share from Crest and Sonangol; 46%), and other investors (4%). The project development timeline depends on regulatory approvals.

Gulf Gateway Deepwater Port, Gulf of Mexico. This is the world's first offshore regasification terminal, 116 miles off the coast of Louisiana. It received its first cargo in March 2005.[79] It is owned by Excelerate Energy, which acquired the rights to the technology from El Paso in 2003. In this innovative approach, an LNG tanker is fitted with vaporization equipment and a mooring system located beneath the forward part of the ship. The equipment connects via a Submerged Turret Loading ™ buoy, which is based on technology used in the North Sea. The facility could deliver 0.5 Bcf/d (5.2 Bcm/y) of gas through lines from the ship to undersea pipelines, which then delivered it onshore. The terminal was closed in 2011 given the market conditions. Excelerate removed the Gulf Gateway buoys, mooring system, and other equipment, and filled, capped, and buried its pipeline in the sea bottom. The components were redeployed at an offshore terminal in Israel. Gulf Gateway received nine cargoes after it was commissioned in March 2005, with the last one in 2008.

Northeast Gateway, Neptune LNG, Massachusetts. Two more deepwater ports were installed in US waters off Massachusetts. Northeast Gateway Deepwater Port, located about 13 miles southeast of Gloucester, was built by Excelerate Energy and opened in 2008, when natural gas prices were hitting $12/MMBtu. It comprises a system of underwater pipelines connected to large buoys that serve as intake valves to receive the gas from shipping tankers. The intake buoys are submerged, and the only visible sign of the terminals are small marker buoys. The port had not received any cargoes since 2010 until one was delivered in early 2015. A similar fate befell Neptune Deepwater Port, with the same system setup, located 10 miles off Gloucester and built by GDF Suez to supplement LNG deliveries into the Boston area through the firm's Everett terminal. The port opened in 2010 and received a few small shipments that year and none after that.

Canada

Canada had an estimated 71.4 Tcf (2 Tcm) of proven natural gas reserves in 2013. The country ranked as one of the world's largest natural gas producers at 5.5 Tcf (155 Bcm) that year, exporting 2.8 Tcf (78.9 Bcm) of natural gas to the United States via pipeline, according to BP Energy Statistics 2014. The Canadian upstream, similar to the United States, is mature, and production of conventional natural gas, mainly in Alberta, is projected to stabilize or even decline. In the middle of the last decade, Canada turned its attention to LNG, mainly in the Atlantic provinces, to meet an anticipated natural gas shortage. Numerous terminals were proposed, but only one, Canaport LNG, was built. With limited local demand, Canaport targeted the northeast US market, but fell victim to the supply revolution south of the border. Imports at Canaport declined to 0.7 MMt in 2013 after peaking at 2.1 MMt in 2011. Concurrently, the potential development of large shale gas fields, largely in British Columbia and northwest Alberta, has spawned a number of project proposals along the country's Pacific coast to export LNG to the Far East.

Canaport LNG, Saint John, New Brunswick. Irving Oil, a large independent refinery owner, and Repsol constructed a 1 Bcf/d (10.3 Bcm/y) terminal, commissioned in June 2009, near Irving Oil's existing refinery and deepwater marine terminal in Saint John, New Brunswick.[80] The terminal's regasified LNG is sold in eastern Canada to Irving's own refinery and to consumers in New England via the Maritimes & Northeast Pipeline. Even though a dozen cargoes were discharged at Canaport in 2013 (from Qatar and Trinidad), the owners are considering adding liquefaction capacity. Plans are in the very early stage of development, and nearby feed gas supply is very limited (table 3–15).

Table 3-15. Canada LNG imports by source (MMt)

	2005	2006	2007	2008	2009	2010	2011	2012	2013
Qatar	0.0	0.0	0.0	0.0	0.1	0.4	1.3	0.6	0.5
Trinidad	0.0	0.0	0.0	0.0	0.7	1.2	0.8	0.8	0.2
Other	0.0	0.0	0.0	0.0	0.1	0.1	0.0	0.0	0.1
Total Imports	–	–	–	–	0.9	1.6	2.1	1.4	0.8

(*Source:* Poten & Partners.)

Mexico

Although Mexico has substantial prospective natural gas resources, particularly in the Burgos Basin near the US border, the national oil company, Petróleos Mexicanos (Pemex), has focused its limited investment capital on oil. As a result, development of the nation's natural gas resources has lagged. The Mexican constitution prohibits foreign ownership of oil and gas resources, and this has further limited development. The country initially made some progress towards opening up its upstream sector through a new contractual arrangement known as the Multiple Service Contract (MSC). Notwithstanding opposition from the public and the Pemex unions, the government passed legislation opening the upstream to foreign participation in 2014, and began a process to auction off exploration rights in 2015. Given the collapse in oil prices, it is too early to tell how successful this will prove to be.

Demand for natural gas in Mexico has been growing, especially in the power and industrial sectors. Residential and commercial consumption have also seen substantial growth due to the 1995 privatization of the transmission and distribution sector that brings natural gas distribution grids into large cities. Mexico's gas consumption registered a 60% increase from 1.8 Tcf (51.4 Bcm) in 2003 to 2.9 Tcf (82.7 Bcm) in 2013.[81] Meanwhile, the nation's domestic production has fallen increasingly short, registering a 36% increase during this period from 41.7 Bcm in 2003 to 56.6 Bcm in 2013. This has resulted in a growing dependence on imports, which are climbing about 20% per year. Pipeline imports from the United States reached 1.1 Tcf (13.6 Bcm), or approximately 70% of all imports. Three pipelines crossing south of the Eagle Ford shale gas field in south Texas account for most of the natural gas imported from the United States. The remaining 30% of imports in 2013 was in the form of LNG.

Mexico's proven gas reserves fell from 70 Tcf (2.0 Tcm) in 1992 to just 14 Tcf (0.4 Tcm) in 2002 according to BP, and to 10 Tcf (0.3 Tcm) in 2013. But the country has a large potential shale gas resource that could be unlocked. The greatest known shale gas potential exists in the portion of the Eagle Ford shale, and extends into Mexico's Burgos basin from south Texas. The EIA estimates that oil- and gas-prone plays extending south from Texas into northern Mexico have an estimated 343 Tcf (9.7 Tcm) of risked, technically recoverable and potential shale gas.[82] The EIA puts the total figure at 545 Tcf (15.4 Tcm) for the country.

Over the long-term, Mexico could become self-sufficient in natural gas under a regulatory framework encouraging foreign investment and access to the upstream technology. In the meantime, Mexico's strongly growing gas demand is being met primarily by pipeline imports from the United States. The EIA estimates that US exports into Mexico reached 2.0 Bcf/d (20 Bcm/y) by the end of 2014.[83] LNG imports play a role as well. Three LNG import terminals have been built in Mexico (fig. 3–20). Together they have a nameplate capacity of 16 MMt/y (2.1 Bcf/d or 22 Bcm/y) of LNG, though actual imports are less than one-third of this with 5.8 MMt (0.8 Bcf/d or 7 Bcm) delivered into Mexico in 2013 (table 3–16).

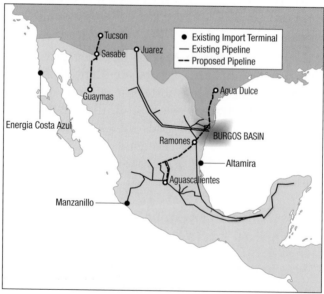

Fig. 3–20. Mexico LNG import terminals. (*Source:* Poten & Partners.)

Table 3-16. Mexico LNG imports by source (MMt)

	2005	2006	2007	2008	2009	2010	2011	2012	2013
Qatar	–	0.1	–	0.1	0.1	0.7	1.3	1.3	1.1
Nigeria	–	0.3	0.5	0.7	2.0	1.5	1.1	0.8	1.1
Peru	–	–	–	0.1	–	0.2	0.4	0.8	1.6
Other	–	0.2	1.6	2.0	0.6	1.9	0.3	0.5	1.5
Total Imports	**0.1**	**0.8**	**2.7**	**3.7**	**2.7**	**4.5**	**3.1**	**3.6**	**5.8**

(*Source:* Poten & Partners.)

Altamira, Tamaulipas State. Built by Shell, Total, and Mitsui, this 4.9 MMt/y (6.7 Bcm/y) LNG facility in the eastern state of Tamaulipas received its first cargo in August 2006.[84] Enagas and Vopak acquired Altamira LNG in 2011. Mexican power generator Comisión Federal de Electricidad (CFE) is the anchor customer at Altamira with a firm supply contract for 3.8 MMt/y (0.5 Bcf/d) of regasified LNG for 15 years from Shell (75%) and Total (25%). The delivered price is at a small premium to the Henry Hub benchmark in Louisiana. In recent years, deliveries into Altamira have ranged from 2.5 to 3.5 MMt/y, primarily from Qatar, Nigeria, and Yemen.

Energía Costa Azul, Baja California. Sempra LNG's Energía Costa Azul terminal in Baja California received its first commissioning cargo in April 2008. Capacity at the 7.5 MMt/y (10.3 Bcm/y) terminal is shared equally between Sempra and Shell.[85] Sempra planned to fill its capacity with LNG from the Tangguh project in Indonesia, while Shell was to rely primarily on deliveries from Russia's Sakhalin II project. It was expected that at least 40% of Energía Costa Azul's initial capacity would be directed to Baja California itself, while the remainder was shipped over the border into the United States. Deliveries into Costa Azul ramped up to a peak of 1.6 MMt from Indonesia in 2010, but have dwindled to three or four cargoes per year as cargoes are diverted to premium-priced outlets in Asia. All the imported LNG is consumed locally. Sempra is now considering adding liquefaction to the facility in partnership with Pemex.

Manzanillo, Michoacan State. This LNG terminal was built by a project consortium comprising South Korean firms Samsung C&T and Korean Gas and Japan's Mitsui Trading under a build, own, and operate (BOO) arrangement.[86] According to the contract, the consortium is entitled to operate the terminal for 20 years after completion of construction, which occurred in 2012. The terminal consists of two 150,000 m³ LNG storage tanks, an LNG regasification and delivery facility with an annual capacity of 3.8 MMt/y, and a jetty to dock and unload LNG tankers. LNG regasified at the terminal will be supplied to the Manzanillo power plant owned and operated by CFE, the Guadalajara independent power plant, and neighboring cities.

CFE entered into a contract for LNG from Repsol Yacimientos Petroliferos Fiscales (YPF) at a price set at a small discount to the Henry Hub benchmark in Louisiana. Repsol had exclusive rights to purchase the production at the Peru LNG project (table 3–16). The CFE contract with Repsol started with the commissioning of the Manzanillo terminal. Approximately 1.6 MMt of LNG was delivered from Peru into Manzanillo in 2013, and deliveries ramped up to 3.2 MMt/y in 2014. Gas demand in southwestern Mexico has been rapidly rising, but the region is inadequately linked to the national gas grid, which contributes to a growing reliance on LNG supplies through the Manzanillo terminal. The contract terms, which are highly favorable to Mexico, have come under heavy criticism from Peruvian authorities. In January 2014, Shell purchased most of Repsol's LNG assets, including participation in Peru LNG and the contract to supply CFE.

Caribbean and South America

The Caribbean and South America regions had 7.7 Tcm (270.9 Tcf) of proven natural gas reserves at the end of 2013, of which 5.6 Tcm (196.8 Tcf) are in Venezuela, according to BP statistics.[87] Yet production at 176.4 Bcm (6.2 Tcf) in 2013 was just one-fifth of that in North America. Regional consumption has increased by 59% from 106.8 Bcm (3.8 Tcf) in 2003 to 168.6 Bcm (6.0 Tcf) in 2013, and LNG imports have surged even as Trinidad and Peru have become significant LNG exporters (fig. 3–21). There is limited connectivity between markets in the region, with pipeline trades primarily from Bolivia to Brazil and Argentina, from Argentina into Chile, and between Colombia and Venezuela.

Once considered a niche market for LNG, the region is gaining prominence. LNG trades reached 13.4 MMt (18.2 Bcm) in 2013. During this year, Argentina and Brazil imported 4.7 MMt (6.4 Bcm) and 4.0 MMt (5.4 Bcm) respectively, followed by Chile at 2.8 MMt (3.8 Bcm) according to Poten & Partners statistics.[88] In the Caribbean, Puerto Rico imported 1.1 MMt (1.5 Bcm) and the Dominican Republic another 0.9 MMt (1.2 Bcm) in 2013. More than half these volumes originated in Trinidad & Tobago, which sold 7.6 MMt (10.3 Bcm), or 55% of its 13.9 MMt (18.9 Bcm) of production in 2013, into these markets (fig. 3–22). Peru was the only South American LNG exporter supplying global markets, with 4.1 MMt (5.6 Bcm) in 2013.

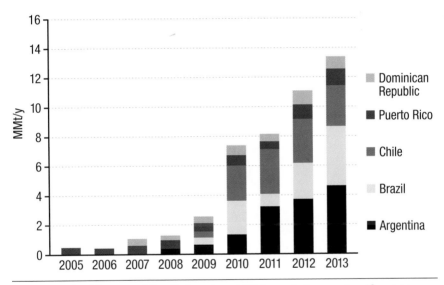

Fig. 3-21. Caribbean and South American LNG imports by country. (*Source:* Poten & Partners.)

Fig. 3–22. Caribbean and South American LNG import terminal map. (*Source:* Poten & Partners.)

Puerto Rico

EcoElectrica, Guayanilla. Enron and its partner Kenetech built the EcoElectrica power plant and adjacent LNG import terminal in Guayanilla, on Puerto Rico's south coast, in 1999.[89] The 500-MW combined cycle plant initially ran on propane delivered into Enron's existing LPG terminal. Upon completion of the new LNG import terminal in July 2000, the facility switched over to LNG supplied under a long-term contract by Cabot LNG (now GDF Suez). Spain's Gas Natural, which holds contracts for Trinidad LNG, purchased

Enron's stake in EcoElectrica in mid-2003. The facility is now owned by International Power/Mitsui (50%), Gas Natural (47.5%), and GE Capital (2.5%). An expansion of the terminal and development of downstream pipelines to other customers on the island is under consideration.

Aguirre Offshore GasPort. Puerto Rico's state-run power authority and US floating LNG specialist Excelerate Energy finalized a deal in early 2014 to develop a second LNG import terminal for the Caribbean island that has struggled to convert its fuel consumption from fuel oil to natural gas.[90] The 15-year agreement for the floating Aguirre Offshore Gasport on the island's south side finalizes a process that has dragged on for years amid regulatory delays and fiscal woes on the island. Aguirre's start-up date is now targeted for 2016, following approval by the FERC in Washington in early 2015.

Dominican Republic

AES Andres. AES followed a business model similar to EcoElectrica in developing the Caribbean's second LNG import terminal in the Dominican Republic. The US-based power generator built an LNG import terminal and associated 310-MW gas-fired power plant, which began operations in 2002.[91] Both the Puerto Rico and Dominican Republic terminals are supplied from Trinidad's Atlantic LNG project, which started up in 1999. BP's contract to supply LNG to the Dominican Republic contained a commercial innovation. While Trinidad will likely be the source of most LNG delivered under this contract, it was not designated as the dedicated source, thus allowing the British firm to claim this as a sale of BP "branded LNG." The terminal was initially extremely underutilized, with just a few cargos delivered in 2003 and 2004, largely because Dominican electric distributors failed to pay for the electricity produced by the power plant. This underutilization in turn forced AES to cancel contracted LNG deliveries from BP. Deliveries were resumed in 2007 and have climbed to nearly 1.0 MMt/y in 2012 and 2013—close to the nameplate capacity of the terminal.

Argentina

Argentina is South America's largest natural gas producer. However, its heavily regulated energy sector includes upstream price controls that, while they shield consumers from rising prices, limit the industry's attractiveness to private investors. Imposed in 2001 to combat inflation and help consumers during the economic crisis, these controls remain in place and cause natural gas to be relatively inexpensive by regional standards. The policy has deterred investment and production, stimulated consumption, and driven the country to rely on growing volumes of imports.

Argentina's proven natural gas reserves have declined from 0.7 Tcm (24.7 Tcf) in 2002 to just 0.3 Tcm (11.1 Tcf) in 2013.[92] Concurrently natural gas production after peaking at 46.1 Bcm in 2006 has fallen by approximately 23% to 35.5 Bcm (1.3 Tcf) in 2013. Meanwhile, consumption has climbed by more than half from 30.3 Bcm (1.1 Tcf) in 2002 to 48.0 Bcm (1.7 Tcf) in 2013. This discrepancy has led to a growing dependence on natural gas imports, which reached 12.1 Bcm in 2013 (with 5.2 Bcm by pipeline from Bolivia and 6.9 Bcm in the form of LNG).

Argentina has suffered severe wintertime shortages of natural gas—reportedly up to 40% of demand. But with the winter peak exactly offsetting the summer drop in northern hemisphere demand, Argentina has been able to utilize LNG to fill the gap—first during the peak demand period and increasingly year-round (table 3–17). Two LNG import terminals using Excelerate Energy's dockside regasification technology, the first at Bahía Blanca and the second at Escobar, were built to facilitate LNG imports, which have increased by a factor of 10 from approximately 500,000 t/y in seasonal operations in 2008 and 2009 to 4.6 MMt in year-round service in 2013.[93]

Table 3–17. Argentina LNG imports by source (MMt)

	2005	2006	2007	2008	2009	2010	2011	2012	2013
Qatar	–	–	–	–	–	0.2	0.6	0.1	0.6
Trinidad	–	–	–	0.3	0.5	1.1	2.1	2.6	2.4
Other	–	–	–	0.1	0.1	–	0.5	1.1	1.7
Total Imports	–	–	–	0.4	0.7	1.3	3.2	3.7	4.6

(*Source:* Poten & Partners.)

In order to leverage Argentina's promising unconventional natural gas resources and revitalize domestic production, the government instituted the Gas Plus program, which entitles companies to sell natural gas from new or unconventional fields at higher prices. Projects that were recently approved under the Gas Plus program will reportedly be allowed to charge around $5/MMBtu for their natural gas, which is double the national average price. So far, this policy has failed to yield significant results, as the government nationalized the assets of Repsol YPF, a major player in Argentina's upstream, in 2012. The ongoing financial crisis in Argentina has also given rise to problems for LNG imports, with severe shortages of hard currency forcing the country to pay for LNG deliveries in advance or else post onerous letters of credit.

Bahía Blanca GasPort. In 2007, when Argentina was faced with a natural gas supply shortage, YPF, S.A., in partnership with Repsol-Stream, approached Excelerate Energy for an economical and timely solution to bring more natural gas to market. In response, Excelerate developed Bahía Blanca GasPort (BBGP), Argentina's first LNG receiving facility, 400 miles south of Buenos Aires in the port city of Bahía Blanca. Utilizing a dockside facility design, BBGP employs an FSRU in conjunction with an articulated high-pressure gas-offloading arm.[94]

At BBGP, an FSRU is moored at the jetty to regasify LNG on board and deliver high-pressure natural gas. The gas comes ashore via an integrated midship gas manifold connected to a high-pressure gas arm. The BBGP is capable of delivering natural gas at a baseload rate of 400 MMcf/d, with peak rates of up to 500 MMcf/d. It is also flexible enough to accommodate increased throughput resulting from the higher vaporization capacity of future FSRUs. Relatively low capital investment and speed of implementation were the primary motivators behind the final investment decision. The facility was completed in less than 10 months from design to completed construction.

GNL Escobar. GNLE, located on the Paraná River about 30 miles outside Buenos Aires City, is Argentina's second LNG receiving facility. It uses the same Excelerate Energy GasPort design that is used at Bahía Blanca. Jointly developed by YPF S.A., ENARSA, and Excelerate Energy, GNLE was commissioned in June 2011, having taken only nine months from construction start to commissioning.[95] GNLE, with a baseload throughput capacity of 500 MMcf/d and peak throughput capacity of 600 MMcf/d, delivers additional natural gas to the Buenos Aires region. Limited depths in the Paraná River restrict the size of ships able to deliver LNG into GNLE and can require complicated LNG lightering and transshipment options at the river's mouth.

Brazil

At the end of 2013, Brazil had 16 Tcf (0.5 Tcm) of proven natural gas reserves according to BP.[96] The Campos, Espírito Santo, and Santos Basins hold the majority of reserves, but sizable reserves also exist in the interior of the country. In 2013, Brazil produced 0.8 Tcf (21.3 Bcm) of dry natural gas, mostly associated with oil. This is more than double the 0.3 Tcf (9.2 Bcm) it had produced a decade earlier. Still, production did not keep pace with consumption, which increased from 0.3 Tcf/y (9.2 Bcm/y) to 1.3Tcf/y (37.6 Bcm/y) during this time frame. The supply shortfall was filled by gas imports. In 2013, Brazil received 0.4 Tcf (10.7 Bcm) of pipeline gas from Bolivia and imported 0.2 Tcf (5.1 Bcm) in the form of LNG.

Hydrocarbon discoveries in Brazil's offshore presalt layer have generated excitement about new gas production. Along with the potential to increase oil production in the country significantly, the presalt areas are estimated to contain sizable natural gas reserves. According to Petrobras, Tupi alone could contain 7 Tcf (0.2 Tcm) of recoverable natural gas, which, if true, could increase Brazil's total natural gas reserves by 50%. However, presalt development promises to be hugely expensive, and plans to monetize presalt gas resources by installing floating LNG production units at the fields have been put on hold.

Natural gas constitutes less than 10% of total energy consumption in Brazil. This reflects the country's 80% reliance on hydroelectricity for power generation. Demand for gas, particularly LNG, varies with rainfall. For example, LNG imports declined from 2.3 MMt in 2010 to just 0.9 MMt in 2011, when there was ample rainfall. Imports rebounded to 2.9 MMt in 2012 and to 4.1 MMt in 2013, as shown in table 3–18.[97] There are three LNG receiving terminals in Brazil, all based on FSRU technology.

Table 3-18. Brazil LNG imports by source (MMt)

	2005	2006	2007	2008	2009	2010	2011	2012	2013
Nigeria	–	–	–	–	0.1	0.7	0.1	0.3	0.7
Trinidad	–	–	–	–	0.4	0.8	0.1	1.1	2.0
Other	–	–	–	–	–	0.8	0.6	1.5	1.4
Total Imports	–	–	–	–	0.5	2.3	0.9	2.9	4.1

(*Source:* Poten & Partners.)

Pecém, Ceara State. Petrobras inaugurated Brazil's first LNG import terminal at the Port of Pecém, in Ceará State in northeastern Brazil in July 2008.[98] The Pecém terminal is capable of regasifying 7 MMm³/d (250 MMcf/d) to be used primarily by power plants in Ceará and Rio Grande do Norte States.

In April 2007, the *Golar Spirit* was chartered by Petrobras for 10 years for service at the Pecém LNG terminal. The conversion to an FSRU started in October 2007 at the Keppel Shipyard in Singapore and was completed in mid-2008. The ship headed to Brazil in June 2008, picking up a cargo in Trinidad on her way to Pecém. The *Golar Spirit* operates at a jetty to which it is moored. LNG is delivered to the FSRU by LNG carriers that berth to the other side of the jetty.

Guanabara Bay, Rio de Janeiro State. Petrobras inaugurated its Guanabara Bay LNG import terminal, which is now capable of regasifying 14 MMm³/d (500 MMcf/d), in March 2009.[99] The technology is essentially the same as that used in Pecém. After testing the installation using the *Golar Spirit* (temporarily relocated from Pecém), the *Golar Winter* was placed in service before being modified and relocated to Bahia. Subsequently, the *Exquisite* assumed duty, before it too was replaced by the *Experience* in 2014. The LNG that is regasified at the Guanabara Bay terminal is transferred over compressed natural gas (CNG) transfer arms to a 15-kilometer gas pipeline gas pipeline (10 km are underwater and 5 km are onshore) connecting the terminal to the southeastern region gas pipeline network. Regasified LNG is delivered to power plants in southeastern Brazil.

Bahia Regasification Terminal. Petrobras commissioned its third LNG import terminal, Bahía Regasification Terminal (TRBA), in January 2014, when gas first flowed into the state gas pipeline network. TRBA is located in the state of Bahía, and has a regasifica-tion capacity of 14 MMm³/d (500 MMcf/d). With the new terminal in operation, Petrobras's natural gas regasification capacity has risen from 21 MMm³/d (740 MMcf/d) to 35 MMm³/d (1,235 MMcf/d). The FSRU *Golar Winter* was stationed at TRBA in October 2013 after being replaced by larger FSRUs at Guanabara Bay.

Chile

The natural gas market in Chile increased significantly to a peak of 8.7 Bcm (0.3 Tcf) in 2004 as direct pipeline connections were built to Argentina. However, it declined to just 2.7 Bcm (0.1 Tcf) in 2008, following a cutoff of deliveries from Argentina. After the introduction of LNG, the market gradually recovered to 5.4 Bcm (0.2 Tcf) in 2012, but fell back to 4.3 Bcm (0.15 Tcf) in 2013, according to BP, of which 3.8 Bcm (2.8 MMt) were in the form of LNG (see table 3–19).[100]

Table 3–19. Chile LNG imports by source (MMt)

	2005	2006	2007	2008	2009	2010	2011	2012	2013
Trinidad	–	–	–	–	0.1	0.6	1.1	2.3	2.3
Other	–	–	–	–	0.3	1.9	1.9	0.7	0.4
Total Imports	–	–	–	–	0.4	2.4	3.0	3.0	2.8

(*Source:* Poten & Partners.)

Chile has disconnected regional natural gas markets. The north of the country is supplied via the Mejillones LNG terminal, and natural gas in that region is used predominantly for electricity generation by the mining industry. In the central metropolitan region (including Santiago), natural gas is primarily supplied by the Quintero LNG terminal. In the far south (Magallanes), natural gas supplies come from local production and feed a methanol plant. The LNG terminals have replaced Argentinian natural gas in the northern and central regions. The methanol plant in the south receives hardly any natural gas, and only one production train out of four is now operating.

GNL Quintero, central region. This LNG import terminal was completed in July 2009.[101] It was the first land-based LNG terminal in South America. The plant is designed with three 5 MMm³/d (175 MMcf/d) vaporizers, one of which functions as a backup. Thus the functional capacity at Quintero is 10 MMm3/d (350 MMcf/d) at peak. A ship usually arrives every two weeks, but stormy conditions in the bay can sometimes disrupt deliveries. BG Group, an original shareholder in Quintero, has negotiated term contracts to supply 2.5 MMt/y of LNG (up from the original 1.7 MMt/y) to three customers at the terminal—Chilean national oil and gas company ENAP, the power company Endesa Chile, and the gas distribution company Metrogas. By 2015, there are plans to incorporate all three vaporizers in regular operation, boosting capacity from 2.5 MMt/y to 3.75 MMt/y. In April 2012, BG Group sold its 40% stake in Quintero to Spain's Enagas, but retained the LNG supply contracts. The three buyers each have 20% stakes in Quintero.

GNL Mejillones, northern region. This regasification terminal, a 50/50 joint venture between state-owned copper company Codelco and GDF Suez, is located in the far north of the country.[102] The terminal initially comprised an offshore floating storage vessel with onshore LNG regasification capable of sending out 5.5 MMcm/d (200 MMcf/d). It began commercial operation in May 2010. In 2014, a 187,000 m³ onshore tank was commissioned, eliminating the need for the floating storage.

Notes

1. IEA, *World Energy Outlook* (Paris: OECD/IEA, 2011).
2. BP, *Statistical Review of World Energy* (London: BP, June 2014), http://www.bp.com/content/dam/bp-country/de_de/PDFs/brochures/BP-statistical-review-of-world-energy-2014-full-report.pdf.
3. Poten & Partners, LNGas Proprietary Service.
4. Ibid.
5. Ibid.
6. Poten & Partners, *Global LNG Outlook* (2014), Q1.
7. Poten & Partners, LNGas Proprietary Service.
8. Poten & Partners, *Global LNG Outlook* (2014), Q1.
9. Poten & Partners, LNGas Proprietary Service.
10. IEA, *World Energy Outlook 2011* (Paris: OECD/IEA, 2011), http://www.worldenergyoutlook.org/weo2011/.
11. Poten & Partners, LNGas Proprietary Service.
12. Poten & Partners, *Global LNG Outlook* (2014), Q1.
13. BP, *Statistical Review of World Energy 2014*.
14. International Energy Agency, *Medium-Term Gas Market Report* (Paris: IEA, 2014).

15. Ibid.

16. Poten & Partners, *Global LNG Outlook* (2014), Q1.

17. BP, *Statistical Review of World Energy 2014*.

18. Poten & Partners, LNGas Proprietary Service.

19. Ibid.

20. Poten & Partners, *Global LNG Outlook* (2014), Q1.

21. Ibid.

22. BP, *Statistical Review of World Energy 2014*.

23. Poten & Partners, LNGas Proprietary Service.

24. Ibid.

25. BP, *Statistical Review of World Energy 2014*.

26. Poten & Partners, LNGas Proprietary Service.

27. BP, *Statistical Review of World Energy 2014*.

28. Ibid.

29. Poten & Partners, LNGas Proprietary Service.

30. Ibid.

31. BP, *Statistical Review of World Energy 2014*.

32. Ibid.

33. Poten & Partners, *Global LNG Outlook* (2014), Q1.

34. Poten & Partners, LNGas Proprietary Service.

35. Poten & Partners, *Global LNG Outlook* (2014), Q1.

36. BP, *Statistical Review of World Energy 2014*.

37. "Drop in 2013 EU Gas Demand Emphasizes Need for Swift Change," *Eurogas.org*, posted March 2014, http://www.eurogas.org/uploads/media/ Eurogas_Press_Release_-_Drop_in_2013_EU_gas_demand_emphasises_need_ for_swift_change.pdf.

38. Ibid.

39. BP, *Statistical Review of World Energy 2014*.

40. Michael Tusiani, "Let's Get Real on Whether Exports of U.S. Gas Can Counter Mr. Putin," *TradewindsNews.com* (March 2014), http://www. tradewindsnews.com/weekly/334460/lets-get-real-on-whether-exports-of- us-gas-can-counter-mr-putin (members-only access).

41. International Energy Agency, *Energy Policies of IEA Countries—The United Kingdom* (Paris: IEA, 2012).

42. BP, *Statistical Review of World Energy 2014*.

43. Shell, "Overview of Ormen Lange Project" (November 7, 2012), http://www. ipt.ntnu.no/~jsg/undervisning/naturgass/lysark/LysarkGupta2012.pdf.

44. Poten & Partners, LNGas Proprietary Service.

45. BP, *Statistical Review of World Energy 2014*.

46. Poten & Partners, LNGas Proprietary Service.

47. BP, *Statistical Review of World Energy 2014*.

48. Ibid.

49. Poten & Partners, LNGas Proprietary Service.

50. BP, *Statistical Review of World Energy 2014*.

51. Poten & Partners, LNGas Proprietary Service.

52. BP, *Statistical Review of World Energy 2014*.

53. Poten & Partners, LNGas Proprietary Service.

54. BP, *Statistical Review of World Energy 2014*.

55. Ibid.

56. Ibid.

57. Poten & Partners, LNGas Proprietary Service.

58. BP, *Statistical Review of World Energy 2014*.

59. Poten & Partners, LNGas Proprietary Service.

60. BP, *Statistical Review of World Energy 2014*.

61. Poten & Partners, LNGas Proprietary Service.

62. BP, *Statistical Review of World Energy 2014*.

63. Ibid.

64. Potential Gas Committee, "Potential Supply of Natural Gas in the United States," Colorado School of Mines (Golden, Colorado: CSM, December 2013).

65. US Energy Information Administration, *Short-term Energy Outlook* (Washington, DC: EIA, June 2014).

66. Ibid.

67. Ibid.

68. Mark Green, "MIT: The Facts on Fracking Methane Emissions," *EnergyTomorrow.org*, posted November 2012, http://www.energytomorrow.org/blog/2012/november/mit-the-facts-on-fracking-methane-emissions.

69. EIA, *Short-term Energy Outlook* (June 2014).

70. GDF Suez NA, "LNG Products and Services," http://www.gdfsuezna.com.

71. Federal Energy Regulatory Commission (FERC), Order Granting Section 3 and Section 7 Authorizations and Approving Abandonment, issued December 17, 2015, http://ferc.gov/whats-new/comm-meet/2015/121715/C-1.pdf.

72. See, for example, http://www.energy.ca.gov/lng/documents/SIGNIFICANT_EVENTS_LNG_HISTORY.PDF.

73. Dominion, "History of Cove Point," https://www.dom.com/corporate/what-we-do/natural-gas/dominion-cove-point/history-of-cove-point.

74. "Cameron LNG," *CameronLNG.com* (c. 2014), http://www.cameronlng.com/.

75. "Sabine Pass LNG Terminal," *Cheniere.com* (c. 2015), http://www.cheniere.com/terminals/sabine-pass/.

76. "Freeport LNG," *FreeportLNG.com* (c. 2015), http://www.freeportlng.com/.

77. "Golden Pass LNG Terminal," *GoldenPassTerminal.com*, http://www.golden-passterminal.com/.

78. "Gulf LNG Clean Energy Project," *Downstream Today* (August 13, 2010), http://www.downstreamtoday.com/projects/project.aspx?project_id=56.

79. National Energy Technology Laboratory, *Liquefied Natural Gas: Understanding the Basic Facts* (Washington, DC: NETL/US DOE, August 2005), http://energy.gov/sites/prod/files/2013/04/f0/LNG_primerupd.pdf.

80. "Canaport LNG," *CanaportLNG.com* (c. 2015), http://www.canaportlng.com/.

81. BP, *Statistical Review of World Energy 2014*.

82. US Energy Information Administration, US Department of Energy, *Mexico Country Brief* (Washington, DC: EIA, April 2014), http://www.eia.gov/beta/international/analysis_includes/countries_long/Mexico/mexico.pdf.

83. Ibid.

84. "Terminals," *Vopak.com* (c. 2015), https://www.vopak.com/terminals.

85. "Energía Costa Azul," *EnergiaCostaAzul.com*, http://energiacostaazul.com.mx/english/.

86. "Samsung C&T Completes Manzanillo LNG Terminal," Samsungcnt.com (March 28, 2012), http://www.samsungcnt.com/EN/trading/ne/501000/articleRead.do?board_id=6&article_id=2209.

87. BP, *Statistical Review of World Energy 2014*.

88. Poten & Partners, LNGas Proprietary Service.

89. "Eco Eléctrica," EcoElectrica.com, http://www.ecoelectrica.com/.

90. "Aguirre Offshore Gasport," *AguirreOffshoreGasport.com*, http://aguirreoffshoregasport.com/

91. Brian Eisentrout, Tim McKinney, Jeff Sipes, and Barbara Weber, "The Dominican Republic LNG Import Terminal: Challenges in Engineering, Procurement and Construction," *LNG Journal* (January/February 2004, 23), http://www.cbi.com/images/uploads/technical_articles/CBI_LNG_Journal_Feb04-lores.pdf; and Governing Board of the Virgin Islands Water and Power Authority, "Energy Production Action Plan" (September 2012, 9), http://www.viwapa.vi/Libraries/PDFs/Energy_Production_Action_Plan.sflb.ashx.

92. BP, *Statistical Review of World Energy 2014*.

93. Poten & Partners, LNGas Proprietary Service.

94. "Bahia Blanca GasPort," *ExcelerateEnergy.com* (c. 2015), http://excelerateenergy.com/project/bahia-blanca-gasport.

95. "GNL Escobar," *ExcelerateEnergy.com* (c. 2015), http://excelerateenergy.com/project/gnl-escobar.

96. BP, *Statistical Review of World Energy 2014*.

97. Poten & Partners, LNGas Proprietary Service.

98. ANP, *Liquefied Natural Gas in Brazil: ANP's Experience in the Implantation of LNG Import Projects* (Rio de Janeiro: ANP, August 19, 2010), http://www.eisourcebook.org/cms/Brazil,%20Liquefied%20Natural%20Gas,%20ANP%20import%20experience.pdf.

99. Ibid.

100. BP, *Statistical Review of World Energy 2014*.

101. "GNL Quintero," *GNLQuintero.com*, http://www.gnlquintero.com/.

102. IEA, *Oil and Gas Security 2012* (Paris: IEA/OECD, 2012, 21), http://www.iea.org/publications/freepublications/publication/chile_2012.pdf.

4

Evolution of the LNG Business Model

Introduction

The high capital costs involved in developing the LNG industry resulted in a cautious and conservative approach during the first 40 years. Nearly all the LNG projects during those years followed an integrated project model, and the output was sold directly to credit-worthy gas or electric utility buyers in countries with high credit ratings. The utilities' credit was underpinned by their captive end-use customers, and the full costs of obtaining the gas were recovered in regulated cost-plus tariffs.

During this period, there were relatively few export and import facilities, and those that existed were tightly controlled, often with significant host country government involvement on the export side and monopoly utilities on the import side. In the absence of contractual commitments between buyers and sellers, potential projects were not developed due to concerns that liquefaction plants, receiving terminals, and/or ships might remain idle for the lack of LNG supply or market demand. Speculative development was not a viable option for participants or financiers, since project economics, then as now, depended on uninterrupted delivery of project volumes. Consequently, long-term contracts that committed most or all of a facility's capacity to creditworthy buyers lay at the heart of the LNG business. Indeed, financial institutions backing a proposed liquefaction project insisted on buyer commitments, as well as certification of upstream reserves and shipping dedicated to the venture, before consenting to project lending.

The Traditional LNG Business Model

Multiyear sale and purchase agreements (SPAs), typically for 20 years, were the basis of the traditional LNG business model. These long-term SPAs minimized project investors' risks by essentially guaranteeing the project's ability to repay its borrowings, while giving the shareholders a reasonable return on their investment. Historically, contract terms were inflexible and did not accommodate radical changes in volume arising from shifts in market demand. Virtually all long-term SPAs contain take-or-pay provisions. As the name indicates, if the buyer failed to take the contracted volumes, it paid for the LNG anyway. Within the plant's capacity limits, the shortfall could be made up later in the contract period when requested by the buyer, and only a price difference from the missed delivery, if any, was paid. There was generally limited offtake flexibility (less than 10% of annual quantities), and no destination flexibility. This meant that the contracted volumes had to be delivered to the import terminal as specified in the SPA. Prices were largely set with reference to crude oil or petroleum products, and there was limited opportunity to review and reopen pricing provisions.[1]

The commercial agreements embedded in the SPAs also reflected an iterative process of infrastructure development coordination between buyers and sellers, and the SPAs were executed before construction started. Tankers were ordered and built for specific trade routes. The number of ships and their sizes were customized to accommodate the specific distances and volumes involved in each trade. Similarly, receiving terminals were built to accommodate the ships and the quality specifications of the natural gas they would be receiving.

The traditional model is conceptually similar to a floating pipeline where gas is transported continuously from one end of the pipeline (the liquefaction facility and loading terminal) to the other end (the import and regasification terminal) by a dedicated shipping fleet. The model of a floating pipeline guaranteed a steady, dependable trade.

Barriers to flexibility in traditional contracts

The traditional model had inherent limitations that discouraged any deviation from the long-term point-to-point trade model. Even if the economics favored diverting contracted volumes, the following factors could prevent that from happening:

Dedicated shipping. SPAs contained restrictions on changing the ships involved in a long-term SPA, a particular concern if the proposed new outlet involved longer shipping distances and therefore could alter the scheduling of the overall fleet. Consultation and approval from all participants in the existing and proposed trades were required to add new ships for an existing trade, or to trade dedicated ships into a new plant or terminal. This was a time-consuming process at best.

Buyer needs. Utility buyers' main concern was security of supply. They insisted on high levels of reliability to ensure security, which was paid for through their contracts. Their high credit ratings underwrote the project. If a ship was diverted from a long-term trade, the reliability was potentially reduced and the risk of shortfall increased. Buyers had no economic incentive to consent to changes in arrangement since they could not, under the utility cost pass-through model, reap any economic benefit from additional value created by such short-term opportunities, but instead had to pass this value on to their customers. On the other side, should the utility come up short of supply, the consequences could be severe, both economically and in terms of relationships with regulators overseeing and approving the utility's investments and rates of return.

Limited LNG markets. Even if sellers had the liquefaction and shipping capacity to produce and transport extra cargoes, there were only a small number of receiving terminals to which such cargoes might be delivered. In 1980, for example, there were just four operating receiving terminals in Europe, two in the United States, and seven in Asia. Moreover, all these terminals had been constructed because their owners had secured long-term supply.

Liquefaction commitments. Project owners did not intentionally build surplus liquefaction capacity for use outside long-term commitments. Although liquefaction trains were not designed with excess capacity, conservative design practices intended to ensure that the facility could meet its guaranteed obligations usually resulted in some spare capacity. As time passed and operations became more efficient, some liquefaction plants developed significant spare capacity, which was available for sale. Again, the terms of the SPAs intervened. Even when other sales opportunities beckoned, the LNG supplier generally had to offer to the existing customers the right of first refusal for additional short- and long-term volumes. These spare volumes were often committed under new or expanded long-term contracts with the original buyers, and the short-term quantities were used by the existing buyers to improve volume flexibility.

Lack of transparency and liquidity. Lack of competition in gas markets meant that there was no transparency and no liquidity. There were no reliable independent price signals to let buyers and sellers know if a given transaction was "in the market." Moreover, natural gas markets entirely based on long-term contracts from the wellhead to the utility left little opportunity to find buyers for offtake short-term volumes, especially with all LNG terminals operating on a proprietary access model. This lack of competitive markets helped entrench the long-term traditional model.

LNG "club rules." The LNG community was held together by long-standing relationships between limited numbers of counterparties. High costs of entry and the need for de facto government approvals on the buyer and seller sides limited the number of international companies that were willing to participate. The result was a small, exclusive club of participants that included state-owned national oil companies, utilities, and energy majors.

In addition to the obstacles identified above, the need for consent between buyer and seller to anything that represented a deviation from the basic SPA created another obstacle to changing the model. The consequences of something going wrong so outweighed the benefits of everything going right that ultimately nothing happened. There was also the unspoken concern that once one party proposed reopening the SPA for new commercial considerations, the other party would also have the right to revisit the provisions of the agreement that they might wish to change. Overlying this was the lack of market transparency that made novel transactions hard to value, and the underlying fear that the party proposing the transaction would disproportionately benefit from the new commercial arrangements.

Early attempts to break the tradition

As long as each new LNG project required mutual long-term commitments, there was no uncommitted capacity in the LNG chain to permit the development of an LNG commodity market via spot or short-term trading of LNG cargoes and transportation. This does not mean that there were no early attempts to inject innovation into the LNG business, and enterprising companies saw the shipping sector as the optimal entry point. In the early days of the industry, several shipowners placed orders for LNG vessels without having first committed them to a dedicated trade. In some cases, the owners of these so-called speculative ships were able to find stable employment for their vessels. Others, instead, met with financial disaster.

For example, the *Gastor* and *Nestor* were built for an Anglo-Dutch group in 1976 and 1977 as a speculative investment. The two vessels were delivered into a shrinking market and remained idle until purchased in 1990 by Nigeria LNG's Bonny Gas Transport at a significant discount to their original cost. Two other ships suffered a similar fate. Built in Sweden in 1981 and 1984, they represented a last-resort effort by the Swedish government to find employment for the Kockums shipyard. Bonny Gas Transport acquired these vessels in 1990 as well.

The US experience was even more painful. El Paso, one of the largest US pipeline companies, undertook a project to import LNG from Algeria to the United States with a fleet of nine dedicated vessels, selling the LNG to the owners of the Cove Point and Elba Island Terminals. However, a dispute arose between the LNG importer, El Paso, and the Algerian seller, Sonatrach, over the price of the LNG. Following the oil price increases that resulted from the Iranian Revolution of 1979, Sonatrach sought much higher prices linked to oil for its LNG sales from all its customers. El Paso asked the DOE to intervene in the pricing dispute, which quickly escalated to a government-to-government negotiation. In 1976, the Canadian government had set a "uniform border price" for natural gas deliveries to the United States based on fuel substitution values tied to US crude oil prices. By 1979, this price had risen to over $4.00/MMBtu, and Sonatrach argued they should be entitled to the same pricing consideration. The DOE refused to accept a price significantly higher than the contract price of $1.30/MMBtu.[2] El Paso's import plans were aborted, and LNG shipments stopped in April 1980. While El Paso had not ordered its ships on a speculative basis, the outcome was the same as if they had: The company could not control the LNG at the export point or in the market through regasification terminals, and so suffered a sad exit from the business.

In February 1981, El Paso announced it was writing off its LNG investment, and it barely avoided a bankruptcy filing. Matters got worse when three LNG tankers ordered by El Paso from Avondale Shipyard failed to pass their trials when new cargo tank insulation broke down under cryogenic temperatures. The ships were declared a total loss and were taken over by the US Maritime Administration (MARAD). Attempts to refit them for other cargo services never proved commercially viable; one vessel grounded and was scrapped, and the two remaining vessels were eventually also scrapped. Three other El Paso LNG vessels were scrapped outright. Three ships were repossessed by MARAD, which had provided financing, and mothballed before being sold in 1990. MARAD took one of the biggest write-offs in its history associated with the financing of the El Paso LNG vessels.

Trunkline's LNG tankers followed El Paso's into MARAD repossession when the trade into Lake Charles was suspended in 1981. Combined with the other idle speculative LNG ships ordered by hopeful LNG shipping players, the laid-up fleet exceeded 20 LNG vessels by the mid-1980s. Seven of the laid-up vessels (including those mentioned above) were secured by Bonny Gas Tranport and Shell for Nigerian LNG service, but did not resume long-term service until 1999. The venture that had proved to be so difficult and expensive for El Paso and the speculative shipowners convinced participants in the LNG industry that the traditional business model should not be tampered with. This mindset continued to dominate the LNG industry for the next decade.

The door to new types of trade opened gradually during the early 1990s when surplus LNG production in Indonesia led state-owned oil and gas company Pertamina to offer innovative sales arrangements in order to sell excess LNG to its Japanese buyers. These deals, known unofficially as the "superdeals," introduced flexibility into individual offtake obligations and used surplus shipping capacity within the Indonesian fleet. Each deal was time- and volume-limited and thus distinct from the respective existing long-term supply contracts. Contracts had no take-or-pay provisions, volumes varied from year to year but were subject to a total number of cargoes to be taken over the life of the contract, and price incentives maximized offtake. The superdeals involved additional sales only to existing contract holders and required buyers' commitments ahead of the contract year so that cargoes could be scheduled. Nonetheless, these contracts represented the first substantial break from traditional contractual arrangements and set the stage for further innovations.

Later in the decade, a series of interlinked events helped to reinforce the growth in nontraditional trades. The Asian financial crisis dampened regional demand at the same time that new liquefaction projects came on line to serve Asian markets. During this period, European countries were supply-constrained because of their growing demand coupled with production problems at the Algerian liquefaction facilities, which dominated European LNG supply. In an unprecedented move, the market responded to these imbalances by redirecting some of the LNG from new Asian projects to Europe and the United States.[3] In December 1986, a cargo of LNG made its way from Bontang, Indonesia, to Boston in one of the first true spot LNG transactions ever undertaken. The two Swedish ships bought by Bonny Gas Transport in 1990 commenced a four-year time charter starting that year, with Pertamina and Malaysia LNG. Several short-term charters followed, before the ships entered service for Nigeria LNG in 2001.

Transformation of the Industry

After about forty years dominated by conventional transactions, recent decades have seen many downstream natural gas and electric power markets transformed by liberalization. The LNG industry has been forced to react to serve the markets that have adopted competition at the wholesale level. The single largest change has been an increase in the number of shorter-term contracts that offer more flexibility in offtake volumes, delivery points, and timing. This increase in flexibility has in turn led to the creation of a growing spot and short-term market for LNG. This transformation is the result of a confluence of multiple factors, discussed below.

Gas and power market liberalization in the United States and Europe

The traditional downstream natural gas industry had, as its primary objective, the obligation to ensure adequate supplies for consumers at reasonable rates. Gas transmission companies and utilities made money by earning a regulated return on their investments in assets, while the cost of natural gas and operating expenses were passed through to consumers. This system did not give incentives to utilities to obtain the best prices or structure the most efficient contracts, as any additional benefits were simply returned to the consumers in the form of lower prices. Under most models for deregulation of the wholesale natural gas market, gas transmission companies either had to exit their merchant (that is, natural gas buying and selling) roles or legally separate the two roles in a process called unbundling. That is, they had to turn into pure transporters under an "open-access" or "third-party access" (TPA) approach that required them to allow other companies to gain access to their transmission systems on a nondiscriminatory basis, and thus open their end customers to competition. The transporters became operators and sellers of transportation capacity. Multiple companies bid on this capacity, and if they were not using it themselves, they resold it. As a consequence, multiple companies might be sending natural gas through a pipeline at the same time. This process began in the United States in the mid-1980s, then spread to the United Kingdom, Canada, and ultimately to the rest of Europe.

By unbundling these functions, the regulators hoped to create an active wholesale market, where natural gas would be freely traded throughout the pipeline grid. As a result, natural gas is actively traded in North America, the United Kingdom, and, to a growing extent, in continental Europe. Such trading is a remote possibility in established Asian LNG markets. Privatization and liberalization in Japan and South Korea's gas and electric industries are developing, although the progress has been slow.[4] Wholesale electric markets, following a similar process, have a "knock-on" effect on natural gas markets, natural gas being the marginal fuel in most competitive wholesale power markets.

In countries that have moved and are moving to competitive markets, the pricing of natural gas is changing with the growth of gas-on-gas competition and the resulting decoupling of natural gas and oil prices. This change is most advanced in the US transmission market where regional trading hubs have been created, benchmark prices are transparent and liquid, and futures, swaps, and basis[5] markets have emerged, along with crosscommodity trading (electricity versus gas) opportunities. Although a true global price for gas may not exist, these regional benchmarks facilitate price transparency, helping LNG buyers and sellers optimize deliveries between North America, Europe, and Asia—the three key importing regions. Rather than a single global price for LNG, there will probably be several marginally correlated regional prices. Since LNG is significantly more expensive to transport than oil, and linkages take years to establish, the LNG price correlation will remain weaker.

Opening wholesale natural gas markets to competition turns the risk–reward equation upside down. Instead of the more stable regulated prices prevalent under long-term contracts, gas prices can become highly volatile, fluctuating widely with swings in demand and supply. There is plentiful opportunity to make and lose vast sums of money very quickly. These opportunities, inherent in the volatility of wholesale gas and electricity markets in North America, drove the rise of the merchant energy traders. Some of the early merchants (Enron, Dynegy, Mirant, and Calpine), who often did not even own assets but lived (and died) on their ability to trade just the wholesale commodity, encountered financial disaster. These markets have also witnessed scandals—the Enron collapse, the California energy crisis, and the demise or crippling of many energy merchant companies.

Increasingly, commercial and investment banks and hedge funds are active in the gas trading markets, seeking to profit from proprietary trading strategies in largely financial products and offering hedging products to asset owners such as gas producers and power generators. The merchants remaining in the physical business are increasingly focused on the retail sector, serving industrial, commercial, and residential consumers in competition with each other and the local utility. Local utilities have often been unbundled as well, though not to the same degree as the wholesale transmission companies.

Before the emergence of shale, this liberalized market in the United States offered LNG suppliers and traders relatively easy access to a liquid and transparent market, and the ability to deliver LNG on very flexible terms (diverting cargoes at the last minute). The United Kingdom offered similar opportunities, and other European LNG terminals have seen growing interest in shorter-term and more flexible deals. However, in the wake of the Fukushima disaster, diversions of volumes went to Japan instead.

With crude oil trading above $100/bbl for several years prior to late 2014, Asian LNG term prices were on the order of $15–$16/MMBtu. At the same time, US natural gas prices rarely went over $4/MMBtu. The result was an opening of wide margins between US Henry Hub–based LNG export prices and Asian prices. This spurred a rush of contracting for US LNG by both Asian utilities and many of the LNG merchant aggregators and trading companies. With new US LNG not subject to destination restrictions, the expectation was that LNG trading could grow dramatically. However, when oil fell back to less than $70/bbl in late 2014 and towards $50/bbl in early 2015, Asian term prices began to follow, as did oil-linked European pipeline supplies. At such low oil prices, if sustained, newly integrated projects appear unlikely to be economic. At the same time, US natural gas prices have not fallen as much as oil prices, leading to narrower or disappearing arbitrage margins between Henry Hub–based US LNG and Asian LNG at oil-linked prices. The resulting uncertainty may make it hard for new projects to advance, although there could be benefits from lower construction costs as contractors and equipment suppliers attempt to gain new business. The future of expanded trading (or at least its profitability) will likewise be challenged if these price levels prevail for an extended period of time.

Competitive financing

Financial institutions are now aggressively competing to fund huge LNG projects. Even with interest rates at historically low levels, lower financing costs are being more than offset by higher investment costs for liquefaction projects, both upstream and downstream, as well as for LNG tankers.

Expansion of the LNG "club"

Throughout the 1980s, only a few integrated major energy companies appeared to possess the project management, technical skills, and financial resources to develop a large new LNG project. Now, there are many international and national energy companies that aspire to this role, including utility and trading companies and private equity funds. The pool of engineering, procurement, and construction (EPC) contractors has expanded but not to the same degree. The dispersion of technological know-how among energy companies, EPC contractors, and equipment manufacturers has helped level the playing field by allowing new entrants to gain access to what was once perceived as proprietary technology and knowledge. Also, many potential exporting countries see benefits to working with companies less likely to dictate terms than the majors. On the market side, simpler technology and regulatory encouragement of competition have opened that segment to many more participants.

Changing upstream strategies

Energy companies have found it increasingly difficult to access international oil prospects. Oil development has become far more competitive and more difficult than gas, and is more tightly controlled by national oil companies who can contract for access to the necessary technology rather than partner with the oil majors. These factors have driven producers to explore for natural gas, often in areas increasingly distant from traditional markets. Given its scale and management challenges, the development of LNG projects lends itself to the super-majors' perceived strengths—namely marshaling the management, commercial, technical, and financial resources needed to execute multibillion dollar energy projects in remote regions of the world. Meanwhile, financial regulators have tightened the definition of proven reserves in financial reporting and are forcing upstream companies to downgrade reserves not under commercial agreement for exploitation. As a consequence, producers are shifting expenditures from exploration to developments connecting known reserves to markets, thus making LNG increasingly a favored option.[6]

Critical and diversified mass

The steady growth and diversification in the global LNG trade is giving participants confidence in their ability to find a buyer or seller within a reasonable period of time should anything happen to upset the initial commercial arrangements, or to begin constructing a project without all its output sold in advance. Examples include the swift redirection of volumes once destined for the Dabhol power project in India, or the redirecting of LNG volumes from the United States and the United Kingdom to Asia, following a reduction in demand or price in the original markets.

What does all this mean for the LNG industry? In some cases it may be more difficult for LNG importers to lock downstream buyers into long-term contracts, because these buyers may be unwilling to take on the long-term risk entailed by an inflexible contract. On the other hand, in a liquid and large gas market, finding long-term buyers may not be necessary and may even be undesirable if arbitrage opportunities between markets are available. With many new participants, including LNG traders, entering the market, finding sufficient creditworthy LNG and downstream buyers to underwrite an entire project is another challenge. At the same time, the increasing number of suppliers and buyers with access to "destination free" LNG allows industry participants to shift their attention from security of supply to price. Buyers and sellers are now willing to consider a wider range of pricing terms and mechanisms, and destination flexibility, as well as less rigid offtake and supply conditions. Competitive market forces, diverging corporate strategies, and higher financial risks are causing the traditional LNG club members to suffer some severe stress, which may lead to less civilized behavior. Commercial innovation may be supplanting technical knowledge as a competitive advantage for industry participants.

New Business Strategies

LNG participants have been forced to rethink their business and asset strategies in order to adapt to industry changes. The following trends represent departures from the traditional model:

- Venture partners are retaining significant volumes from their LNG export projects for their trading operations via equity LNG or sale and purchase agreements.

- Downstream integration by LNG producers as well as upstream integration by LNG buyers, seeking to improve financial returns through what each perceives to be a more lucrative segment of the business.

- Tolling business models where liquefaction capacity is being disaggregated from the gas supplies, and the LNG is being acquired by the capacity holders in the tolling facilities. Similar trends are developing in open-access import terminals, which largely operate with tolling agreements, and ownership of the LNG and natural gas is separated from ownership of the terminal.

- Ordering ships that are not dedicated to specific trades or projects is in practice, as well as extending the life of aging LNG carriers by shifting them into trading activities.

- Technological innovation that has opened up unconventional gas, deepwater reserves, and floating LNG technology, both for receiving terminals and for liquefaction.

- The emergence of a North American business model of "disintegrated" LNG based on gas supplies acquired in a freely traded market and LNG supplies which can be traded anywhere in the world.

Venture partners as offtakers

Some of the larger LNG players are developing portfolio approaches to LNG. Through LNG project participation, they have contracted for their own dedicated LNG supplies. They have ordered their own ships and are gaining access to markets through SPAs that allow them to supply LNG from different sources, and/or they are establishing dedicated regasification capacity in multiple markets. Examples of companies pursuing this approach include Shell, BP, and BG.

Integration along the LNG chain

Despite the growth in the number of shorter-term contracts, securing external financing for a project still requires finding long-term markets for the gas. An emerging strategy by liquefaction participants has been to contract for import terminal capacity, or build import terminals to provide a primary sales channel. Such integration helps ensure access to a market, mitigates some of the risk of excess liquefaction capacity, and allows the LNG seller to take advantage of price arbitrage opportunities across different markets. The ExxonMobil–Qatar Petroleum partnership built three large LNG import terminals—Golden Pass on the US Gulf Coast, South Hook in Wales, and Rovigo in Italy—as part of their megatrain strategy. They also built super-large LNG carriers to implement a floating pipeline concept.

Conversely, large buyers have opted to participate in upstream resource development and/or liquefaction plants to ensure access to LNG via SPA, or equity LNG via the acquisition of LNG linked to their shareholding in the project. They often opt to own their own LNG carriers, which provides them with destination flexibility and opens up the potential for arbitrage profits through trading. The result is that buyers can dispose of excess cargoes when their local demand is weak, and they are more able to ensure adequate LNG when local demand is strong. This strategy has been actively exploited in Australia to ensure market access as well as to help finance ever more costly projects. Project sponsors in Canada's British Columbia are also adopting this strategy to strengthen their prospects. They are all seeking strong Asian buyers as equity partners and offtakers.

The irony is that while the same results could be achieved through arms-length commercial arrangements, the respective players perceive the others' business segments as potentially more lucrative and offering higher returns than their own core businesses. It seems unlikely that both sides can be correct in their assumptions.

Tolling business model

The tolling business model was first implemented in Trinidad and then in Egypt. It allowed project participants to process their gas through the LNG plant and to market the gas themselves. This model has evolved in the United States, where terminal owners spotted the opportunity to add liquefaction trains and to sell capacity to third parties who did not necessarily own any upstream reserves. Few US upstream companies have taken capacity in the US liquefaction plants, perhaps reflecting the more diverse nature of the US gas production sector. In contrast, overseas buyers, anxious to acquire LNG at Henry Hub–linked prices, have actively signed capacity agreements. Most of them plan to purchase feed gas from the pipeline grid, though some are also tying up upstream resources. The terminal owners, often pipeline infrastructure companies, were comfortable with a cost-of-service model that encouraged them to adopt the tolling model. Moreover, a guaranteed cost-of-service fee embedded in the tolling agreements supports a funding process that promises to make the sponsor's banks comfortable. It is more an infrastructure play than a commodity one.

Shipping

Destination flexibility and spare shipping capacity are increasingly prized by buyers and sellers alike. If LNG buyers are to continue to accept rigid offtake obligations in the face of increased domestic market uncertainty, they must have control of shipping and destination rights to manage their supply/demand balance through merchant trading activities. Conversely, if sellers accept flexible offtake obligations, they must have access to shipping so as to move discretionary volumes to alternative markets. Both approaches will likely require that the parties have access to spare shipping capacity to maximize their opportunities in diverse markets. In contrast, the spare shipping capacity typical of the classic early project models was maintained solely as a means of ensuring highly reliable deliveries.

Spot and short-term trades have also helped transform the LNG shipping industry. For the most part, new LNG ships are still being ordered in response to specific requirements arising from long-term SPAs, and this will be the case for the majority of the LNG fleet. However, a surprising number of new ships are ordered without long-term charter arrangements. Even though the shipowner nearly always wants a secure revenue flow from a long-term charter, the ability to charter out the vessel spot or short-term lessens the financial risk until a long-term charterer can be found. With wide divergence in gas prices between geographically disparate markets, spot charter rates have varied from as low as $25,000 per day, when numerous ships are idle, to over $120,000 per day when the market is tight. By comparison, long-term charter rates had settled in the range of $70,000 per day for modern tonnage at the end of 2014.

The logistics of multiple buyers and sellers operating at export plants and import terminals creates new issues. In the traditional LNG project model, ships were committed to specific trades with a predictable schedule of loadings and discharges. But if there are several buyers each controlling their own ships—often of different cargo capacities and with the potential to discharge at different terminals on short notice—the scheduling process becomes more complex and can cause a loss of LNG production due to the plant filling its storage.

These logistical challenges may also be duplicated at the LNG import terminal. After all, it is becoming more common for receiving facilities that are located in countries with open natural gas markets to have multiple capacity holders, and for those capacity holders' separate LNG supply and shipping arrangements to inevitably conflict with one another. Given that the LNG importer may have entered into downstream pipeline commitments and/or strict natural gas sales agreements with end users at the terminal's tailgate, the terminal operator must carefully schedule each ship arrival and the sendout of LNG. The conflict between different users' cargo discharge and sendout objectives invariably requires complex contractual undertakings and almost always results in a diminution of the terminal's effective throughput capacity. The situation becomes even more complex if the original capacity holders are permitted to assign, lease, or sell their terminal capacity rights to third parties.

New technology

The most significant recent influence on natural gas supply has been the breakthroughs in developing unconventional gas—notably shale gas—through the combination of horizontal drilling and multistage hydraulic fracturing (or fracking). According to the

International Energy Agency (IEA), technically recoverable remaining global natural gas reserves totaled 810.0 Tcm at year-end 2012, representing a reserve-to-production (R/P) ratio of 238 years based on natural gas production for the year that ended December 31, 2011.

Another technological innovation, deepwater drilling, often coupled with subsea completion and production, was initially developed to exploit oil and gas fields in the US Gulf of Mexico, and is now opening new gas basins. Two in particular have significant LNG implications. The main operators off Mozambique, Anadarko and Eni, claim to have found 175 Tcf of gas reserves in deep water and are planning to build an onshore plant to process this gas for export. While exploration activity is in an earlier stage in Tanzanian waters, sufficient reserves have also been found there to encourage upstream operators to start talking about a second East African LNG project. Meanwhile, Noble Energy has identified 29 Tcf of gas in Israeli waters and another 7 Tcf offshore Cyprus. Both nations are now examining LNG export possibilities.

Finally, technology has advanced in the LNG industry, through the increasing maturity of floating regasification, and the initial commitments to floating liquefaction. Small-scale LNG projects are developing rapidly in response to niche markets for transportation fuel and marine fuel, or to permit the exploitation of remote gas reserves and markets.

"Disintegrated LNG"

The transformation of the North American supply picture, which has come about as a result of the shale gas revolution, has also transformed the potential future of the LNG industry. In contrast to the more traditional integrated projects based on stranded gas reserves, most North American LNG export projects are based on accessing a deep, growing, and liquid gas market, positioning the liquefaction plant just like any other large consumer of natural gas. In this model, the upstream may be entirely decoupled from liquefaction, as may be the transmission system (other than the last few miles to tie the liquefaction plant to the transmission grid). Liquefaction plant owners are often completely independent of and insulated from risks for gas supply and LNG sales and shipping. The LNG produced by these projects is almost all "destination free," promising to dramatically expand the short- and medium-term nature of the industry. The impact on the global LNG market and especially on global gas prices is the subject of much speculation and debate.

Effect of the transformation of the industry

For those companies electing to pursue it, the shift from the floating pipeline model to a network model with highly flexible trading arrangements is often characterized as an improvement. Optimization of assets is commonly cited as a positive aspect of the new model, but it depends on what part of the chain is being optimized. Although liquefaction may be more effectively used in the new model, since excess LNG can be sold in multiple locations, such optimization comes at a cost. It requires flexibility in shipping and import terminals, and excess capacity in both, and this may not represent the most efficient utilization of these assets. However, given the much larger capital commitments to liquefaction, these costs are easily absorbed if they lead to higher sales volumes or access to more lucrative markets. While the Qatar projects are now making limited use of the terminals they developed in the Atlantic, and have idled several of their LNG carriers, they have more than offset these costs by exploiting the increased demand and higher prices in the Asian markets. These markets are more available to them than they were during the early development of their projects.

The rise of spot and short-term LNG trades

Over the past decade, spot and short-term LNG trades grew from virtually nothing to nearly a third of the global LNG market today. In the LNG industry, spot and short-term trades comprise single and multicargo transactions as well as term deals up to four years in duration (fig. 4–1). Although frequently used, this vocabulary carries a different connotation in the LNG industry than in other commodity markets. Normally, spot trading refers to a cash sale and immediate delivery of a commodity, without restrictions on subsequent sales. In most commodity markets, spot sales dominate commercial transactions. However, with LNG, because of the planning required, the lack of a globally liquid gas market, and the relatively concentrated assets in the industry, sales of cargoes for immediate delivery comprise a very small share of spot LNG trades. It is also still rare to have cargoes change hands on the water, as is the case with the oil industry.

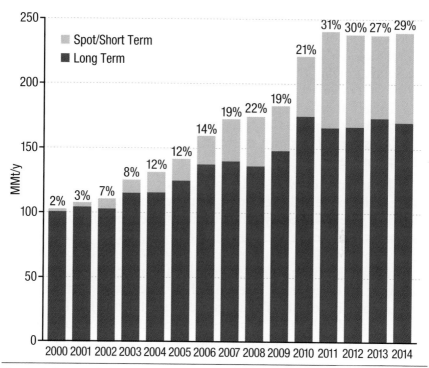

Fig. 4–1. Spot and short-term trade as a percentage of global LNG. (*Source: Poten & Partners.*)

A decade ago, when spot and short-term LNG trades accounted for a marginal share of the LNG business—less than 10 MMt/y of LNG and less than 10% of total trade—many major LNG sellers considered the volume too small to be worth pursuing. Furthermore, these trades added complexity to carefully crafted shipping logistics. Buyers were only interested if spot trades helped satisfy an immediate supply shortfall. This type of trade was never deemed part of a long-term supply strategy even to meet seasonal demand peaks. Given that almost all buyers were utilities unable to profit from trading the commodity, there was little motivation to pursue spot and short-term trades. This has all changed. In 2011, around 70 MMt of LNG was sold under these deals, accounting for about 30% of global LNG trade, according to Poten & Partners. The flexibility and profitability provided by these trades is now a critical component of the business for both buyers and sellers, as they adjust to unforeseen events such as nuclear shutdowns in Japan and varying price differentials between regional gas markets.

Korea Gas (Kogas), one of the two largest LNG buyers along with Japan's Tokyo Electric Power Co., was among the first buyers to embrace spot and short-term trades. In advance of every winter, the South Korean firm would enter the market with large purchasing programs geared to meet the peak winter demand. Kogas also found itself short of LNG year-round owing to the government's reluctance to approve the company's long-term purchasing plans, and Kogas turned to short- and medium-term LNG purchases to fill this gap as well.

In Japan, utilities were at first reluctant to rely too heavily on spot trades. They preferred the supply security that long-term purchases provided. They cautiously entered into some free-on-board (FOB) purchases, valuing the flexibility that control of shipping and destination provided. That reluctance waned in 2007 when the Kashiwazaki-Kariwa nuclear power plant shut down for 21 months following an earthquake. The shift then accelerated with the shutdown of all of the nation's nuclear plants following the Fukushima disaster in March 2011. Japanese utilities had no choice but to seek cargoes in the market on both a spot and short-term basis, frequently from their term suppliers. Qatar played a critical role in meeting Japan's needs by diverting cargoes away from Europe and North America on relatively short notice.

Buyers have realized the value of FOB purchases. They can contract for FOB LNG to ensure supply to their domestic markets, and with shipping control, can divert cargoes elsewhere when they are not needed at home. In recent years, European buyers, experiencing depressed gas demand as a result of economic recession, have encouraged cargo diversions and the reexport of cargoes from storage at their import terminals. European Union energy liberalization policy has also forced LNG and gas suppliers to drop the destination clauses in their SPAs, furthering this trend.

Meanwhile, new importers are relying heavily on spot and short-term supplies. Argentina goes to the market every year with large tenders for its two floating import terminals, and Brazil uses the spot market to address the unpredictability of rainfall affecting hydroelectric generation. New importers in the Middle East, South America, and Asia are adopting short-term purchasing strategies because of local demand uncertainties.

Regional price arbitrage had encouraged a large flow of cargoes from the Atlantic Basin to Asia, and many companies were chasing these profitable trades, attracted by the premium prices that cargoes commanded in Asia during the winter months. However, with the fall in oil prices in the second half of 2014 and early 2015, the arbitrage between Europe and Asia disappeared in early 2015. Still, many major LNG firms that viewed themselves primarily as upstream players are now building portfolios of FOB LNG supply not only to meet term sales obligations but also for trading. Spot and short-term trades are providing commodity trading groups a cost-effective entrée into the LNG business as well, though so far with varying success.

From a supply perspective, the use of spot and short-term trading was facilitated by LNG production volumes, which exceeded long-term SPA commitments resulting from:

Conservative liquefaction plant design. A focus on reliability of deliveries led to the overdesign of liquefaction trains and to conservative operating assumptions, with the result that actual plant capacity is often well above design capacity. Subsequent efforts to debottleneck and improve initial plant production also create extra volumes. The surplus volumes can be made available for spot trades, if they are not acquired by the existing buyers on a short- or long-term basis by being added to the existing long-term contracts.

Wedge volumes. This is capacity created when a new liquefaction train is brought on line and LNG is available for sale before buyers have achieved their long-term contract plateau volume offtakes. A liquefaction train requires very little buildup time before reaching full capacity, whereas SPAs very often include a buildup period—sometimes lasting for years—before the contract plateau level is reached. The buildup period is designed to allow the buyers time to build market demand, which usually materializes at a modest pace. LNG contracts also usually factor in extra time between the start of the plant and that of firm contractual deliveries, to provide a buffer in the event of construction delays. During this period, the buyers are rarely obliged to take early volumes. The dramatic increase in train sizes and output has lengthened the duration of the ramp-up period, leaving significant excess capacity available for spot sales.

Contract and operational flexibility. Spot capacity can arise when a buyer in a long-term contract exercises whatever volume flexibility is permitted under the terms of the contract. While these flexibilities may be limited to 10% or less, nevertheless they represent a source of spot volumes. Another source of spot volumes will arise as LNG suppliers offer seasonal deliveries to temperature-sensitive markets, as has been the case in recent term purchases by Kogas. The off-peak volumes can be marketed elsewhere. Liquefaction plants can often ramp up production for short terms for a variety of reasons. For example, the Atlantic LNG plant in Trinidad can increase production by as much as 5% during rainy periods.

Expiration of long-term contracts. Historically, before a long-term contract ended, the original buyer and seller renegotiated the terms and extended—and often expanded—the volumes under contract, maintaining the volume flow without interruption. This is no longer always the case. The increased number of buyers and sellers, regulatory changes, declining reserves, production behind liquefaction plants, and evolving corporate strategies often result in volumes under term contracts being reduced or not renewed at all.

Contract failure. In some cases, unforeseen circumstances will lead to the failure and early termination of an SPA. In one well-known example, India's Dabhol project, which was never completed because of a pricing dispute, reneged on its SPAs with Oman, forcing Oman to find new markets. However, in contrast to Sonatrach in the 1980s, which was left with a surplus of LNG when deals with the United States fell through, the Oman sellers were able to make use of the spot market to move volumes while seeking long-term buyers.

The use of spot and short-term volumes has moved well beyond these events, with firms acting as intermediaries by acquiring LNG portfolio supplies which can then be offered to buyers on a term basis or traded in the spot market. Access to shipping is an essential part of any LNG trading strategy. When an export venture has excess LNG to sell, it will tender it out on a delivered basis if it has sufficient shipping at its disposal, or FOB if it does not. Only firms with access to shipping can bid on the FOB cargoes. Those without available tonnage are frozen out of the tender. This is typically the case during the winter months when shipping is seasonally tight and arbitrage opportunities abound. Ship availability, no matter the season, has been a significant hurdle to entry of financial and other traders into this potentially very profitable business. Control of a sizeable LNG fleet is also a requirement for a successful portfolio LNG player.

Now, US liquefaction projects are breaking the traditional model completely. Not only will firms have to lease liquefaction capacity on a long-term basis, they will need to secure their own feed gas and, arrange its transport to the plant, and they will have to have shipping in order to transport the LNG to market. But they will be free to sell the LNG to any buyer and under whatever conditions they can negotiate. Many nontraditional players in the long-term LNG market are betting that Henry Hub–linked LNG will prove to be a good arbitrage play delivered to premium Asian and possibly European markets, and they are therefore queuing up to take part in this exciting new business opportunity. This development would expand LNG supply in general and spot and short-term trades in particular. While the arbitrage plays looked to be very lucrative through the middle of 2014, the subsequent collapse in oil prices may place the profitability and stability of these capacity arrangements in jeopardy.

Companies are using surplus capacity and LNG supply portfolios to arbitrage price disparities between different markets. However, simply because the price is higher in one market than in another does not mean it is high enough to justify selling the gas into the higher priced market. For example, assume that a Middle Eastern supplier finds ready buyers in both the United States and Spain. A single ship traveling from the Middle East can deliver only two cargoes to the United States in the same amount of time it would take to deliver three cargoes to Spain. In these circumstances, the costs of acquiring incremental short-term shipping would be a key factor in determining which market was the preferred destination.

Spot trading becomes even more complex when cargo diversions are taken into account. Most diversion arrangements call for the sharing of incremental profits between the buyer and seller of the diverted cargo. However, defining incremental revenues and costs is not always an easy matter, especially when the cargo is to be diverted to a market where pricing is not transparent, or where each of the parties may have taken different financial hedging positions which may require unwinding. As complicated as it is to agree on value sharing between two counterparties to a cargo diversion, the idea of accomplishing such agreement between three or four parties creates an almost insurmountable obstacle. By the time any agreement might be concluded, it is entirely possible the opportunity will have vanished. Diversion transactions are a function of destination restrictions and may slowly disappear from the commercial landscape.

In Asia, where most spot and short-term sales are made, the spot and short-term market is driven primarily by the physical requirements of the customers because of the lack of other sources of natural gas to create market liquidity. Because long-standing relationships are extremely important (driven by the potential for additional long-term sales), spot deliveries are often agreed between existing sellers and buyers at the underlying term contract price. Only if such sales cannot be made will the parties go outside their existing contractual arrangements. This is considered a part of the service for captive customers. This practice has eroded somewhat as buyers have used the spot market to obtain additional supplies on terms not readily matched by their existing relationships. Examples would include spot purchases by Japanese electric utilities to cover nuclear power production problems; by Kogas to cover peak winter season requirements; or by Indonesia's Pertamina to cover shortfalls in its own LNG production levels.

When Pacific buyers seek spot LNG, price is not usually the primary concern. Assurance of delivery or flexibility is paramount, as even very high-priced volumes will be averaged in with long-term contract purchases and the price differential passed on to the end customers. Conversely, offering low-priced cargoes to existing utility buyers may not attract any interest as the utility's customers have little or no ability to increase their consumption to take advantage of the lower price. Nor is there any mechanism to avoid the price averaging across all customers. Unlike the conditions which apply in North America, and to a lesser but growing extent in Europe, the Asian buyers do not operate open-access facilities (LNG terminals or downstream pipelines), do not have access to an integrated gas transmission grid, and do not have other sources of gas that can be substituted freely for LNG. Absent these elements, it appears unlikely that Asia will see the development of a traded LNG market in the sense it has occurred elsewhere. In spite of that, Singapore has declared its ambitions to become an LNG trading hub as well as provider of cooldown services for LNG ships—which may prove challenging, especially as its own market is very limited in size.

Reaching the negotiation stage for a spot LNG transaction requires finding a supplier with uncommitted capacity or finding potential buyers with available import capacity, and locating a vessel. Negotiating a spot cargo can be extremely time and resource intensive, and there is no guarantee that the deal will be brought to fruition, even in the face of compelling economics. Arranging a single cargo often takes almost as much time as arranging a term trade. To address this drawback, active participants in short-term markets have moved to more standardized "master" SPAs where the majority of the terms and conditions are established in advance. Individual term sheets addressing price, schedule, and other cargo specific aspects are then executed as cargoes become available.

Even with master contracts, these deals may be tricky to arrange. If a ship is required outside the seller's or buyer's fleet, it has to be vetted and must meet the requirements of the loading and unloading facilities. Also, scheduling additional cargo loadings and discharges at busy terminals can create further obstacles to successful transactions.

Among the emerging topics of debate is whether LNG can facilitate a global marketplace for natural gas, and when and how such a market will materialize. As this book goes to print, there is no global benchmark price for natural gas, although there are a number of regional benchmarks. As LNG increases its share of the global gas market, and spot trades increase their share of the LNG market, the market will become more sophisticated in arbitraging price disparities between regions. This should increasingly link various regional prices together and could move the natural gas industry towards global pricing benchmarks. The spot and short-term market will also drive gas-on-gas competition in markets where gas pricing has traditionally been linked to other fuels. Still, the absence of a physically liquid and price-transparent market in Asia, and the presence of a constrained one in Europe, will continue to create barriers to the creation of a global gas market.

Notes

1. Alaska and Brunei began LNG sales on a fixed-price basis. Indonesian sales, which began in 1977, were indexed to oil. Alaska (and probably Brunei) changed their contracts to oil parity in 1981. Abu Dhabi LNG sales started in 1977, had a fixed price, and changed to Murban Crude parity-plus-shipping in 1980. Early Asian contracts had meet-and-discuss clauses for price, but no requirement to agree to anything. However, prices rose after oil-indexed Indonesian prices began. When Brunei and Alaska contracts shifted to oil-indexed pricing, there were more frequent price reviews.

2. The base purchase price stated in the LNG sales contract between El Paso and Sonatrach was $1.30/MMBtu, subject to semiannual escalation based on the New York Harbor prices of No. 2 and No. 6 fuel oils. See US Energy Regulatory Administration, "Application to Import LNG from Algeria," ERA Docket No. 77-006-LNG (N.p.: ERA, December 21, 1978).

3. Abu Dhabi's ADGAS signed short-term ex-ship deals with Gaz de France and Belgium's Distrigaz to help fill Algerian supply shortfalls in 1995 and during the first quarter of 1996. The sales to Gaz de France and Distrigaz were followed by a short-term FOB contract with Spain's Enagas after the successful sale of two trial cargoes in early 1995. This contract was progressively extended through the end of 1998. ADGAS also sold several FOB cargoes to Cabot's Distrigas terminal in Boston in 1994 and 1995.

4. Japan and Korea are different markets. Japan has isolated markets with dedicated import terminals and distribution systems. Electric power utilities are also quite isolated, with their own import terminals, power plants, and transmission systems with limited connectivity to others. Kogas imports LNG and supplies natural gas to 12 city gas or electric power generation companies. It is Kogas's natural gas market dominance and the government's role in Kogas that delays liberalization.

5. Basis is the difference in price between two locations and can be seen as a proxy for the market value of transportation between these locations, even when the transportation service is provided by pipelines under a regulated tariff (which rarely matches the basis differential).

6. In 2005, 95% of ExxonMobil's worldwide gas reserve additions were associated with LNG projects in Qatar.

5

LNG Projects

Introduction

An LNG project traditionally comprises an integrated chain of facilities from the upstream to the downstream; however, the word "project" in this chapter refers to the liquefaction facility to which the upstream and shipping components are connected. This chapter contains a brief description of all liquefaction facilities that were operating, under construction, or in advanced permitting in 2014. It is impossible to predict which prospective projects will ultimately be completed, since circumstances constantly change. Therefore, we have not covered the more speculative ones. Indeed, the complete list of proposed projects is far too long to be covered in this book. For example, more than 30 liquefaction projects have been proposed for the United States, but details on only the five most advanced from a regulatory perspective are included here.

From 2000 through 2013, the number of countries producing LNG increased from 12 to 17. LNG exports more than doubled from 103.3 MMt (140.5 Bcm) in 2000 to 246 MMt (334 Bcm) in 2013, driven in large part by Qatar. When the decade began, Qatar was exporting approximately 10 MMt/y (13.6 Bcm/y), or approximately 10% of the global total; in 2013, this Middle Eastern nation exported approximately 77 MMt (104.7 Bcm), and its share had climbed to nearly 33%.

Building on the momentum established since 2000, the number of liquefaction projects has mushroomed beyond what anyone could have foreseen a few years ago, and now numerous projects and prospective schemes exist for Australia, North America, East

Africa, West Africa, the eastern Mediterranean, Russia, Indonesia, and Malaysia (fig. 5–1). Poten & Partners projects that global LNG production will reach 360 MMt/y (490 Bcm/y) by 2020, based primarily on Australian developments, and 450 MMt/y (612 Bcm/y) by 2030, driven by exploitation of North America gas for LNG export as well as exploitation of new natural gas resource basins such as East Africa.[1]

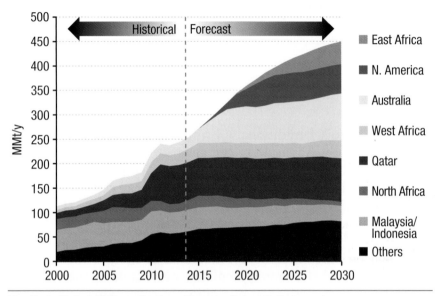

Fig. 5–1. Global LNG production. (*Source:* Poten & Partners.)

The Pacific Basin

Asia-Pacific

The Asia-Pacific region accounted for 36% of global LNG production, or 83 MMt (113 Bcm), in 2013 (fig. 5–2). This left a large regional net supply deficit that was supplied from the Middle East (primarily Qatar) and the Atlantic basin. (Asia-Pacific demand reached 178 MMt [242 Bcm] in 2013.) There has been a growing reliance on regional imports since the middle of the first decade of the 21st century, when this market was predominantly supplied by Indonesia and Malaysia. In 2005, regional LNG production was 65 MMt/y (88 Bcm/y) and demand totaled 92 MMt (124 Bcm/y). At the time, Indonesia and Malaysia were the two largest regional and global LNG exporters at 23.4 MMt (31.8 Bcm) and 21.6 MMt (29.4 Bcm), respectively. In that year, Qatar was third at 20.0 MMt (27.2 Bcm), and much of its production was delivered to markets east of Suez.

By 2013, Malaysian production had increased only modestly to 24.9 MMt (33.9 Bcm), while Indonesian output declined to 18.0 MMt (24.5 Bcm) (fig. 5–3). Meanwhile, Australian production more than doubled from 11.4 MMt (15.5 Bcm) in 2005 to 23.4 MMt (31.8 Bcm) in 2013. Concurrently, Qatar became the top supplier to the Asia-Pacific region—of its 77.1 MMt (104.9 Bcm) of production in 2013, 55.0 MMt (74.8 Bcm) were delivered to Asia-Pacific buyers.

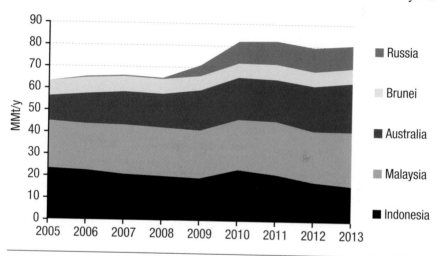

Fig. 5–2. Asia-Pacific LNG exports. (*Source:* Poten & Partners.)

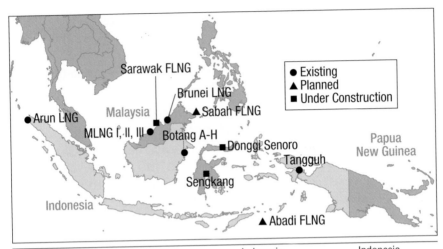

Country	Brunei	Indonesia	Indonesia
Project	Brunei LNG	Bontang A,B	Arun NGL
Location	Lumut	Bontang	Lhokseumawe
Trains	5	2	3
Capacity (MMt/y)	7.2	4.5	5.8
Status	Operating	Operating	Converted to Import
Start-up	1972	1977	1978
Shareholders	Brunei Government, Mitsubishi Corp., Shell	Pertamina	ExxonMobil, Japan Indonesia LNG, Pertamina
Liquefaction	C3MR	C3MR	C3MR

Country	Malaysia	Indonesia	Indonesia
Project	MLNG I (Satu)	Bontang C,D	Arun Phase I Expansion
Location	Bintulu	Bontang	Lhokseumawe
Trains	3	2	2
Capacity (MMt/y)	8.1	4.5	3.9
Status	Operating	Operating	Converted to Import
Start-up	1983	1983	1985
Shareholders	Mitsubishi Corp., Petronas, Sarawak State	Pertamina	ExxonMobil, Japan Indonesia LNG, Pertamina
Liquefaction	C3MR	C3MR	C3MR

Country	Indonesia	Indonesia	Indonesia
Project	Arun Phase II Expansion	Bontang E	Bontang F
Location	Lhokseumawe	Bontang	Bontang
Trains	1	1	1
Capacity (MMt/y)	1.9	2.3	2.53
Status	Converted to Import	Operating	Operating
Start-up	1986	1989	1993
Shareholders	ExxonMobil, Japan Indonesia LNG, Pertamina	Pertamina	Pertamina
Liquefaction	C3MR	C3MR	C3MR

Country	Malaysia	Indonesia	Indonesia
Project	MLNG II (Dua)	Bontang G	Bontang H
Location	Bintulu	Bontang	Bontang
Trains	3	1	1
Capacity (MMt/y)	7.8	2.8	3
Status	Operating	Operating	Operating
Start-up	1994	1996	1996
Shareholders	Mitsubishi Corp., Shell, Petronas, Sarawak State	Pertamina	Pertamina
Liquefaction	C3MR	C3MR	C3MR
Country	Malaysia	Indonesia	Indonesia
Project	MLNG III (Tiga)	Tangguh LNG	Donggi Senoro
Location	Bintulu	Berau Bay	Central Sulawesi
Trains	2	2	1
Capacity (MMt/y)	6.8	7.6	2.1
Status	Operating	Operating	Under Construction
Start-up	2003	2009	2015
Shareholders	Mitsubishi Corp., Nippon Oil, Petronas, Shell, Sarawak State	BP, CNOOC, Inpex, JNOC, LNG Japan, Mitsubishi Corp., Mitsui & Co, Nippon Oil, Talisman	Pertamina, PT Medco Energi Intl., Sulawesi LNG Development
Liquefaction	C3MR	C3MR	C3MR
Country	Indonesia	Malaysia	Malaysia
Project	Sengkang LNG	Petronas FLNG (Sarawak)	MLNG T9
Location	Sulawesi	Malaysia (FLNG)	Bintulu
Trains	4	1	1
Capacity (MMt/y)	2	1.2	3.6
Status	Under Construction	Under Construction	Under Construction
Start-up	2015	2016	2017
Shareholders	EWC	Petronas	Petronas
Liquefaction	IPSMR	Nitrogen Expander	C3MR
Country	Malaysia	Indonesia	Indonesia
Project	Petronas FLNG (Sabah)	Tangguh LNG T3	Abadi FLNG
Location	Sabah	Berau Bay	Timor Sea
Trains	1	1	1
Capacity (MMt/y)	1.5	3.8	2.5
Status	Planned	Planned	Prospective
Start-up	2017	2019	—
Shareholders	Petronas	BP, CNOOC, Inpex, JNOC, LNG Japan, Mitsubishi Corp., Mitsui & Co, Nippon Oil, Talisman	Indonesian Government, Inpex, Shell
Liquefaction	Nitrogen Expander	C3MR	—

Fig. 5–3. Southeast Asia LNG infrastructure and export project profiles. (*Source:* Poten & Partners.)

Southeast Asia.

Brunei. Brunei LNG Sendirian Berhad's (BLNG's) five-train lique-faction facility was the first large-scale liquefaction plant in Asia. Commissioned in 1972, it was also Shell's first foray into LNG.[2] BLNG is owned 50% by the state of Brunei with the other 50% split equally between Shell and Mitsubishi. Brunei Shell Petroleum SB, which produces feed gas in four offshore gas fields to supply the plant, is owned equally by the state of Brunei and Shell. BLNG sent its first shipments to Japan in 1972, and the plant has consistently produced close to its 7.2 MMt/y (9.8 Bcm/y) capacity (table 5–1). In 2013, Brunei LNG produced 6.7 MMt (9.1 Bcm). Japan is the primary market, buying 5.1 MMt (6.9 Bcm) in 2013; Korean Gas, which began purchasing LNG under long-term contract in 1994, received 1.1 MMt (1.5 Bcm).

Table 5–1. Brunei LNG export volumes (MMt)

Destination	2005	2006	2007	2008	2009	2010	2011	2012	2013
Japan	6.3	6.5	6.4	6.1	6.1	5.8	6.3	5.9	5.1
South Korea	0.6	0.9	0.6	0.7	0.5	0.7	0.8	0.8	1.1
Other	–	–	–	–	–	–	–	–	0.5
Total Exports	6.9	7.4	7.0	6.8	6.6	6.6	7.1	6.7	6.7

(*Source:* Poten & Partners.)

The view of Brunei as a declining gas player may be changing with a surge in discoveries over the last couple of years. Brunei has secured enough gas reserves to underpin LNG exports under new contracts through 2023. Recent drilling success and improved coop-eration with Malaysia's Petronas have raised speculation that plans for a long-delayed sixth train could be revived.

Indonesia. Indonesia was once the world's largest supplier of LNG. Production at the Arun and Bontang facilities, which are owned 100% by state-owned Pertamina, peaked at 28.7 MMt/y (39.0 Bcm/y) in 1999, when Indonesia accounted for 31% of the global total (table 5–2). Two separate operating companies, majority owned by Pertamina (55%), manage each facility. Because of feed gas shortages, production trains at Arun have been progressively shut down. As Arun operations were scaled back, Indonesian LNG production slid to 19.4 MMt (26.5 Bcm) in 2009, and the country's export ranking dropped to third behind Qatar and Malaysia.

A third project, the Tangguh LNG venture, began operating in 2009. Tangguh contributed to a bump-up in Indonesian LNG production to 23.5 MMt (32.0 Bcm) in 2010. Tangguh is owned and operated solely by foreign companies. Pertamina plays no role in the project. Feed gas supply problems at Bontang caused Indonesian production to slide to 16.5 MMt (22.4 Bcm) in 2013. Several small-scale liquefaction projects are being built, and a potential third train at Tangguh promises a modest boost in Indonesian production. But not all of this new production will be exported, as some will supply Indonesian regasification terminals.

Table 5-2. Indonesia LNG export volumes (MMt)

Destination	2005	2006	2007	2008	2009	2010	2011	2012	2013
Japan	14.3	14.4	13.6	14.1	13.0	12.8	9.3	6.2	6.3
South Korea	5.5	5.0	3.8	3.0	3.0	5.5	7.9	7.4	5.6
Other	3.6	3.2	3.5	3.1	3.5	5.2	4.2	4.5	4.6
Total Exports	23.4	22.7	20.9	20.2	19.4	23.5	21.4	18.1	16.5

(*Source*: Poten & Partners.)

Discovered in 1971, the Arun field was the first major natural gas discovery in Indonesia.[3] It contained 17 Tcf (500 Bcm) of gas reserves. The state owns all hydrocarbons in Indonesia and oversees production sharing contracts (PSCs) with producers. ExxonMobil operated the natural gas field for the Arun LNG complex, located in Aceh Province in the northern part of the island of Sumatra. The first three trains at the Arun complex were completed for the 1978 start-up. Two additional trains were added in 1984, and a final sixth train was added in 1986. With all six trains in operation, Arun had a production capacity of 12.5 MMt/y (17.0 Bcm/y).

Supply declines in the Arun field forced the early shutdown of two of the original six trains in April 2000, and supply contracts were switched to Bontang. Despite added production from smaller fields to supplement Arun, feed gas supply continued to decline. By 2013, only a single train was still operating, and this train was shut down in 2014, when Arun was converted into an LNG hub and regasification terminal. The import facility capitalizes on existing storage, which includes five LNG tanks with 636,000 cubic meters of capacity as well as LPG and condensate storage and three jetties, two for LNG and one for LPG.

Bontang LNG has a liquefaction capacity of 22.5 MMt/y (30.6 Bcm/y) from eight trains.[4] The facility, one of the largest in the world, is located in East Kalimantan on the island of Borneo's east coast. The Badak gas reservoirs were discovered in 1972, along with a series of nearby discoveries that were to underpin the Bontang LNG project. Shipments began in 1977 from the first two trains, A and B, with an aggregate capacity of 4.5 MMt/y. As new buyers were added, additional trains were built. Trains C and D were added in 1983, E in 1990, F in 1993, G in 1997, and H in 1999. In 2005, Pertamina announced tentative plans for a ninth train (I), which was never built because of faltering feed gas supplies.

The Mahakam block in East Kalimantan—under the management of Total (the operator) and INPEX—is the primary gas supply for Bontang LNG. Mahakam production peaked in 2001, when 3.3 Bcf/d was supplied to the plant to produce 21.4 MMt (29.1 Bcm) of LNG and 1.2 MMt/y of LPG. Since then, Mahakam natural gas output has declined to just 1.7 Bcf/d in 2013, producing approximately 10.7 MMt (14.6 Bcm) of LNG. With significant underutilization of productive capacity, there are plans to channel new supplies from other fields into the Bontang complex, including Chevron's 1.1 Bcf/d Indonesia Deepwater Development (IDD) project and Eni's 0.45 Bcf/d Jangkrik field development, but not until late in this decade.

The majority of Arun and Bontang's LNG production is exported to Japan, with smaller trades with South Korea and Taiwan. Beginning in 2003, Pertamina was underdelivering contract quantities, and this problem worsened with the decline in feed gas supplies to the two LNG complexes. Pertamina shifted some sales commitments from Arun to Bontang and is scaling back term sales. In 2008, Pertamina completed contract renegotiations with its Japanese buyers, which significantly reduced Bontang contract volumes and settled issues related to underdelivered cargoes. A total supply of 25 MMt over 10 years was agreed with the Japanese buyers. This included 3 MMt/y from 2011 through 2015 and 2 MMt/y from 2016 through 2020. The former Bontang sale and purchase agreements (SPAs) with these buyers included 8.45 MMt/y under the 1973 ex-ship contract and 3.63 MMt/y under the 1981 FOB contract. In the future, third-party suppliers of gas into Bontang LNG will market their own LNG production.

The fully integrated Tangguh LNG project is located in Western Papua, within the Papua Province of Indonesia.[5] Tangguh is operated by BP Indonesia as a PSC contractor to BP Migas (now SKK Migas), Indonesia's upstream regulatory authority. The Tangguh LNG project, with two trains each of 3.8 MMt/y (5.2 Bcm/y) production capacity, was completed in 2009. Feed gas is supplied from six gas fields in the Wiriagar, Berau, and Muturi PSCs in Bintuni Bay in close proximity to the plant. The fields were discovered in the mid-1990s. While the government owns the LNG plant and related infrastructure, BP holds a 37.16% stake in the PSCs and operates all aspects of the project. Other upstream PSC partners include Mitsubishi/INPEX (17.71%), CNOOC (16.96%), Nippon Oil (13.45%), JNOC/Mitsui (7.37%), and LNG Japan (7.35%). The six PSCs have a combined proven and certified gas reserve base of 14.4 Tcf (408.1 Bcm).

BP was also appointed by BP Migas as Tangguh's seller of record, meaning that BP led the LNG marketing effort on behalf of the project. This is a departure from Indonesia's other two LNG export ventures, in which Pertamina was the seller of record. Tangguh LNG landed its first term sale in September 2002 when CNOOC agreed to purchase 2.6 MMt/y for 25 years for delivery to its Fujian terminal. In July 2003, the venture secured a second sale to South Korea's Posco and SK Corporation (now K-Power). The deal was finalized in 2004 for 1.35 MMt/y split between the two buyers but delivered into Posco's terminal at Gwangyang. The terms of the sales were unfavorable to the seller. The final term sale was to San Diego–based Sempra Energy for 3.7 MMt/y for 20 years into the California firm's Costa Azul import terminal, located 60 miles south of the US border in Baja, Mexico. The contract was finalized in October 2004 and was linked to US gas prices. With North American gas prices falling to well below global LNG prices, these volumes have been diverted to premium Asian buyers.

With sufficient reserves to support a third train, FID is likely in 2015, enabling commissioning in 2019. Marketing activities for the expansion train, which will be jointly managed by BP and SKK Migas, commenced in 2012, with a focus on buyers in Japan and South Korea. About 40% of natural gas production, mostly in the form of LNG, has been earmarked for the domestic market.

The Donggi-Senoro LNG (DSLNG) project consists of a one-train 2 MMt/y (2.7 Bcm/y) LNG plant near Luwuk, Central Sulawesi Province, Indonesia.[6] DSLNG is the first Indonesian LNG project developed as a "downstream business activity" under Law No. 22 of 2001, which shifted the investment and the associated risks of the LNG plant from the government of Indonesia to a private venture. DSLNG is also the first LNG export project in which Mitsubishi has taken the lead and is the operator. Until DSLNG, Mitsubishi had taken minority interests in projects spearheaded by international oil majors. A special purpose company, Sulawesi LNG, set up by Mitsubishi with a 75% stake and Korea Gas with 25%, has a 59.9% stake in the liquefaction plant. State-owned Pertamina holds 29%, and another Indonesian firm, Medco, holds 11.1%.

DSLNG is to be completed in 2015. Feed gas will be supplied from the Senoro-Toili and Matindok fields, located in close proximity to the plant. Pertamina holds a 50% interest in Senoro-Toili and 100% in Matindock. The other 50% in Senoro-Toili is split between Medco with 30% and Mitsubishi with 20%. Senoro-Toili will supply the plant with 250 MMcf/d, with Matindok supplying 85 MMcf/d. In February 2009, DSLNG signed preliminary agreements with Japan's Kansai Electric and Chubu Electric to buy the entire planned output of 2 MMt/y for 15 years with an expected start in 2012. However, the Indonesian government was pushing for a significant local market obligation, and Kansai allowed the tentative deal to lapse in August 2009. This contributed to a two-year delay until the government agreed to reduce the domestic requirement to 25% and replacement buyers were lined up. Chubu eventually firmed up its 1 MMt/y purchase, but the DSLNG partners had to scramble to find replacement buyers for the Kansai volume, eventually signing deals with Kyushu Electric for 0.3 MMt/y of LNG and Kogas for 0.7 MMt/y as part of an equity deal.

Japan's INPEX Corporation, with a 65% operating interest, and Shell, with a 35% shareholding, are proposing a 2.5 MMt/y floating LNG (FLNG) production project on the Abadi field in the Masela Block in Indonesia's Arafura Sea.[7] INPEX is the operator with a 65% interest in the Masela Block, discovered in 2000. In December 2010, the Indonesian government approved a phased Plan of Development (POD) for Abadi, including an FLNG unit able to produce 2.5 MMt/y of LNG and 8,400 barrels per day of condensates in the first phase.

Commissioning of the FLNG unit is penciled in for 2018, though it could slip from that date since no order has yet been placed. Field reserve estimates have climbed to 18.4 Tcf of reserves, and Phase 2 plans for a second 2.5 MMt/y FLNG unit have expanded and could include two larger scale FLNG units or an onshore complex on Tanimbar Island.

After INPEX selected Shell as the strategic partner, the Anglo/ Dutch firm acquired a 30% stake in the integrated project in July 2011. INPEX continues to be the project operator, but Shell adds expertise in offshore production, gas liquefaction, LNG shipping, and FLNG technology. PT EMP Energi Indonesia subsequently transferred its 10% shareholding to INPEX and Shell, raising their respective stakes to 65% and 35%. Traditional Asian LNG buyers, particularly in Japan, are likely to be the target market. Given the relatively modest scale of Phase 1, Japan's electric and gas utilities could easily absorb the volumes. INPEX itself could take some of the volumes for its 1.5 MMt/y (2.0 Bcm/y) import facility at Naoetsu in northwestern Japan, which was commissioned in 2013. The government of Indonesia may enforce a Domestic Market Obligation of 25% to as much as 40% on full Abadi development to supply domestic regasification terminals.

Malaysia. Malaysia is one of the world's largest exporters of LNG and a major supplier to the Asian market (table 5–3). There are three separate liquefaction facilities located at the port of Bintulu: MLNG Satu, MLNG Dua, and MLNG Tiga. MLNG Satu first shipped LNG in 1983[8] and has three 2.8 MMt/y (3.8 Bcm/y) liquefaction trains. MLNG Dua became the second operational facility in 1995 when the first of three 3.2 MMt/y (4.4 Bcm/y) liquefaction trains came on line. Trains 4 and 5 at MLNG Dua started producing LNG in 1995, followed by Train 6 in 1996, together adding 9.6 MMt/y (13.1 Bcm/y) of production capacity. The two-train 7.7 MMt/y (10.5 Bcm/y) MLNG Tiga project shipped out its first cargo in 2003. The existing eight trains at the complex can produce up to 25.7 MMt/y (35.0 Bcm/y).

Table 5–3. Malaysia LNG export volumes (MMt)

Destination	2005	2006	2007	2008	2009	2010	2011	2012	2013
Japan	13.6	12.0	13.3	13.1	12.6	14.0	15.0	14.6	14.9
South Korea	4.7	5.7	6.1	6.3	5.8	4.8	4.1	4.1	4.3
Taiwan	3.0	3.3	3.0	2.7	2.6	2.9	3.3	2.8	2.9
Other	0.3	0.1	0.1	–	0.8	1.3	2.0	1.9	2.7
Total Exports	21.6	21.1	22.5	22.1	21.8	22.9	24.3	23.4	24.9

(*Source*: Poten & Partners.)

Malaysian LNG production reached 21.6 MMt (29.4 Bcm) in 2005 after the completion of MLNG Tiga. It has subsequently increased to 24.9 MMt (33.9 Bcm) in 2013, for a capacity utilization of 97%. After struggling to maintain production levels, a string of upstream discoveries has buoyed production and encouraged the state-owned firm to approve a ninth LNG train at the Malaysia LNG complex. Train 9, which will utilize existing storage and marine infrastructure, is scheduled for completion at the end of 2015, ramping up to capacity in 2016. When the new train is commissioned, the Bintulu complex will have a nameplate capacity of 29.3 MMt/y (39.9 Bcm/y), solidifying its position as one of the world's largest LNG production complexes.

Of the 24.9 MMt (33.9 Bcm) of LNG produced at the Bintulu complex in 2013, 14.9 MMt (20.3 Bcm) was sold to buyers in Japan, followed by 4.3 MMt (5.9 Bcm) into South Korea, 2.9 MMt (4.1 Bcm) into Taiwan, and 2.7 MMt (3.7 Bcm) into China. Satu volumes are sold primarily to Tokyo Electric and Tokyo Gas. Dua production goes mainly to Japanese buyers, with Korea Gas taking 2.0 MMt/y (2.7 Bcm/y) and Taiwan's CPC getting 2.25 MMt/y (3.1 Bcm/y). Japanese buyers dominate the buyer list at Tiga as well, with Korea Gas also taking 2.0 MMt/y (2.7 Bcm/y) and China's CNOOC getting 3.03 MMt/y (4.1 Bcm/y). Petronas is expected to target Asian buyers for Train 9 volumes as well, building upon established relationships. The new Bintulu train and the smaller floating LNG projects could also serve the domestic market. Currently, LNG is imported on a short-term basis into the 3.8 MMt/y Sungai Udang LNG regasification plant in Malaka State. The import facility was commissioned in 2013.

Petronas plans to monetize stranded gas fields in Malaysian waters using two moderately sized floating LNG production units.[9] The 1.2 MMt/y (1.6 Bcm/y) Petronas Floating LNG 1 (PFLNG-1) will be stationed at the Kanowit gas field off Sarawak, Malaysia, and could become the world's first FLNG unit in operation when completed in the fourth quarter of 2015. Petronas is simultaneously advancing a 1.5 MMt/y (2.0 Bcm/y) FLNG unit, Petronas FLNG 2 (PFLNG-2), to monetize the Rotan discovery off Sabah, owned jointly by Petronas and Murphy Oil. PFLNG-2 was sanctioned in January 2014 and is tentatively scheduled for a 2018 start. Construction of the two PFLNG units and MLNG Train 9 at the Bintulu complex would boost Malaysia's LNG capacity to 32 MMt/y (43.5 Bcm/y). LNG from the two floating LNG projects is likely to be pooled within Petronas's growing portfolio of domestically produced LNG and supplies procured from overseas projects to serve domestic markets.

Australia. By 2020, Australia is expected to be one of the top two LNG producers globally (probably passing Qatar) and the top supplier into the Asia-Pacific market. By then, as many as 22 trains may be operational with nearly 80 MMt/y (108.4 Bcm/y) of production capacity, based upon existing projects and those under construction, compared to production of 22.2 MMt/y (30.2 Bcm/y) in 2013 (table 5–4). In addition to seven conventional LNG projects and one of the world's first floating liquefaction plants off the Western Australia coast, there are three LNG projects in the eastern province of Queensland designed to exploit coalbed methane reserves (fig. 5–4).

Table 5–4. Australia LNG exports (MMt)

Destination	2005	2006	2007	2008	2009	2010	2011	2012	2013
China	–	0.7	2.5	2.7	3.5	3.9	3.6	3.6	3.6
Japan	10.2	12.2	12.1	12.0	11.9	13.3	14.0	15.9	17.9
South Korea	0.7	0.6	0.4	0.5	1.3	1.0	0.8	0.8	0.6
Other	0.4	0.2	0.1	0.1	1.5	1.0	0.6	0.3	0.1
Total Exports	11.3	13.7	15.1	15.3	18.3	19.2	19.0	20.6	22.2

(*Source:* Poten & Partners.)

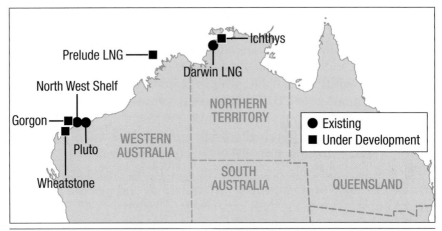

Country	Australia	Australia	Australia
Project	North West Shelf	North West Shelf T3	North West Shelf T4
Location	Burrup Peninsula	Burrup Peninsula	Burrup Peninsula
Trains	2	1	1
Capacity (MMt/y)	5	2.5	4.2
Status	Operating	Operating	Operating
Start-up	1989	1993	2004
Shareholders	BHP Billiton, BP, Chevron, MIMI, Shell, Woodside	BHP Billiton, BP, Chevron, MIMI, Shell, Woodside	BHP Billiton, BP, Chevron, MIMI, Shell, Woodside
Liquefaction	C3MR	C3MR	PMR

Country	Australia	Australia	Australia
Project	Darwin LNG	North West Shelf TS	Pluto LNG T1
Location	Wickham Point, Darwin	Burrup Peninsula	Burrup Peninsula
Trains	1	1	1
Capacity (MMt/y)	3.5	4.4	4.3
Status	Operating	Operating	Operating
Start-up	2006	2008	2012
Shareholders	ConocoPhillips, ENI Gas & Power, Inpex, Santos, Tepco/Tokyo Gas	BHP Billiton, BP, Chevron, MIMI, Shell, Woodside	Kansai Electric, Tokyo Gas, Woodside
Liquefaction	Opt. Cascade	PMR	DMR

Country	Australia	Australia	Australia
Project	Gorgon LNG	Ichthys	Prelude FLNG
Location	Barrow Island	Blaydin Point, Darwin	Browse Basin (FLNG)
Trains	3	2	1
Capacity (MMt/y)	15.6	8.4	3.6
Status	Under Construction	Under Construction	Under Construction
Start-up	2015	2016	2016
Shareholders	Chevron, Chubu Electric, ExxonMobil, Osaka Gas, Shell, Tokyo Gas	Chubu Electric, CPC, Inpex Osaka Gas, Toho Gas, Tokyo Gas, Total	CPC, Inpex, KOGAS, Shell
Liquefaction	C3MR	C3MR	DMR

Fig. 5–4. Western Australia LNG infrastructure and export project profiles. (*Source:* Poten & Partners.)

Country	Australia
Project	Wheatstone
Location	Ashburton North, WA
Trains	2
Capacity (MMt/y)	8.9
Status	Under Construction
Start-up	2016
Shareholders	Apache, Chevron, KUFPEC, Kyushu Oil
Liquefaction	Opt. Cascade

Fig. 5–4. (Continued)

Western Australia. The foundation of the Australian LNG industry is the North West Shelf (NWS) project in Western Australia, which was developed by a Woodside Petroleum–led joint venture.[10] Following the Domestic Gas project (Phase 1), which started producing gas in 1986, Phase 2, the initial LNG phase comprising two 2.2 MMt/y (3.0 Bcm/y) LNG trains, was commissioned in 1989. The third 2.2 MMt/y (3.0 Bcm/y) train increased LNG production capacity to 6.6 MMt/y (9.0 Bcm/y) in 1993. After a relatively low-cost debottlenecking in 1995, each of these train's capacities was raised to 2.5 MMt/y (3.5 Bcm/y). A fourth 4.2 MMt/y (5.9 Bcm/y) unit was added in 2005. After a fifth 4.4 MMt/y (6.2 Bcm/y) train came into service in 2008, the project's total capacity (Trains 1 through 5) was 16.1 MMt/y (21.9 Bcm/y). The shareholders are now focusing venture investment in developing additional feed gas supplies to maintain production with existing trains rather than with new ones.

The NWS project is an integrated joint venture of six international firms, each with a 16.67% stake. In addition to Woodside, the partners are BHP Billiton, BP, Chevron, MIMI (Mitsubishi/Mitsui 50/50), and Shell. China's CNOOC participates in the project but does not share ownership of NWS infrastructure. Woodside Offshore Petroleum Ltd., a subsidiary of Woodside, is the operator of the entire project. LNG marketing efforts are carried out by North West Shelf Australia LNG, which again comprises the six members of the NWS venture itself.

Long-term contracts with Japanese utilities totaling more than 9 MMt/y (12.2 Bcm/y) make Japan by far the largest market for NWS LNG. The venture's marketing arm has extended its sales beyond Japan. In a complex negotiation that included the sale of an upstream position to CNOOC, a deal was concluded with the Chinese company for the sale of 3.7 MMt/y (5.0 Bcm/y) on FOB terms from 2006 through 2031. NWS has also entered into a number of medium-term transactions. Uncommitted cargoes are sold via spot and short-term contracts.

The ConocoPhillips-operated Darwin LNG project began operating in early 2006.[11] Darwin LNG is a single-train, 3.3 MMt/y (4.6 Bcm/y) facility. The ConocoPhillips-led partners, which include Italy's ENI, Australia's Santos, and Japan's INPEX, Tokyo Electric, and Tokyo Gas, would like to expand Darwin LNG and already have environmental approvals from Australian authorities for up to 10 MMt/y (14 Bcm/y). It is supplied from the Bayu-Undan field, located in the Timor Sea's Joint Petroleum Development Area about 500 kilometers northwest of Darwin and 250 kilometers south of Timor-Leste. The Bayu-Undan field contains an estimated 3.4 Tcf (96.4 Bcm) of gas reserves. This low reserve level (by LNG project standards) has led to somewhat unusual 17-year SPAs that commit essentially 100% of the field's reserves to Tokyo Electric Power and Tokyo Gas for almost all Darwin LNG's annual production. Additional gas reserves are required for a second train. ConocoPhillips has touted its Browse Basin drilling program as a potential feed gas source. ConocoPhillips and partners Santos and SK Group could also supply the plant from the Caldita and Barossa fields in the Timor Sea, either for an expansion train or to backfill Bayu-Undan when it comes off plateau production levels.

Burrup Park, the home of Pluto LNG, comprises a single 4.3 MMt/y (5.9 Bcm/y) LNG train and is 90% owned by Woodside with Tokyo Gas and Kansai Electric holding minority interests.[12] Construction commenced in the fourth quarter of 2007, with start-up planned for late 2010. After several delays, the train was commissioned in the second quarter of 2012. The complex, which is next to the port of Dampier, occupies a 200-hectare site, providing space at the site for up to four liquefaction trains. The owner and developer, Woodside, opted to build an LNG processing facility comprising many modules constructed overseas in an attempt to overcome Australia's high labor costs.

Australia's Woodside Petroleum discovered the Pluto field in April 2005 and the Xena field in September 2006 in exploration permit WA-350-P, which is located in the Carnarvon Basin in Western Australian waters. The gas fields can supply Train 1 for 20 years.

Woodside has failed to find adequate gas reserves through its own drilling program or to tie up third-party gas to sanction expansion trains. In addition to becoming minority partners in Pluto LNG, Japanese buyers Tokyo Gas and Kansai Electric are the anchor buyers at the first train, taking nearly 90% of the train's nameplate capacity for 15 years from the start of production.

Chevron Australia, which has retained a 64.14% shareholding in the venture, began construction of the Wheatstone LNG Project, located 12 kilometers west of Onslow on Western Australia's Pilbara coast, in late 2011. The foundation project consists of two LNG trains with a combined capacity of 8.9 MMt/y (12.1 Bcm/y), plus a domestic gas plant.[13] First LNG production is expected in 2016. Wheatstone has permission to expand to 25 MMt/y (34.0 Bcm/y). The Wheatstone and Iago fields operated by Chevron will supply 80% of the feed gas. The remaining 20% will be supplied from the Woodside and Kuwait Foreign Petroleum Exploration Company (KUFPEC) Julimar and Brunello fields, also located in the Carnarvon Basin. The Wheatstone Project is a joint venture between Australian subsidiaries of Chevron, KUFPEC, Woodside, and Japan's Kyushu Electric, together with PE Wheatstone Pty Ltd., owned in part by TEPCO.

Chevron has adopted equity liftings rather than joint sales at Wheatstone. Tokyo Electric Power Company is by far the largest offtaker from Wheatstone at 4.2 MMt/y (5.7 Bcm/y). Another Japanese utility, Kyushu Electric, is taking 0.8 MMt/y (1.1 Bcm/y), while Chubu Electric and Tohoku Electric have each purchased 1.0 MMt/y (1.4 Bcm/y) under term sales deals. Shell sold its 0.57 MMt/y (0.8 Bcm/y) in equity LNG to its Asian subsidiary Shell Eastern, prior to relinquishing its position in the project to Chevron via an asset swap.

Gorgon LNG Phase 1 includes three 5.2 MMt/y (7.1 Bcm/y) liquefaction trains, with the first train scheduled for commissioning in 2015.[14] The project, which is located on environmentally sensitive Barrow Island, will be supplied from the Greater Gorgon Area, a prolific natural gas province located in the Carnarvon basin. A modular construction strategy is being employed to minimize the environmental impact on the island, a Class-A nature preserve, and to overcome labor shortages. The island is located 50 kilometers off the Pilbara coast. Feed gas from the Gorgon and Jansz-Io fields will be processed at the three-train 15.6 MMt/y (21.2 Bcm/y) Barrow Island facility. Adding to project complexity and expense has been the need to separate a very high proportion of CO_2 (15%) from the feed gas stream and reinject it underground. With ample feed gas in the Greater Gorgon fields, Chevron, the project operator with a 47.3% stake, has touted a fourth train at the Barrow Island complex, but ExxonMobil and Shell with 25% each (the rest is held by Japanese buyers) are reluctant to sanction an expansion train until market conditions sort out and investment costs for the first three trains are finalized. Reportedly budgeted at $37 billion, the project is now estimated to cost around $54 billion, making it the most expensive private construction project in history.

Each partner assumed responsibility for its equity share of the LNG sales, with sales going to China, India, Japan, and South Korea under oil-linked deals. Chevron Australia has executed term sales deals totaling 4.75 MMt/y (6.5 Bcm/y). Of this, about 4.5 MMt/y (6.1 Bcm/y) are with Japanese buyers, and South Korea's GS Energy is taking 0.25 MMt/y (0.3 Bcm/y). In January 2015, Chevron announced it had reached an agreement with the LNG trading arm of South Korea's SK Group to sell 4.15 MMt during a five-year period, reflecting the difficulty of securing long-term contracts in a challenging market environment. As part of price renegotiations, several Japanese buyers took minority interests in the venture, giving them destination-free equity LNG. ExxonMobil signed long-term SPAs with India's Petronet and PetroChina for 1.5 MMt/y and 2.25 MMt/y respectively, essentially selling out all of its equity LNG. Shell also tapped PetroChina for a 2 MMt/y (2.7 Bcm/y) term sale, with another 0.5 MMt/y (0.7 Bcm/y) going to BP. The firm held the rest back for its trading operations and to supply its own import terminals.

The Ichthys joint venture led by Japan's INPEX is building a two-train 8.4 MMt/y (11.4 Bcm/y) liquefaction plant at Blaydin Point within Darwin's Middle Arm Peninsula industrial area.[15] A fully modular construction strategy, which will minimize onshore construction at the site, is being employed with the first train scheduled for commissioning in 2017; the second will follow nine months later. Ichthys LNG is the first major LNG project sponsored by Japan Inc. Japanese firms, led by INPEX, hold 76% of the equity in the project, and about 70% of the output is destined for a consortium of five Japanese buyers: Tokyo Electric, Tokyo Gas, Kansai Electric, Osaka

Gas, and Kyushu Electric. Taiwan's CPC is an important anchor buyer as well. The project is being built by a Japanese-led EPC group and is financed by Japanese government and private bank loans.

Prelude FLNG will be the largest floating object in the world when it is installed and commissioned in 2016.[16] The FLNG unit will be anchored in approximately 250 meters of water, with its subsea production systems radiating 2 to 5 kilometers outwards from the vessel. The vessel is a double-hulled steel design, measuring 488 meters in length and 74 meters in width. When fully loaded, it will weigh around 600,000 tons, about six times the weight of the world's largest aircraft carrier. The vessel is being constructed at Samsung Heavy Industries' Geoje Island shipyard in South Korea.

There are compelling reasons why these fields are good candidates for FLNG development. First, the fields are 100% owned by Shell, providing the company full control of project development. Shell, which is one of the firms most advanced in developing FLNG technology and in a strong position to spur the project forward on its own, was anxious to implement its first FLNG project and prove the concept for further deployment. Shell, with its vast financial assets, could fund the project with its balance sheet. Moreover, conventional development options are limited because of the small resource base and long distance to shore, which undermine onshore liquefaction project viability.

Shell overcame initial buyer resistance to signing purchase contracts from Prelude FLNG by using its globally traded LNG portfolio to backstop these contracts. The Korea Gas SPA started in 2013 with an initial 1 MMt/y (1.4 Bcm/y) supplied from Shell's global portfolio. Deliveries at contract plateau volumes of 3.6 MMt/y (4.9 Bcm/y) will shift to Prelude FLNG once it starts up and will continue to 2035. Shell believes Prelude LNG production could be higher than 3.6 MMt/y, which would enable some deliveries to Taiwan's CPC and Japan's Kyushu Electric, who have portfolio supply deals with Shell.

The large-scale exploitation of coalbed methane (CBM) has opened up a new LNG export province in eastern Australia (fig. 5–5). CBM is naturally occurring methane gas in coal seams. It is also referred to as coal seam methane (CSM) and coal seam gas (CSG). CBM forms by either biological or thermal processes. During the earliest stage of coalification (the process that turns plants into coal), biogenic methane is generated as a by-product of microbial action (similar to the mechanism that generates methane in landfills). Biogenic methane is generally found in near-surface, low-rank coals. The methane produced is absorbed onto micropore surfaces and stored in cleats, fractures, and other openings in the coals. It can occur also in groundwaters within the coal beds. CBM is held in place by water pressure and does not require a sealed trap as do conventional gas accumulations. The coal acts as a source and reservoir for the methane gas, while the water is the seal.

Drilling CBM wells only began around the year 2000 in Australia, but activity has picked up dramatically. In 2013, 1,300 CBM wells were drilled in Queensland, bringing the total to 6,000. It is estimated that 25,000 wells, at a cost of approximately $1 million each, will have to be drilled over the life of the CBM-to-LNG projects. The most prospective CBM fields are located in the Bowen and Surat Basins. Together, the two fields hold 35.1 Tcf (994 Bcm) of proved and probable (2P) reserves at the end of 2013, according to Trade and Investment Queensland. The Surat Basin is the larger of the two.

The CBM bounty has spawned a number of CBM-to-LNG projects, of which three are under construction with a combined capacity of 25.3 MMt/y (34.4 Bcm/y). They are all being built by Bechtel using the ConocoPhillips Optimized Cascade process technology on Curtis Island, off Gladstone in central Queensland. CBM will be transported from the Surat and Bowen Basins by separate project pipelines. Given the vast number of wells that must be drilled and uncertainties surrounding production profiles at CBM wells, there are concerns that feedstock supply difficulties could prevent the facilities ramping up output as fast as sponsors would like.

The Queensland Curtis LNG Project (QCLNG), sponsored by BG, started production at the first train at the end of 2014, with the second train coming on stream in July 2015. QCLNG is the first project to convert coalbed methane, or any unconventional gas for that matter, into LNG for export. The foundation two-train 8.5 MMt/y (11.6 Bcm/y) project was sanctioned in October 2010.[17] As the natural gas extracted from the Surat Basin is about 98% methane, minimal processing beyond water removal and treatment is required before it is piped to Curtis Island via a 540-kilometer pipeline. The site can accommodate an expansion to 12 MMt/y (16.3 Bcm/y). One challenge with coalbed methane is that the wells need to be dewatered within 6 to 12 months before they reach peak gas production, and water disposal could be a long-term concern. To facilitate the ramp up, BG has contracted gas from third-party suppliers, which could make up some 10% to 20% of supply to the plant during the commissioning phase. As many as 6,000 wells will have to be drilled over the life of the project.

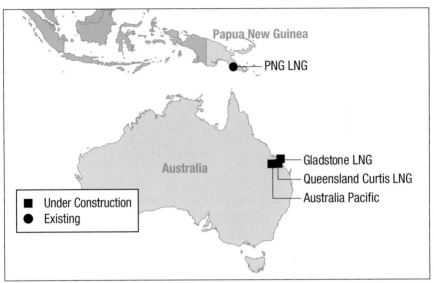

Country	Papua New Guinea	Australia
Project	PNG LNG	Queensland Curtis LNG
Location	Port Moresby	Gladstone, Queensland
Trains	2	2
Capacity (MMt/y)	6.9	8.5
Status	Operating	Under Construction
Start-up	2014	2015
Shareholders	ExxonMobil, Nippon Oil, Oil Search, Petromin, PNG Government, PNG	BG, CNOOC, Tokyo Gas
Liquefaction	C3MR	Opt. Cascade

Country	Australia	Australia
Project	Australia Pacific LNG	Gladstone LNG
Location	Gladstone	Curtis Island
Trains	2	2
Capacity (MMt/y)	9	7.8
Status	Under Construction	Under Construction
Start-up	2016	2015
Shareholders	ConocoPhillips, Origin Energy, Sinopec	KOGAS, Petronas, Santos, Total
Liquefaction	Opt. Cascade	Opt. Cascade

Fig. 5–5. Queensland, Australia, and Papua New Guinea LNG infrastructure and export project profiles. (*Source:* Poten & Partners.)

At FID in 2010, QCLNG was anchored by nearly 5.2 MMt/y (7.1 Bcm/y) of long-term sales to buyers in China and Japan, which included CNOOC, Chubu Electric, and Tokyo Gas. Additional supply from QCLNG was intended to fulfill BG's obligations to Singapore LNG and GNL Quintero (Chile) under the company's portfolio supply arrangements. In May 2013, China's CNOOC agreed to buy an additional 5 MMt/y (6.8 Bcm/y) of LNG from BG's flexible portfolio supply. BG is expected to supply CNOOC from its share of QCLNG uncommitted production, or potentially from a third expansion train of 4.25 MMt/y (5.8 Bcm/y). CNOOC also agreed to increase its equity shareholding in the two-train project to 25%, and a similar shareholding in the upstream. CNOOC would also get an option for a similar stake in an expansion train. BG has dropped expansion plans at the moment, but these plans could be revisited following the acquisition by Shell in 2016. BG's original 100% shareholding has been diluted to 73.75% via equity sales of 25% to CNOOC and 1.25% to Tokyo Gas.

Gladstone LNG (GLNG) became the second CBM-to-LNG project (after BG's QCLNG) to reach FID when Santos and its partners sanctioned the 7.8 MMt/y (10.6 Bcm/y) venture in January 2011.[18] The GLNG project includes construction of a 420-kilometer gas transmission pipeline from the Bowen and Surat Basins to Curtis Island. Approval for GLNG followed a binding Heads of Agreement for the purchase of 3.5 MMt/y (4.8 Bcm/y) of LNG signed by Korea Gas (Kogas) in December 2010, building upon an existing 3.5 MMt/y (4.8 Bcm/y) sale to Malaysia's Petronas. GLNG Train 1 delivered its first cargo in October 2015. Australia's Santos is the lead partner with a 30% stake in the venture, followed by Malaysia's Petronas and France's Total with 27.5% each and Kogas with 15%.

The LNG plant will be supplied with feed gas from CBM fields located along a north–south coal seam trend in Queensland's Bowen and Surat basins, which includes the Fairview, Scotia, and Roma fields among others. The partners must upgrade the resource base to ensure they have sufficient feed gas for the first two trains. As commissioning approaches, the project has begun seeking third-party gas to meet its feed gas requirements.

The Australia Pacific LNG Project (APLNG), which comprises two liquefaction trains with combined production capacity of 9.0 MMt/y (12.2 Bcm/y), was the last of the three CBM-to-LNG projects to be sanctioned. Venture partners led by ConocoPhillips and the Origin Group[19] expect the first train to be commissioned in early 2016 and the second later that year. Feed gas will be supplied from the Surat and Bowen Basins, where Australia's Origin Energy has established a strong resource base. In October 2008, ConocoPhillips joined Origin to launch APLNG, the partners each owning a 50% interest. China's Sinopec later joined the venture with a 25% stake, reducing the partners' shareholdings to 37.5%. APLNG comprises Origin, ConocoPhillips, and Sinopec.

APLNG has the largest portfolio of certified CBM resources in Australia. Sinopec has agreed to purchase 7.6 MMt/y (10.3 Bcm/y) of LNG, which also facilitated the sanctioning of Train 2, lowering unit costs since the infrastructure investment was front-loaded on Train 1. Only 1,015 wells will be drilled prior to project start, with 10,415 wells drilled over the project life.

Papua New Guinea.

PNG LNG. The first cargo from the 6.9 MMt/y (9.4 Bcm/y) PNG LNG project was shipped in May 2014. The two-train liquefaction complex located northwest of Port Moresby on the Gulf of Papua receives feed gas from the Southern Highlands and Western Provinces of Papua New Guinea.[20] Over 700 kilometers (450 miles) of pipelines deliver the feed gas from the fields over difficult terrain to the liquefaction plant (fig. 5–5). Flooding, minimal preexisting infrastructure, and extremely steep slopes were among obstacles that were overcome after construction began in 2010. Pipe had to be airlifted in some areas because the soil could not support heavy machinery, and lack of infrastructure required construction of supplemental roads, communication lines, and a new airfield. Partners in the PNG LNG venture are ExxonMobil with a 33.2% stake, Oil Search with 29%, Santos with 13.5%, Nippon Oil with 4.7%, and PNG with 19.6%.

Nonassociated and associated gas fields dedicated to PNG LNG provide enough feed gas for the first two trains. Nonassociated gas comprises approximately 80% of the total. Hides is the largest nonassociated gas field, and Kutubu is the largest associated resource. ExxonMobil and its partners are considering adding expansion trains, but have not yet identified reserves to underpin an expansion.

ExxonMobil and Oil Search are now focusing their efforts on the P'nyang field in their search for feed gas for an expansion train.

Long-term sales total 6.5 MMt/y (8.8 Bcm/y), essentially selling out both trains. Two Japanese buyers, Tepco and Osaka Gas, are purchasing 1.8 MMt/y (2.5 Bcm/y) and 1.5 MMt/y (2.0 Bcm/y), respectively; China's Sinopec is taking 2.0 MMt/y (2.7 Bcm/y); and Taiwan's CPC has signed up for 1.2 MMt/y (1.6 Bcm/y). The Sinopec and CPC purchases are DAP or CIF, while the Japanese utilities opted for a mix of ex-ship and FOB.

Russia

Sakhalin II LNG. In spite of Russia's vast natural gas resources and huge pipeline natural gas exports, the nation's first foray into the LNG business did not occur until 2009, when the Sakhalin II LNG plant was commissioned.[21] The complex, which is located in Prigorodnoye in Aniva Bay, 13 kilometers east of Korsakov on Russia's Sakhalin Island, has two trains and an LNG nameplate production capacity of 9.6 MMt/y (13.1 Bcm/y). The project has performed significantly above this level, however, producing 10.5 MMt (14.3 Bcm) in 2011, 10.9 MMt (14.8 Bcm) in 2012, and 10.5 MMt (14.3 Bcm) in 2013. It is part of a larger Sakhalin Island oil and gas project which includes development of the Piltun-Astokhskoye oil field and the Lunskoye natural gas field, offshore Sakhalin Island in the Okhotsk Sea, together with the associated onshore infrastructure. In addition to LNG, Sakhalin II produces nearly 400,000 b/d of oil. The project is managed and operated by Sakhalin Energy Investment Company Ltd.

Under the original Sakhalin Energy venture, Shell was the operator with a 55% ownership, with Japanese partners Mitsui (25%) and Mitsubishi (20%). Under legal and political pressure from Moscow following large cost overruns that were essentially passed through to the Russian government under the terms of the PSC, the partners in 2007 sold a 50%-plus-one-share stake in Sakhalin Energy to Russia's Gazprom, leaving Shell with 27.5%, Mitsubishi with 12.5%, and Mitsui with 10.0%. The foreign partners in Sakhalin Energy, led by Shell, intend to build a third LNG train. While brownfield economics are compelling, majority partner Gazprom is reluctant to commit the necessary gas resources to underpin another approximately 5

MMt/y (6.8 Bcm/y) train. Instead, the Russian firm is prioritizing the domestic market and a new LNG project at Vladivostok on the mainland (fig. 5–6).

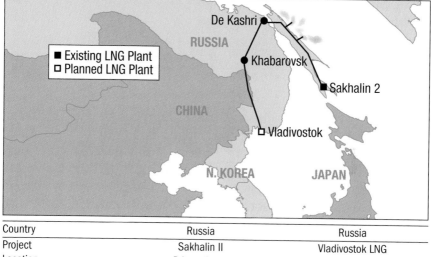

Country	Russia	Russia
Project	Sakhalin II	Vladivostok LNG
Location	Prigorodnoye	Vladivostok
Trains	2	2
Capacity (MMt/y)	9.6	10
Status	Operating	Prospective
Start-up	2009	2018
Shareholders	Gazprom, Mitsubishi, Mitsui, Shell	TBD
Liquefaction	DMR	–

Fig. 5–6. Russia Far East LNG infrastructure and export project profiles. (*Source:* Poten & Partners.)

Japanese buyers hold term contracts for 4.7 MMt/y (6.4 Bcm/y), while Korea Gas purchases 1.5 MMt/y (2.0 Bcm/y). Shell and Gazprom have each retained 1 MMt/y (1.4 Bcm/y) for their trading operations. Gazprom sells the remaining production into the spot market. The venture stepped up supplies to Japan to make LNG available to power generators following the Fukushima disaster in March 2011 (table 5–5). In 2013, Sakhalin II delivered 8.6 MMt (11.7 Bcm) to buyers in Japan and 1.8 MMt (2.5 Bcm) to Kogas.

Table 5–5. Russia LNG export volumes (MMt)

Destination	2005	2006	2007	2008	2009	2010	2011	2012	2013
China	–	–	–	–	0.2	0.4	0.3	0.4	–
Japan	–	–	–	–	2.8	6.0	7.1	8.3	8.6
South Korea	–	–	–	–	1.0	2.9	2.9	2.2	1.8
Other	–	–	–	–	1.0	0.6	0.3	–	0.1
Total	–	–	–	–	4.9	9.9	10.5	10.9	10.5

(*Source:* Poten & Partners.)

Vladivostok LNG. Criticized by its Russian rivals after its Shtokman LNG project in the Barents Sea failed, Gazprom is determined to develop its own LNG export project. It has selected a site near Vladivostok in Russia's Far East Primorye Territory.[22] The Vladivostok LNG project will have a capacity of at least 15 MMt/y (20.4 Bcm/y) across three trains. The first train is to be commissioned in 2018, with the second train following in 2020. Additional trains will be added as this strategic energy and LNG hub develops. A special-purpose company, Gazprom LNG Vladivostok, was registered in September 2013. Gazprom wants to reduce its equity stake in the venture to 51%, with Japanese firms being likely investors. Japan Far East Gas Company (JFG), comprising Itochu, Japan Petroleum Exploration, Marubeni, and INPEX, signed a memorandum of understanding (MOU) in June 2013, outlining plans to cooperate in developing Vladivostok LNG.

The first two trains will be supplied with gas from the Sakhalin Island gas production center via the Sakhalin–Khabarovsk–Vladivostok gas transmission system, already constructed by Gazprom. Additional trains will be supplied via the planned Power of Siberia gas pipeline, which will link large, new Siberian gas production centers at the Yakutia and Irkutsk fields to Vladivostok. Gazprom is targeting Asian buyers for sales of Vladivostok LNG, including Japan, China, and South Korea, which are all in close proximity to the proposed plant. Gazprom and JFG will jointly market LNG to Japanese buyers, who are expected to be major purchasers of Vladivostok LNG. Further development may be curtailed as Russia has succeeded in breaking into the Chinese market with at least one massive pipeline project now under development between eastern Siberia and northern China.

North America

Canada. Canada and the United States have emerged as potentially large LNG exporters based upon their vast unconventional gas resources (fig. 5–7). A late 2013 report from the British Columbia government claims that British Columbia's total natural gas reserves are around 3,000 Tcf (83 Tcm)—enough to support development and LNG export operations for more than 150 years.[23]

The recent estimate is based on analysis of the Montney Formation, which spans northeast British Columbia south of Fort Nelson into the Fort St. John area and across the border into Alberta. In November 2013, Canada's National Energy Board (NEB) boosted its Montney unconventional energy reserve estimate significantly. The new assessment places expected marketable, unconventional gas reserves in the Montney at 449 Tcf (13 Tcm), ranging from a low of 316 Tcf (9 Tcm) to a high of 645 Tcf (18 Tcm). These figures are just 10% to 12% of estimated in-place reserves, pointing to a huge potential in the basin.

This resource base, centered on the Montney and Horn River shale gas fields, has encouraged as many as 18 LNG export proposals, of which 9 have received export licenses from the NEB. The construction and operation of LNG terminals generally fall under provincial regulation, while the export of LNG is federally regulated through the NEB. The NEB reviews export license applications to ensure that the proposed volume of gas to be exported is surplus to Canadian requirements. Both provincial and federal authorities have been supportive of LNG export projects.

The BC LNG export proposals are structured more like traditional greenfield LNG export projects than the developments that predominate in the United States. The proposals mostly involve dedicated upstream developments with considerable up-front investment, rather than purchasing feed gas from the grid. In addition, they have higher investment costs because they do not use existing infrastructure (such as storage tanks and marine facilities at existing LNG import terminals). Finally, building pipelines to deliver feed gas from fields in eastern British Columbia to the LNG export plant sites on the coast will add billions of dollars to each project's costs.

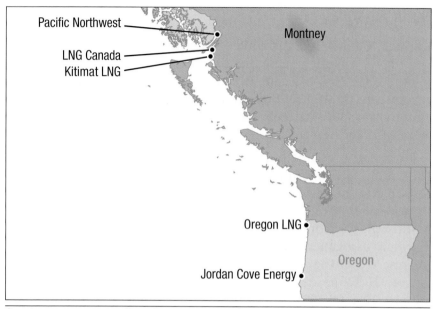

Country	Canada	Canada	Canada
Project	Kitimat LNG	Pacific Northwest LNG	LNG Canada
Location	Bish Cove, BC	Lelu Island, Prince Rupert, BC	Kitimat, BC
Trains	2	2	2
Capacity (MMt/y)	11	12	12
Status	Approved	Planned	Approved
Start-up	2018	2018	2019
Shareholders	Apache, Chevron	JAPEX, Petronas	KOGAS, Mitsubishi, PetroChina, Shell
Liquefaction	C3MR	–	DMR

Country	USA	USA
Project	Jordan Cove Energy Project	Oregon LNG
Location	Coos Bay, OR	Warrenton, OR
Trains	4	2
Capacity (MMt/y)	6	9.6
Status	Under Regulatory Review	Under Regulatory Review
Start-up	2019	2019
Shareholders	Veresen	LNG Development Co.
Liquefaction	–	–

Fig. 5–7. Pacific Coast North America LNG infrastructure and export project profiles. (*Source:* Poten & Partners.)

These cost disadvantages are partially offset by lower shipping costs to East Asian markets and by greater buyer comfort with the traditional business model, which is sweetened by participation in the upstream. It remains to be seen whether buyers will be willing to commit to the oil-linked pricing terms that will be needed if these greenfield projects are to be realized, when Henry Hub–linked prices are available from US projects. Poten & Partners, while reluctant to pick winners, expects that the first BC LNG project will be commissioned shortly after 2020, with production reaching 15.0 MMt/y (20.0 Bcm/y) by 2025. Another two LNG export schemes could be built in Oregon based on accessing BC and Rocky Mountain gas reserves through the existing transmission network at lower pipeline costs than BC projects. Many BC projects need to negotiate permission from the First Nations tribes who have historic treaty rights to the project sites and the lands through which the pipelines would run, and this process of consultation with these groups has proved lengthy and contentious. The Oregon projects have also had their share of local opposition.

Kitimat LNG. Kitimat LNG (KM LNG) was the first BC liquefaction project to receive LNG export authorization from Canada's NEB.[24] Owned by Chevron and Woodside, KM LNG comprises a two-train 10.0 MMt/y (13.6 Bcm/y) LNG project at Bish Cove, near Kitimat. Feed gas for KM LNG would be delivered from the Horn River and Liard shale gas fields over the Spectra pipeline system and the proposed 460-kilometer Pacific Trail Pipeline (PTP) to KM LNG. PTP, which is wholly owned by KM LNG, would add $1.5 to $3 billion to project costs. Following entry into KM LNG in December 2012, a Chevron executive said the decision to build KM LNG was unlikely before late 2014 or early 2015, which would facilitate a 2018 or 2019 start-up. However, Chevron cautioned that FID would depend upon locking in long-term sales agreements for at least 60% to 70% of the gas at robust price terms, which continues to elude the partners.

Galveston LNG, a privately owned Canadian energy development company, first promoted KM LNG in 2006 as an LNG import project. It had garnered all necessary federal and provincial permits. But surging North American unconventional gas production eliminated the need for LNG imports into British Columbia. Galveston identified the LNG export opportunity presented by shale gas and announced in September 2008 that it had replaced its LNG import scheme with an LNG export project. Galveston sold its position to Apache, EOG, and Encana. However, the venture failed to sign up any buyers. In December 2012, Chevron joined the venture, buying out EOG and Encana, and bringing strong Asian LNG marketing relationships to the venture. Apache has now sold its LNG interests (along with its share of Wheatstone) to Woodside.

LNG Canada. LNG Canada, a Shell Canada–led joint venture, is proposing to build a four-train LNG export project in Kitimat.[25] It is accessible via British Columbia's largest coastal fjord—the Douglas Channel. Phase 1 will comprise two 6 MMt/y (8.2 Bcm/y) trains. The NEB has approved the export of up to 24 MMt/y (32.6 Bcm/y), the full four-train build-out capacity of the plant. In May 2012, Shell Canada, Korea Gas, Mitsubishi, and PetroChina announced they would jointly develop the proposed LNG Canada project. Shell holds a 40% working interest in the integrated venture, with Kogas, Mitsubishi, and PetroChina each holding 20%. The partners will get LNG based upon their equity stakes, giving Shell 4.8 MMt/y (6.5 Bcm/y) of Phase 1 volumes, while the other partners would each get 2.4 MMt/y (3.3 Bcm/y). Pending receipt of necessary regulatory approvals, construction of LNG Canada Phase 1 could start in 2016, which would place it in line for a 2019–2020 start-up.

LNG Canada would be supplied with feed gas from partner acreage in the Horn River and Montney shale basins as well as the Cordova Embayment. LNG Canada Phase 1 will require feed gas of 1.85 Bcf/d. Shell and PetroChina hope to supply more than half of Phase 1 from Groundbirch, while Kogas plans to supply its capacity from reserves it holds jointly with Encana in the Montney and Horn River basins. LNG Canada has selected TransCanada Corporation to design, build, own, and operate the proposed multibillion dollar Coastal GasLink pipeline that will transport shale gas from near Dawson Creek in northeast British Columbia to the plant.

Pacific NorthWest LNG. Progress Energy, which is owned 100% by Malaysia's Petronas, plans to build an LNG export facility at Lelu Island near Prince Rupert.[26] The proposed Pacific NorthWest LNG (PNW LNG) facility will comprise an initial two 6 MMt/y (8.2 Bcm/y) trains with a third train of the same size later boosting plant capacity to 18 MMt/y (24.5 Bcm/y). China's Sinopec, Japan's JAPEX, Indian Oil Corporation, and Petroleum Brunei have taken shareholdings in Pacific NorthWest LNG and the associated natural gas supply. They also get an equivalent amount of LNG offtake. The plant would liquefy and export natural gas produced in northeastern British Columbia.

Petronas's Canadian subsidiary, Progress Energy, holds 1.3 million net undeveloped acres of leases in northeast British Columbia and northwest Alberta. Approximately 820,000 acres are in the Montney basin. Progress Energy plans to supply the LNG plant from this acreage. Up to 3 Bcf/d of feed gas at full build-out is to be delivered through a TransCanada-owned, 900-kilometer pipeline system running from North Montney to Prince Rupert, which will be developed and owned by TransCanada Corporation.

The Malaysian firm has indicated an interest in reducing its ownership to as little as 50%, with partners getting equity LNG commensurate with their shareholdings. Depending upon the firm's ability to sell down its equity, Petronas would retain at least 6 MMt/y (8.2 Bcm/y) of Phase 1 LNG production and 9 MMt/y (12.2 Bcm/y) at full build-out for its own portfolio. The Malaysian firm plans to direct this lean LNG preferentially into its home market, freeing up its Malaysia-produced BTU-rich LNG for buyers in Asia, particularly those in Japan, that favor high liquids content product. This would facilitate extension deals for expiring Malaysia LNG contracts at oil-linked price terms. Under the venture timeline, FID was to have occurred before year-end 2014, construction would begin in 2015 and Train 1 production would start at the end of 2018 or early 2019. In late 2014, Petronas engaged in a high-visibility negotiation with the BC government over the terms of the province's fiscal regime. The proposed tax rates were reduced, but in December, Petronas announced it was indefinitely postponing the project in the face of high costs and declining energy prices.

Notwithstanding the Petronas decision, ExxonMobil and Imperial Oil (ExxonMobil's Canadian affiliate) filed for regulatory approval for their proposed 15 MMt/y export project in mid-January 2015, and most other proposed projects appear to be moving ahead.

United States. Two Oregon LNG export projects—Jordan Cove LNG and Oregon LNG—face considerable local opposition, but continue to make progress. They have both received conditional approval to export LNG to non-FTA countries and are advanced in the FERC review process. Canada's NEB has also authorized both projects to source feed gas from Canada, which means they could operate with little or no impact on the price or supply of US natural gas. Similar to BC LNG projects, they will benefit from transportation costs to markets in Asia that are significantly below those from projects located on the US Gulf of Mexico. Moreover, these projects will benefit from the existing pipelines that deliver natural gas to buyers on the US West Coast, though both will require new connections to those existing lines.

Project sponsors—Veresen Inc. for Jordan Cove and Leucadia National Corp. for Oregon LNG—are not experienced in the LNG business. They intend to implement a tolling business model under which the tolling company will lease liquefaction capacity, secure its own feed gas, and market the LNG. It is likely that tolling clients will have the opportunity to take shareholdings in the ventures.

Jordan Cove LNG. In March 2014, Veresen Inc.'s 6 MMt/y (8.2 Bcm/y) Jordan Cove LNG became the first West Coast project to receive conditional approval from the US DOE to export LNG to non-FTA countries. Veresen (formerly Fort Chicago Energy L.P.) is a Calgary-based company active in the energy infrastructure sector. The project is located within the Port of Coos Bay on the North Spit of lower Coos Bay, on an undeveloped site zoned for industrial development. The site "is approximately 7 nautical miles from the entrance of the federally controlled and maintained navigation channel. Jordan Cove LNG and the South Dunes Power Plant sit on 500 acres of privately purchased land designated an Enterprise Zone by the State of Oregon."[27] The Jordan Cove project still needs approval from the Federal Energy Regulatory Commission (FERC) to begin construction, and has received a favorable draft environmental impact statement from the FERC.

Oregon LNG. In July 2014, Oregon LNG became the second West Coast project to receive authorization from the US DOE to export LNG to non-FTA countries.[28] LNG Development Company, LLC, which is a subsidiary of Leucadia National Corporation, proposes to build a 9.6 MMt/y (13.1 Bcm/y) LNG export plant near the mouth of the Columbia River, on the Skipanon Peninsula in Warrenton, Oregon. The project has filed an application at FERC and is currently under National Environmental Policy Act (NEPA) review. Final approval by FERC and DOE, as well as other approvals, are necessary before the project can commence construction.

United States, Alaska.

Kenai LNG. Located at Nikiski on the Kenai Peninsula, Alaska is home to the world's second-oldest liquefaction facility, which started operations in 1969.[29] Kenai LNG was the first project to deliver LNG to Japan, which remained the destination for the vast majority of its LNG. Production at approximately 1.25 MMt/y until 2005 was close to its capacity. With a feed gas shortfall in Cook Inlet fields, the plant was mothballed during the winter of 2011. Kenai LNG is owned by ConocoPhillips (70%) and Marathon (30%).

The gas supply situation improved because of new drilling, to the point that a surplus is available for export during summer months. ConocoPhillips restarted the plant in early 2014, and shipments of LNG resumed in May 2014, with five shipments planned in 2014. However, gas production is reserved for local utilities during the winter months (table 5–6).

Table 5–6. Alaska LNG export volumes (MMt)

Destination	2005	2006	2007	2008	2009	2010	2011	2012	2013
Japan	1.2	1.2	0.9	0.7	0.6	0.6	0.3	0.3	–
Other	–	–	–	–	–	–	–	–	–
Total Exports	1.2	1.2	0.9	0.7	0.6	0.6	0.3	0.3	–

(*Source*: Poten & Partners.)

Alaska LNG. In early 2015, the Alaska LNG project, a joint venture of ExxonMobil, ConocoPhillips, BP, and the state of Alaska began the DOE and FERC permitting process for a nominal three train, 20 MMt/y LNG export plant to be located at Nikiski on the Cook Inlet, not far from the Kenai LNG plant. The feed gas for this project would be the gas reserves associated with the giant Prudhoe Bay oilfield on the North Slope of Alaska, which would be delivered via a new 800-mile large-diameter transmission line. These reserves are generally assessed to be as much as 24 Tcf and have been the subject of numerous monetization proposals over the past 30 years. Ownership by the state and agreements on fiscal and other state incentives may provide more momentum for the project than in previous incarnations. The capital cost of the project is estimated at between $45 and $65 billion.

South America

Peru LNG. Peru LNG, a 4.4 MMt/y single-train LNG plant, was commissioned in 2010 with the first cargo dispatched from the project in June of that year.[30] Natural gas is delivered to the plant, at Pampa Melchorita 169 kilometers south of Lima, through a pipeline system that starts at the Camisea gas fields in the Peruvian Highlands and crosses the Andes Mountains and the Atacama Desert. Peru LNG, South America's first LNG export plant, is owned by a consortium of Hunt Oil (operator with a 50% interest), South Korea's SK Corporation (20%), Shell (previously Repsol, 20%), and Japan's Marubeni (10%). (See fig. 5–18.)

Repsol (now Shell) has exclusive rights to market the entire output of Peru LNG for 18 years from start of operations (table 5–7). Repsol resold a large share of this offtake to Mexico's state-owned power generator Comisión Federal de Electricidad (CFE) at a price linked to the US benchmark price at Henry Hub, Louisiana. With the surge in US shale gas production, Henry Hub trades at just a third to a half of globally traded LNG. This yields a very low netback to Peru, drawing criticism from Peru's president regarding the depressed tax revenues generated by the project. In January 2014, Shell acquired Repsol's Peruvian assets, giving the firm the exclusive offtake rights at Peru LNG as well as interests in the Camisea fields, the feed gas pipeline, and the LNG plant. It also marks the return of Shell—the discoverer of Camisea—to Peru.

Table 5-7. Peru LNG export volumes (MMt)

Destination	2005	2006	2007	2008	2009	2010	2011	2012	2013
Japan	–	–	–	–	–	–	0.4	0.8	0.7
Mexico	–	–	–	0.1	–	0.2	0.4	0.8	1.6
South Korea	–	–	–	–	–	–	0.8	–	0.5
Spain	–	–	–	–	–	0.4	1.3	2.0	1.1
Other	–	–	–	–	–	0.6	0.8	0.4	–
Total Exports	–	–	–	0.1	–	1.2	3.6	4.0	4.0

(*Source:* Poten & Partners.)

The Middle East and East Africa

The Middle East is home to the world's largest oil and gas reserves, while East African waters have emerged as a promising natural gas basin (fig. 5–8). The Middle East was not a large LNG exporting region until the middle of the last decade, when Qatar built a series of megatrains (fig. 5–9). In 2000, regional production was just 17.7 MMt (24.1 Bcm), accounting for just 17% of global trade. By 2005, regional production increased by 85% to 32.6 MMt/y (44.3 Bcm/y), but still represented only 23% of global supply. Primarily because of the completion of six Qatari megatrains of 7.8 MMt/y (10.6 Bcm/y) each, regional production trebled to 98.8 MMt/y (134.4 Bcm/y) by 2013–including 20.8 MMt/y (28.3 Bcm/y) from the United Arab Emirates, Oman, and Yemen. The regional share of global production surged to 41%. Meanwhile, East Africa has emerged as a focal point for LNG export project investment following exploration successes in waters off Mozambique and Tanzania.

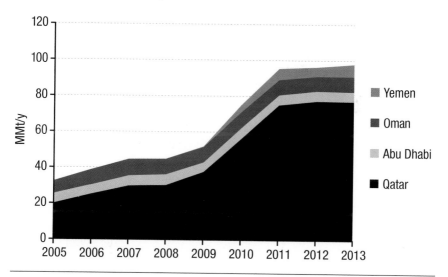

Fig. 5–8. Middle East LNG exports. (*Source:* Poten & Partners.)

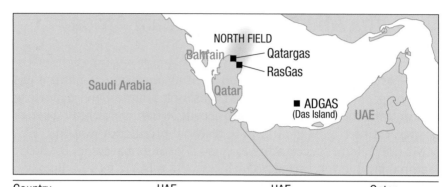

Country	UAE	UAE	Qatar
Project	ADGAS 1	ADGAS 2	Qatargas 1
Location	Das Island	Das Island	Ras Laffan
Trains	2	1	3
Capacity (MMt/y)	2.5	3.3	10.0
Status	Operating	Operating	Operating
Start-up	1977	1994	1996
Shareholders	ADNOC, BP, Mitsui & Co, Total	ADNOC, BP, Mitsui & Co, Total	ExxonMobil, Marubeni, Mitsui & Co, QP, Total
Liquefaction	C3MR	C3MR	C3MR

Country	Qatar	Qatar	Qatar
Project	RasGas	RasGas 2	RasGas 3
Location	Ras Laffan	Ras Laffan	Ras Laffan
Trains	2	3	2
Capacity (MMt/y)	6.6	13.1	15.6
Status	Operating	Operating	Operating
Start-up	1999	2004	2009
Shareholders	ExxonMobil, Itochu Corp., Korea LNG, LNG Japan, QP	ExxonMobil, QP	ExxonMobil, QP
Liquefaction	C3MR	C3MR	C3MR

Country	Qatar	Qatar	Qatar
Project	RasGas 2 T5	Qatargas 2 T4	Qatargas 2 T5
Location	Ras Laffan	Ras Laffan	Ras Laffan
Trains	1	1	1
Capacity (MMt/y)	4.7	7.8	7.8
Status	Operating	Operating	Operating
Start-up	2007	2009	2009
Shareholders	ExxonMobil, QP	ExxonMobil, QP	ExxonMobil, QP, Total
Liquefaction	C3MR	APX	APX

Fig. 5–9. Arabian Gulf LNG infrastructure and export project profiles. (Source: Poten & Partners.)

Country	Qatar	Qatar	Qatar
Project	RasGas 3 T6	Qatargas 3 T6	RasGas 3T7
Location	Ras Laffan	Ras Laffan	Ras Laffan
Trains	1	1	1
Capacity (MMt/y)	7.8	7.8	7.8
Status	Operating	Operating	Operating
Start-up	2009	2010	2010
Shareholders	ExxonMobil, QP	ExxonMobil, Mitsui & Co, QP	ExxonMobil, QP
Liquefaction	APX	APX	APX

Country	Qatar
Project	Qatargas 4T7
Location	Ras Laffan
Trains	1
Capacity (MMt/y)	7.8
Status	Operating
Start-up	2011
Shareholders	QP, Shell
Liquefaction	APX

Fig. 5–9. (Continued)

Arabian Gulf

Qatar. Qatar is home to two LNG export complexes, built and operated by separate project companies, Qatargas and RasGas. Qatargas began operations in 1996, and RasGas followed in 1999. The early Qatari LNG export ventures were initially conceived to supply traditional LNG markets east of Suez. However, the subsequent rise of Atlantic LNG market prospects encouraged both Qatargas and RasGas to seek sales west of Suez. Because of large natural gas reserves, ample space at Ras Laffan for development of liquefaction plant and storage and loading facilities, and the availability of liquid and transparent markets in the Atlantic basin, the Qatari ventures took advantage of economies of scale to develop six megatrains with capacities of around 7.8 MMt/y (10.6 Bcm/y) each. By 2012, the country's nameplate LNG production capacity had reached 77.1 MMt/y (104.9 Bcm/y), more than tripling the 2005 capacity total of 22.8 MMt/y (31.0 Bcm/y) as shown in table 5–8.

Table 5–8. Qatar LNG export volumes (MMt)

Destination	2005	2006	2007	2008	2009	2010	2011	2012	2013
Belgium	–	0.3	1.7	2.2	4.7	4.3	4.5	3.3	2.4
China	–	–	–	–	0.4	1.2	2.3	5.0	6.8
India	3.9	5.2	6.9	6.3	6.6	7.7	9.8	11.8	11.1
Japan	6.3	7.5	8.2	8.2	7.7	7.6	11.9	15.7	16.1
South Korea	6.2	6.6	8.1	9.0	6.9	7.5	8.2	10.3	13.5
Spain	3.5	4.7	3.9	3.7	3.4	4.0	3.5	2.9	2.7
United Kingdom	–	0.1	0.2	0.1	4.5	10.3	15.7	10.3	6.2
Other	0.1	0.8	0.8	1.0	3.6	13.7	19.3	18.2	18.6
Total Exports	**20.0**	**25.1**	**29.8**	**30.3**	**37.8**	**56.2**	**75.2**	**77.4**	**77.2**

(*Source*: Poten & Partners.)

The Qatargas ventures operate seven LNG trains at Ras Laffan Industrial City.[31] The seven Qatargas trains were developed through varying partnerships with international firms including ExxonMobil, Total, ConocoPhillips, Shell, and Mitsui. Qatargas 1, Qatar's maiden LNG project, was commissioned in 1996. Following debottlenecking in 2005, the three-train facility had a nameplate production capacity of 10.0 MMt/y (13.6 Bcm/y). Subsequently, four megatrains boosted production capacity by early 2011 to the current 42.0 MMt/y (57.1 Bcm/y).

Qatar Petroleum's RasGas venture also operates seven LNG trains at Ras Laffan with a total nameplate production capacity of 36.3 MMt/y (49.4 Bcm/y).[32] While the initial two trains have capacities of 3.3 MMt/y (4.5 Bcm/y) each, the subsequent three trains were 40% larger at 4.7 MMt/y (6.4 Bcm/y) each, followed by Trains 6 and 7 at 7.8 MMt/y (10.6 Bcm/y) each. Along with Qatargas's four similarly sized units, dubbed megatrains, they set a new size benchmark for the LNG industry. The seven RasGas trains, commissioned between 1999 and 2010, were developed in partnership with ExxonMobil, Itochu, Korea LNG, and LNG Japan.

Qatar's 14 LNG trains, 7 at Qatargas and 7 at RasGas, are supplied with feed gas from the North Field, which holds approximately 900 Tcf (25 Tcm) of gas. Qatar has placed a moratorium on further LNG expansion while the field's long-term production capability is evaluated. The Qatargas and RasGas ventures ordered a fleet of 47 Q-Flex and Q-Max LNG carriers that are sized from 210,000 m³ to

266,000 m^3 to provide transportation services for the megatrains. Qatar Petroleum and ExxonMobil also built large-scale LNG import terminals in the United States (Golden Pass) and in the United Kingdom (South Hook) as part of their megatrain projects, with the capability to accommodate the Q-Flex and Q-Max vessels, which initially were too large to berth at most LNG import terminals.

When major Atlantic Basin markets faltered, the Qatari ventures refocused their marketing efforts to Asia. As a result, Qatar's top five markets in 2013 were in Asia. Japan topped the list by importing 16.1 MMt (21.9 Bcm), followed by South Korea with 13.4 MMt (18.2 Bcm). India was number three, buying 10.3 MMt (14.0 Bcm), with China number four at 6.8 MMt (9.3 Bcm). Taiwan rounded out the top five buyers with 6.4 MMt (8.7 Bcm) in 2013. During the year, Qatar exported 77.2 MMt (105.0 Bcm) and accounted for one-third of global trade. Of the total, 56.3 MMt (76.6 Bcm), fully three-quarters of Qatari LNG production, was delivered east of Suez.

United Arab Emirates (UAE). The Abu Dhabi Gas Liquefaction Company (ADGAS) became the first LNG supplier in the Middle East with the 1977 start-up of its two-train liquefaction facility located on Das Island.[33] ADGAS, a subsidiary of Abu Dhabi National Oil Company (ADNOC), added a 3.3 MMt/y (4.5 Bcm/y) train to the two 1.25 MMt/y (1.7 Bcm/y) trains in 1994. In recent years, LNG exports have ranged from a low of 5.3 MMt (7.2 Bcm) in 2009 to a high of 6.0 MMt (8.2 Bcm) in 2010. In 2013, Abu Dhabi exported 5.4 MMt (7.3 Bcm), all to Japan's Tokyo Electric Power Company (TEPCO), the sole long-term buyer of ADGAS LNG and LPG production (table 5–9).

Table 5–9. UAE LNG exports (MMt)

Destination	2005	2006	2007	2008	2009	2010	2011	2012	2013
Japan	5.1	5.2	5.6	5.6	5.1	5.2	5.5	5.5	5.4
Other	0.3	0.0	0.1	0.3	0.2	0.8	0.1	0.1	0.0
Total Exports	5.4	5.2	5.7	5.9	5.3	6.0	5.6	5.6	5.4

(*Source:* Poten & Partners.)

ADNOC has carried out a preliminary evaluation of adding a new 3.3 MMt/y (4.5 Bcm/y) train to replace the two older units. But this is unlikely to happen until the TEPCO contract expires in 2019. By then, the two older trains will be more than 40 years old, making them prime candidates for replacement. However, if Abu Dhabi is still short on gas, it will be hard to justify a new train or to even extend the TEPCO contract past 2019. Feed gas for the LNG complex could be diverted to the domestic market to meet peak demand rather than importing expensive LNG. Abu Dhabi currently imports pipeline gas from Qatar to meet gas demand for power generation and is planning to import LNG as well.

Other Middle East.

Oman. Gas discoveries made in 1989 led to the Oman LNG project at Qalhat. The resource was far in excess of local demand, and provided the opportunity for the government to diversify away from oil export revenue. The base two-train project, which first exported gas in 2000, has a production capacity of 7.4 MMt/y (10.2 Bcm/y). A third, 3.7 MMt/y (5.1 Bcm/y) liquefaction train with a slightly different ownership structure, known as Qalhat LNG, was commissioned in November 2005 (table 5–10). Oman LNG and Qalhat LNG merged in September 2013.[34]

Table 5-10. Oman LNG exports (MMt)

Destination	2005	2006	2007	2008	2009	2010	2011	2012	2013
Japan	1.0	2.4	3.6	3.2	2.6	2.9	3.8	4.0	4.0
South Korea	4.2	5.2	5.1	4.8	4.1	4.6	4.2	4.1	4.3
Spain	1.5	0.4	–	0.4	1.5	0.4	0.1	–	0.1
Other	0.3	0.5	0.5	0.3	0.4	1.1	0.3	0.1	–
Total Exports	7.1	8.5	9.2	8.7	8.6	9.0	8.4	8.2	8.5

(*Source:* Poten & Partners.)

Both companies were highly integrated prior to the merger, with Oman LNG operating all three trains. These companies occupy adjacent sites on the outskirts of the eastern town of Sur, and both are supplied with feed gas from central Oman gas fields by Petroleum Development Oman (PDO). The government of Oman is the largest shareholder in both companies, with direct stakes of 51.0% in Oman LNG and 46.8% in Qalhat LNG. Shell is the second largest shareholder in Oman LNG with 30.0%. Oman LNG owns 36.8% of Qalhat LNG. Between direct and indirect interests, the government's shareholding in Qalhat LNG is 65.6%.

Gas supplies are limited, and the two LNG projects have been getting only enough feed gas to meet their contractual commitments.

The two-train Oman LNG plant is particularly underutilized, with approximately 1.5 MMt/y (2.0 Bcm/y) of unfilled capacity. Qalhat LNG's single train is almost full, with its three long-term contracts accounting for essentially all of its capacity. Each of the three trains could produce 3.7 MMt/y (5.0 Bcm/y) if extra gas were available. Oman produced 8.5 MMt (11.6 Bcm) of LNG in 2013, delivered almost exclusively to buyers in Japan and South Korea.

Yemen. The Yemen LNG (YLNG) project is located in Balhaf, about 400 kilometers east of Aden.[35] The plant consists of two LNG trains with a total capacity of 6.7 MMt/y (9.1 Bcm/y) as shown in table 5–11. The first train started producing LNG in October 2009; the second train started in April 2010. Reserves dedicated to the project lie in the Marib-Jawf gas fields northeast of Sanaa. YLNG was developed by a consortium led by France's Total (with a 39.62% shareholding), along with US company Hunt Oil (17.22%), Yemen Gas Company (16.73%), South Korea's SK Corp (9.55%), Kogas (6%), Hyundai Corporation (5.88%), and the General Authority for Social Security & Pensions of Yemen (5%).

Table 5–11. Yemen LNG export volumes (MMt)

Destination	2005	2006	2007	2008	2009	2010	2011	2012	2013
China	–	–	–	–	–	0.5	0.8	0.6	1.1
Japan	–	–	–	–	–	0.1	0.2	0.3	0.5
South Korea	–	–	–	–	0.2	1.7	2.7	2.6	3.6
Other	–	–	–	–	0.1	1.7	2.6	1.7	1.7
Total Exports	–	–	–	–	0.3	4.0	6.3	5.2	6.9

(*Source:* Poten & Partners.)

Although YLNG suffered repeated delays since it was first proposed in 1995, it has performed surprisingly well since its commissioning in late 2009. In spite of operating in a civil war zone with repeated attacks on the feed gas pipeline and facing a shipping threat from Somali pirates, YLNG produced a record 6.8 MMt (9.3 Bcm) in 2013. Even though the French buyers had intended to sell most of their purchases into the United States, 85% of 2013 production was sold into Asia, with half of YLNG production delivered to South Korea. As of this writing, Yemen was plunging into civil war, the YLNG plant was shut down, and the foreign partners had withdrawn their employees.

East Africa

Mozambique. Anadarko and Eni are jointly developing the Afungi LNG Park near Palma, Cabo Delgado Province, Mozambique.[36] They were ambitiously targeting an FID on Phase 1 development for early 2014 to allow first exports in 2018. But the government has not yet established a regulatory framework, and this timeline for FID has passed, delaying first LNG until post-2020. The two operators have allocated 12 Tcf into a common pool to support the initial four 5 MMt/y (6.8 Bcm/y) trains, with Trains 1 and 2 supplied from Anadarko's Area 1 reserves, and Trains 3 and 4 from Eni's adjoining Area 4 block. Afungi LNG will be developed as a common facility and could eventually host as many as 10 trains, with total capacity of a massive 50 MMt/y (68 Bcm/y). Anadarko and Eni have agreed in principle to sell the LNG from their respective trains separately, with the partners in Area 1 jointly marketing the 10 MMt/y produced at Trains 1 and 2 and the participants in Area 4 selling an equivalent amount from Trains 3 and 4. Anadarko and Eni are separately exploring FLNG options to exploit reserves outside of Prosperidade-Mamba not dedicated to the onshore liquefaction project.

Tanzania. While LNG export plans in Tanzania are lagging those in Mozambique, upstream players in the nation's offshore fields are pushing hard to catch up. However, as in Mozambique, the government has not yet put a definitive set of regulations in place, which appears unlikely until after constitutional reforms are decided and elections slated for late 2015 take place. But the government is already working with upstream firms to develop a joint LNG export plant. The four proponents—BG, Ophir, Statoil, and ExxonMobil—have yet to form a joint venture to carry out this project. But the selected site in southern Tanzania has been approved by the government, and BG will lead the pre-FEED, followed by FEED in 2015 and a potential project sanction of an initial two-train LNG project with capacity of 10 MMt/y (13.6 Bcm/y) in 2016 (fig. 5–10).

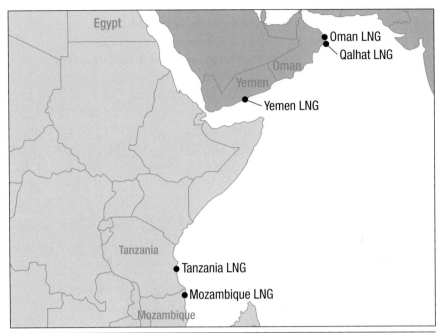

Country	Oman	Oman	Yemen
Project	Oman LNG	Qalhat LNG	Yemen LNG
Location	Qalhat	Qalhat	Balhaf
Trains	2	1	2
Capacity (MMt/y)	6.6	3.3	6.7
Status	Operating	Operating	Suspended
Start-up	2000	2005	2009
Shareholders	Itochu, Korea LNG, Mitsubishi, Mitsui. Oman, Shell, Total	Itochu, Mitsubishi, Oman Oman LNG, Osaka, UFG	GASSP, Hunt Oil, Hyundai SK Corp, Total, Yemen Gas
Liquefaction	C3MR	C3MR	C3MR

Country	Mozambique	Tanzania
Project	Mozambique LNG	Tanzania LNG
Location	Palma	—
Trains	2	2
Capacity (MMt/y)	10	10
Status	Planned	Prospective
Start-up	2019	—
Shareholders	Anadarko, BPRL, Cove Energy, ENI, Mitsui, Videocon	BG, Ophir
Liquefaction	—	—

Fig. 5–10. Other Middle East and East Africa LNG infrastructure and export project profiles. (*Source:* Poten & Partners.)

The Atlantic Basin

The Atlantic and Mediterranean region, sometimes referred to as the Atlantic basin, was home to the world's first LNG project in Algeria. Atlantic Basin production has expanded at a moderate pace when compared to the Middle East and Asia-Pacific. Nevertheless, as of mid-2014, six countries in the region were exporting LNG from seven projects: two production centers in Algeria; and one each in Angola, Equatorial Guinea, Nigeria, Norway, and Trinidad and Tobago. Two liquefaction projects in Egypt have been shuttered, at least temporarily, because of a lack of feed gas, and the aging plant in Libya was destroyed during the 2012 civil war (fig. 5–11).

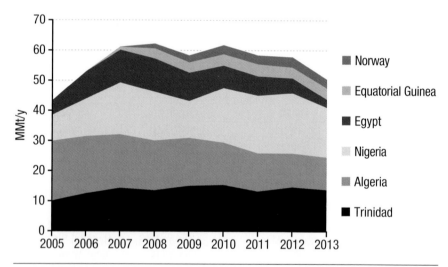

Fig. 5–11. Atlantic Basin LNG exports. (*Source:* Poten & Partners.)

Production in the region peaked at 61.7 MMt (83.9 Bcm) in 2010 and has since declined by 13% to 54 MMt (73 Bcm) in 2013 as feed gas shortages in several countries (including Algeria and Egypt) and civil unrest (Libya and Nigeria) hampered operations. Meanwhile, the Snøhvit project in Norway has struggled with technical problems, and Angola LNG has encountered prolonged start-up issues.

Algeria, the founding Atlantic Basin LNG producer, expanded and added additional liquefaction projects during the late 1970s to become a global leader in LNG production. However, after the Libyan liquefaction facility came on line in 1970, no Atlantic Basin country developed a greenfield LNG export project until 1999 when Nigeria and Trinidad both started production. Oversupply in the European and US markets was the main cause of this inactivity.

However, declines in regional pipeline supplies and continued natural gas demand growth, coupled with falling costs led to a rebirth of Atlantic Basin LNG prospects. Expansion trains were added in Nigeria and Trinidad between 2002 and 2008, and two projects in Egypt went on line in 2005, followed by projects in Equatorial Guinea and Norway in 2007. Another lull in capacity additions (largely due to the collapse in US LNG demand) followed until 2013, when a replacement train in Skikda, Algeria, was commissioned to replace smaller trains destroyed by an explosion. A greenfield LNG plant in Angola followed. Algeria's third liquefaction center (Gassi Touil) is currently under construction with completion scheduled for 2015.

The surge in unconventional natural gas production in the United States has injected new life in the Atlantic Basin liquefaction business. Even as demand for LNG in the United States dwindled, idling many US import terminals, it opened up a new business opportunity, adding liquefaction trains at these vastly underutilized facilities. Based upon a conservative outlook that only the five most advanced proposals will be realized, Poten & Partners forecasts US LNG exports to approach 34 MMt/y (45 Bcm/y) by 2020 and 60 MMt/y (82 Bcm/y) by 2025.[37]

By the end of 2014, four of these liquefaction projects—Sabine Pass, Freeport LNG, Cameron, and Cove Point LNG—were already under construction, and, when commissioned, they will propel the United States to the top rank among Atlantic Basin exporters. A fifth, Corpus Christi LNG, has advanced rapidly toward an FID and construction began in mid-2015. However, the United States would still lag behind Qatar, the top Middle Eastern LNG exporter with 105 Bcm/y, and Australia, which would hold the top ranking in the Asia-Pacific with nearly 80 MMt/y (109 Bcm/y) by 2025. Nevertheless, the United States would climb to third in the global rankings, perhaps even higher if more proposed projects are realized. The United States promises to propel Atlantic Basin LNG production to 133 MMt/y (180 Bcm/y) by 2025, more than double the 2013 level.

Africa

On Africa's Mediterranean coast, Algeria and Libya were the first Atlantic Basin LNG exporters supplying markets in Europe and the United States. Algeria first exported LNG in 1964 and Libya in 1971. Both countries have struggled to meet their export capabilities because of poor maintenance at the plants and, more recently, feed gas shortages in Algeria. The Libyan civil war permanently shuttered the country's small, outdated facility. Egypt opened two LNG export plants in 2005, but production has been hampered by feed gas shortages, which have closed both plants. North African LNG production peaked at 29.3 MMt (39.8 Bcm) in 2007 and has since collapsed to just 13.5 MMt (18.4 Bcm) in 2013, with declines in all three countries (fig. 5–12).

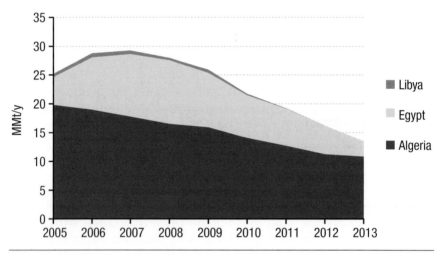

Fig. 5–12. North Africa LNG exports. (*Source:* Poten & Partners.)

Algeria. Algeria has provided the historical foundation for Atlantic LNG trades. The country was the world's first commercial exporter, starting with shipments to the United Kingdom in 1964, and dominated the Atlantic Basin trade for decades until competing export ventures started up in Nigeria, Trinidad, Egypt, and the Middle East. Algeria's greatest advantage is its proximity to European markets. However, LNG exports peaked at 21.4 MMt (29.1 Bcm) in 2003 and declined to just 10.8 MMt (15.0 Bcm) in 2013, due to a shortage of feed gas and weakening demand in Europe as well as in the United

States.[38] Liquefaction capacity utilization has fallen to less than 50%. Besides LNG, Algeria is a leading exporter of pipeline gas to Europe, which (along with gas demand for power generation) contributes to the lack of feed gas for liquefaction.

Algeria has two liquefaction centers: one located in Arzew/Bethouia and one in Skikda. Complexes are 100% owned by Sonatrach, the state-owned energy company (fig. 5–13). The oldest plant, GL4-Z (Camel) at Arzew, which was commissioned in 1964 and had a capacity of 1.5 MMt/y (2.0 Bcm/y), was closed in 2010. At Bethioua, the six trains of GL1-Z and six of GL2-Z, after being refurbished and upgraded, have a total nameplate capacity of 21 MMt/y (28 Bcm/y), though there is always a debate concerning actual production capacity. At Skikda, after three trains were destroyed in an accident in January 2004, the three remaining trains had an aggregate capacity of 3.35 MMt/y (4.6 Bcm/y).

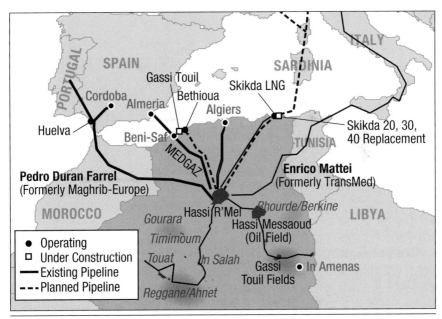

Fig. 5–13. Algeria natural gas/LNG infrastructure and project profiles. (*Source:* Poten & Partners.)

Country	Algeria	Algeria	Algeria
Project	Arzew GL4-Z (Camel)	Skikda GL1-K T10	Skikda GL1-K T20, 30
Location	Arzew	Skikda	Skikda
Trains	3	1	2
Capacity (MMt/y)	1.5	0.85	1.7
Status	Shuttered	Operating	Destroyed
Start-up	1964	1972	1972
Shareholders	Sonatrach	Sonatrach	Sonatrach
Liquefaction	Cascade	Technip	Teal
Country	Algeria	Algeria	Algeria
Project	Bethioua GL1-Z	Bethioua GL2-Z	Skikda GL2-K T40
Location	Bethioua	Bethioua	Skikda
Trains	6	6	1
Capacity (MMt/y)	10.5	10.5	0.85
Status	Operating	Operating	Destroyed
Start-up	1978	1981	1981
Shareholders	Sonatrach	Sonatrach	Sonatrach
Liquefaction	C3-MR	C3-MR	Prico
Country	Algeria	Algeria	Algeria
Project	Skikda GL2-K T5P, 6P	Skikda 20, 30, 40 replacement	Gassi Touil GL3-Z
Location	Skikda	Skikda	Arzew
Trains	2	1	1
Capacity (MMt/y)	2.5	4.5	4.7
Status	Operating	Operating	Under Construction
Start-up	1981	2013	2014
Shareholders	Sonatrach	Sonatrach	Sonatrach
Liquefaction	Prico	C3MR	C3MR

Fig. 5–13. (Continued)

Commissioning of the 4.5 MMt/y (6.1 Bcm/y) Skikda replacement train in March 2013 boosted Algeria's LNG production capacity to 25.5 MMt/y (34.7 Bcm/y). Liquefaction capacity will increase by another 4.7 MMt/y (6.4 Bcm/y) when the Gassi Touil project is commissioned in 2015. While it is located between the GL1-Z and GL2-Z plants, it is a greenfield development and does not share common facilities. The project, based upon the Gassi Touil gas field, was originally sponsored by a joint venture comprising Sonatrach and two Spanish firms, Repsol and Gas Natural. But after a dispute, Sonatrach became the sole shareholder. It is the only facility not supplied by the giant Hassi R'Mel gas field.

Algeria exports most of its LNG under long-term contracts to European buyers, but has been slowly increasing sales of LNG under spot and short-term deals (table 5–12). In 2013, Sonatrach sold 92%, approximately 9.8 MMt (13.3 Bcm), of its LNG into Europe under long-term contracts, including 3.9 MMt (5.3Bcm) into France, 2.8 MMt (3.8 Bcm) into Turkey, 2.5 MMt (3.4 Bcm) into Spain, and 0.5 MMt (0.7 Bcm) into Greece (table 5–12). Approximately 0.2 MMt (0.3 Bcm) were delivered into the United Kingdom, where Sonatrach has leased terminal capacity. Another 0.9 MMt (1.2 Bcm) were sold under spot and short-term arrangements predominantly into outlets east of Suez, including 0.4 MMt (0.5 Bcm) into Japan.

Table 5–12. Algeria LNG export volumes (MMt)

Destination	2005	2006	2007	2008	2009	2010	2011	2012	2013
France	5.8	6.3	6.0	5.8	6.0	4.9	4.5	3.5	3.9
Italy	1.9	2.2	1.7	1.1	0.9	1.3	1.3	0.7	–
Spain	4.1	2.6	3.5	3.8	4.0	3.7	2.9	2.9	2.5
Turkey	3.1	2.5	2.2	3.1	3.2	2.5	2.9	3.0	2.8
Other	4.9	5.5	4.3	2.7	1.9	1.7	1.1	1.1	1.6
Total Exports	19.8	19.0	17.8	16.5	15.9	14.1	12.7	11.2	10.8

(*Source*: Poten & Partners.)

Libya. Libya's National Oil Corporation (NOC) is the sole owner of a plant at Marsa el Brega completed in 1971. Original design capacity of the single-train plant was 2.3 MMt/y (3.2 Bcm/y). Exports to its two customers in Spain and Italy increased for the next few years, reaching a peak of 2.6 MMt/y (3.6 Bcm/y) in 1977. After Libya nationalized the plant in 1980, LNG exports went into decline. Unlike other plants, this facility was designed to process heavy natural gas liquids along with the natural gas, which forced buyers to extract liquids at their import sites. Few regasification terminals at that time were equipped to handle liquids extraction. In a 1990 refurbishment, Italy's sole terminal at La Spezia was modified and lost its ability to accept liquids-rich Libyan LNG. After that, Spain was the only offtaker of Libyan LNG. In 2010, the last full year of operation, exports totaled just 250,000 tons (table 5–13). In 2011, Marsa el Brega was destroyed in the Libyan civil war, and plans to refurbish the plant have been shelved. Libya still exports natural gas via the 8 Bcm/y (0.8 Bcf/d) Greenstream pipeline to Italy.

Table 5–13. Libya LNG export volumes (MMt)

Destination	2005	2006	2007	2008	2009	2010	2011	2012	2013
Spain	0.6	0.7	0.6	0.4	0.6	0.3	0.1	–	–
Other	–	–	–	–	–	–	–	–	–
Total Exports	0.6	0.7	0.6	0.4	0.6	0.3	0.1	–	–

(*Source*: Poten & Partners.)

Egypt. Egypt built two LNG export projects to capitalize on its natural gas resource base. The first liquefaction complex located at Damietta, 60 kilometers west of Port Said, started operations in September 2004.[39] The second project, located at Idku, 50 kilometers east of Alexandria, began production in May 2005.[40] Combined, the two projects—the single-train 5.5 MMt/y (7.5 Bcm/y) Damietta unit and the two-train 7.2 MMt/y (9.8 Bcm/y) Idku plant—have a nameplate production capacity of 12.7 MMt/y (17.3 Bcm/y). Both projects are located in the Nile Delta on Egypt's Mediterranean coast, and each has provisions to toll third-party gas through their facilities. Expansion plans have been shelved at both projects because of feed gas supply issues, and neither is currently producing LNG (fig. 5–14).

The Spanish Egyptian Gas Company (SEGAS) LNG complex in Damietta was supplied with surplus feed gas from the Egyptian grid rather than from dedicated upstream reserves. Surging Egyptian gas demand has eliminated this surplus, and the complex no longer produces LNG. The operating company, SEGAS, is controlled by Union Fenosa Gas (UFG) (80%) and two state-owned Egyptian companies—Egyptian Natural Gas Holding Company (EGAS) (10%) and Egyptian General Petroleum Corporation (EGPC) (10%). UFG is equally owned by Union Fenosa of Spain and Eni of Italy. The owners of the gas resources in offshore Israel are reportedly considering delivering a portion of these reserves to Damietta for processing into LNG.

The Egyptian LNG (ELNG) plant, located at Idku, was developed and operated by a partnership of BG, Petronas, EGAS, EGPC, and GDF Suez. It operates as a tolling facility that provides liquefaction service to the upstream West Delta Deep Marine (WDDM) partners who retain title to the gas through LNG production and who secure LNG buyers. BG and Petronas are the 50:50 concession holders for WDDM. However, in 2013, two-thirds of WDDM natural gas production was delivered to the domestic market. Even though diversion to the domestic market is constrained by pipeline infrastructure, LNG production at Idku dropped to 2.7 MMt (3.7 Bcm) in 2013, representing a capacity utilization of less than 40%. In 2014, all WDDM natural gas production was diverted to the domestic market, and LNG production came to a halt.

Country	Egypt
Project	Egyptian LNG
Location	Idku
Trains	2
Capacity (MMt/y)	7.2
Status	Mothballed
Start-up	2005
Shareholders	BG, EGAS, EGPC, GDF SUEZ, Petronas
Liquefaction	Opt. Cascade
Country	Egypt
Project	SEGAS LNG T1
Location	Damietta
Trains	1
Capacity (MMt/y)	5.5
Status	Mothballed
Start-up	2005
Shareholders	EGAS, EGPC, Union Fenosa Gas
Liquefaction	C3-MR

Fig. 5–14. Egypt LNG infrastructure and project profiles. (*Source:* Poten & Partners.)

Egyptian LNG production had ramped up to a peak of 11.1 MMt (15.1 Bcm) in 2008. Spain was the largest destination at 3.8 MMt (5.2 Bcm), followed by Japan and South Korea at 1.7 MMt (2.3 Bcm) each. By 2013, Egyptian LNG, primarily from ELNG, totaled just 2.7 MMt (3.7 Bcm). Of this, only 0.3 MMt (0.4 Bcm) stayed in the Atlantic Basin, and 2.4 MMt (3.3 Bcm) were shipped to markets east of Suez. Japan and South Korea were the top buyers, each importing 0.6 MMt (0.8 Bcm) of LNG, followed by China and India at 0.4 MMt (0.5 Bcm) each (table 5–14).

Table 5–14. Egypt LNG export volumes (MMt)

Destination	2005	2006	2007	2008	2009	2010	2011	2012	2013
Japan	–	0.5	1.2	1.7	0.2	0.4	0.7	1.0	0.6
South Korea	0.3	0.9	1.1	1.7	0.2	0.7	0.5	0.6	0.6
Spain	2.7	3.6	3.4	3.8	3.2	2.3	1.7	0.5	–
USA	1.5	2.1	2.3	1.2	3.4	1.5	0.7	0.1	–
Other	0.4	2.0	2.8	2.7	2.5	2.4	2.9	2.8	1.5
Total Exports	**4.8**	**9.1**	**10.9**	**11.1**	**9.4**	**7.4**	**6.5**	**5.0**	**2.7**

(*Source*: Poten & Partners.)

Nigeria became the first West African country to export LNG in 1990. The Nigeria LNG complex now comprises six trains capable of producing 22 MMt/y (29.9 Bcm/y) of LNG, though this capacity has never been fully utilized (fig. 5–15). Other Nigerian proposals—a megatrain expansion at Nigeria LNG as well as two greenfield projects, Brass LNG and OK LNG—have not been built, but export projects have started up in Equatorial Guinea (2007) and Angola (2012) as shown in figure 5–16.

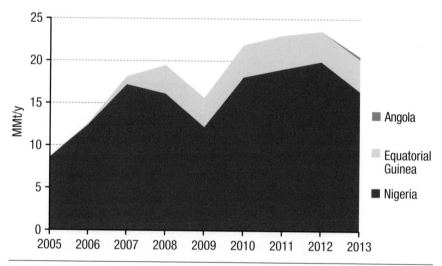

Fig. 5–15. West Africa LNG exports. (*Source:* Poten & Partners.)

Country	Nigeria	Nigeria	Nigeria
Project	NLNG 1,2	NLNG 3	NLNG-Plus (T 4,5)
Location	Bonny Island	Bonny Island	Bonny Island
Trains	2	1	2
Capacity (MMt/y)	6.6	3.3	8
Status	Operating	Operating	Operating
Start-up	1999	2002	2005
Shareholders	ENI Gas & Power, NNPC, Shell, Total	ENI Gas & Power, NNPC, Shell, Total	ENI Gas & Power, NNPC, Shell, Total
Liquefaction	C3MR	C3MR	C3-MR
Country	Equatorial Guinea	Nigeria	Angola
Project	Equatorial Guinea LNG	NLNG 6	Angola LNG
Location	Bioko Island	Bonny Island	Soyo
Trains	1	1	1
Capacity (MMt/y)	3.7	4	5.2
Status	Operating	Operating	Operating
Start-up	2007	2008	2012
Shareholders	Marathon, Marubeni, Mitsui & Co, Sonagas	ENI Gas & Power, NNPC, Shell, Total	BP, Chevron, ENI Gas & Power, Sonangol, Total
Liquefaction	Opt. Cascade	C3MR	Opt. Cascade

Fig. 5–16. West Africa LNG infrastructure and project profiles. (*Source:* Poten & Partners.)

Nigeria. With six trains completed since 1999, Nigeria LNG (NLNG) in Finima, Bonny Island, in the Niger Delta region is capable of producing 22 MMt/y (29.9 Bcm/y) of LNG and 5 MMt/y of LPG and condensates.[41] The first two trains, each with 3.3 MMt/y (4.5 Bcm/y) of production capacity, were commissioned in October 1999 and February 2000. The similar-sized Train 3 was completed in November 2002. Train 4 came on stream in November 2005 and Train 5 in February 2006. The size of these later so-called NLNGPlus trains increased to 4 MMt/y (5.4 Bcm/y) each. Train 6, NLNGSix, was commissioned in December 2007, boosting NLNG to its current size. NLNG is jointly owned by Nigerian National Petroleum Corporation (NNPC) 49%, Shell 25.6%, Total 15%, and Eni 10.4%.

Plans to add a seventh train, a megatrain, to boost NLNG LNG production capacity to 30 MMt/y (40.8 Bcm/y) have stalled in part because of an unstable social environment and regulatory uncertainties stemming from the long-pending Nigerian Petroleum Investment Bill (PIB), which have undermined investor confidence. Two greenfield projects (Brass LNG located in Brass River, Niger Delta and OK-LNG located in Olokola, a free-zone port on the border between Ondo and Ogun states) have also stalled.

Sabotage of gas pipelines often limits feed gas supply below the 3.5 Bcf/d the NLNG plant is designed to process. An uneasy truce with Niger Delta militant groups offered a respite from disruptions, and the facility produced 19.1 MMt (26.0 Bcm) of LNG in 2011 and 20.0 MMt (27.2 Bcm) in 2012. However, production fell back to 16.5 MMt (22.4 Bcm) in 2013 when NLNG declared force majeure on exports, given the sabotage on Shell's Trans-Niger pipeline and a tax dispute with the Nigerian Maritime Administration and Safety Agency.

NLNG cargoes have increasingly been diverted to east of Suez markets. By 2012, half or more of NLNG exports went to these outlets. Prior to 2012, Spain was the top destination for NLNG cargoes. In 2013, Japan was the top destination with 3.8 MMt (5.2 Bcm), followed by South Korea at 2.8 MMt (3.8 Bcm), and Spain, which imported 2.5 MMt (3.4 Bcm). Mexico and South America have also developed into important buyers of Nigerian LNG (table 5–15).

Table 5–15. Nigeria LNG export volumes (MMt)

Destination	2005	2006	2007	2008	2009	2010	2011	2012	2013
France	2.9	3.0	3.0	2.7	1.8	2.8	2.8	2.2	0.8
Japan	–	0.2	0.7	1.8	0.6	0.6	1.9	4.8	3.8
Portugal	1.1	1.1	2.0	1.9	1.6	2.1	1.8	1.4	0.9
South Korea	–	0.1	0.2	0.3	0.2	0.9	1.2	1.8	2.8
Spain	3.7	5.8	6.8	5.8	3.8	6.0	5.0	3.9	2.5
Other	1.0	2.2	4.5	3.7	4.2	5.7	6.3	5.8	5.7
Total Exports	8.6	12.4	17.2	16.1	12.2	18.1	19.1	20.0	16.5

(*Source*: Poten & Partners.)

Equatorial Guinea. The Equatorial Guinea LNG (EG LNG) plant is located on the northwest side of Bioko Island at Punta Europa, near the capital city of Malabo.[42] The first cargo from the 3.7 MMt/y (5.1 Bcm/y) plant was exported in June 2007. Feed gas for the project is purchased from the Alba gas field participants, Marathon Oil, Noble Energy, and GEPetrol. Marathon Oil controls 60% of EG LNG with Sonagas, the national oil company, holding a 25% interest. Two Japanese companies, Mitsui and Marubeni, hold 8.5% and 6.5%, respectively. EGLNG wants to establish an LNG export hub for regional gas, but the venture has not secured feed gas for an expansion train, and the existing train could be short of feed gas in the long term.

In June 2004, BG signed a deal to purchase 3.4 MMt/y (4.6 Bcm/y) from EG LNG. The primary destination for the LNG was to be the Lake Charles import terminal in Louisiana, but with destination flexibility. The FOB purchase price was linked to the Henry Hub Index. When surging US shale gas production curbed LNG demand at Lake Charles and depressed Henry Hub prices, BG sold EG LNG cargoes into premium outlets elsewhere, generating considerable arbitrage profits. Of the 3.8 MMt (5.2 Bcm) produced in 2013, BG delivered 2.2 MMt (3.0 Bcm) to Japan, 0.6 MMt (0.8 Bcm) to Singapore, and 0.4 MMt (0.5 Bcm) to China and Taiwan each (table 5–16). No cargoes were delivered into the Lake Charles terminal.

Table 5-16. Equatorial Guinea LNG export volumes (MMt)

Destination	2005	2006	2007	2008	2009	2010	2011	2012	2013
China	–	–	–	0.1	0.1	0.1	0.1	–	0.4
Japan	–	–	0.3	1.2	1.3	0.5	1.5	2.8	2.2
Singapore	–	–	–	–	–	–	–	–	–
South Korea	–	–	–	0.9	1.1	1.2	0.8	0.4	0.1
Taiwan	–	–	0.3	0.7	0.6	0.3	0.6	0.1	0.4
Other	–	–	0.7	1.3	1.1	2.0	1.7	0.4	1.4
Total Exports	–	–	**1.0**	**3.4**	**3.4**	**3.8**	**4.0**	**3.6**	**3.8**

(*Source:* Poten & Partners.)

Angola. Angola LNG, the latest LNG export project to be commissioned in the Atlantic Basin, is a single-train 5.2 MMt/y (7.1 Bcm/y) facility built near the town of Soyo in Zaire Province, northern Angola.[43] It was commissioned in mid-2013. The plant is supplied from offshore associated natural gas fields on Blocks 14, 15, 17, and 18, and will also be supplied from nonassociated natural gas fields Quiluma, Atum, Polvo, and Enguia. The project is characterized as being environmentally friendly, because most of the associated gas produced with crude oil had previously been flared. The facility will be supplied with 1 Bcf/d of feed gas from the offshore blocks. The project also supplies 125 MMcf/d of gas to the domestic market. As the oil fields mature and associated gas production declines, the nonassociated gas fields in Blocks 1 and 2 will be developed to supplement the associated gas supply. Under Angolan law, Sonangol is the owner of all associated and nonassociated gas discovered in the blocks. Chevron has a 36.4% stake in Angola LNG, followed by Sonangol at 22.8%, and BP, Total, and Eni at 13.6% each.

The original plan was to deliver Angola LNG into the Gulf LNG Clean Energy terminal in Mississippi. However, the boom in shale gas production made exports to this terminal unattractive. To compensate for the lack of market in the United States, Angola LNG set up London-based Angola LNG Marketing Ltd in November 2011. It is led by Sonangol and sells cargoes originally destined for the US Gulf Coast. In 2013, only five cargoes were produced. Of these, four went to buyers in East Asia and one to Brazil. In early 2014, the plant was taken out of service for extensive redesign and refurbishment, due to a variety of problems. It could be out of service until 2016.

Arctic

Building an LNG export project in the Arctic is a very challenging undertaking owing to scarce local labor and the harsh weather conditions, which severely limit available work time at the site. To counter these conditions, the partners in the Snøhvit project decided to build a liquefaction barge in Spain and float it to an island near Hammerfest, north of the Arctic Circle, where production started at the 4.3 MMt/y (5.9 Bcm/y) plant in 2007. All expansion plans have been put on hold as the plant operator deals with ongoing operating glitches at the technologically complex facility.

Now a Russian company, Novatek, is building a second, much larger Arctic project on the even more challenging Yamal Peninsula with the full support of the Russian government. Train 1 of the three-train 16.5 MMt/y (22.4 Bcm/y) project is scheduled for commissioning in 2018. The project has one of the most challenging LNG shipping programs ever undertaken, requiring a fleet of ice-breaking LNG tankers with plans to transit the Northern Passage to Asia during the summer months (fig. 5–17).

Country	Norway	Russia
Project	Snøhvit	Yamal LNG
Location	Melkoya Island	Yamal Peninsula
Trains	1	3
Capacity (MMt/y)	4.3	16.5
Status	Operating	Under Construction
Start-up	2007	2018
Shareholders	GDF Suez, Petoro, RWE, Statoil, Total	CNPC, Novatek, Total
Liquefaction	MR Cascade	C3MR

Fig. 5–17. Atlantic Basin Arctic LNG infrastructure and project profiles. (*Source:* Poten & Partners.)

Norway. The single-train 4.3 MMt/y (5.9 Bcm/y) Snøhvit LNG project is Europe's sole LNG export plant and the northernmost ever constructed.[44] Located on the remote island of Melkøya north of the Arctic Circle, it was built to monetize fields in the Barents Sea, too far north to be connected to Norway's North Sea pipeline infrastructure. It was also intended to provide socioeconomic benefits to the 1.5% of the Norwegian population living in the Hammerfest area. After exploring joint venture possibilities with existing liquefaction process licensees to adapt their process to Melkøya's harsh Arctic conditions, project-lead Statoil formed an alliance with LNG newcomer Linde AG to develop the Mixed Fluid Cascade (MFC) liquefaction process employed at Snøhvit LNG.

Owing to the difficult construction environment in northern Norway, the majority of the liquefaction unit was assembled on a barge, constructed in the Dragados shipyard in Cadiz, Spain. It was floated to Melkøya Island, where it was integrated into a preconstructed dock. Statoil pushed the technological envelope on the upstream as well, which adversely impacted an operating performance that had been troubled since the plant was commissioned in October 2007. Statoil is the operator with a 36.79% interest, state-owned Petoro owns 30.0%, France's Total and GDF Suez own 18.4% and 12.0% respectively, and RWE owns 2.81%.

In 2008, the first full year of operation, Snøhvit LNG struggled to ramp up to capacity, producing just 1.6 MMt (2.2 Bcm) of LNG. The plant has continued to experience operational difficulties, producing 3.0 MMt/y (4.1 Bcm/y) on average since 2010, still only 70% of nameplate capacity. In 2013, Norway exported 2.9 MMt (3.9 Bcm) of LNG, including 0.9 MMt (1.2 Bcm) to Spain. Five countries (Brazil, Japan, Mexico, the Netherlands, and the United Kingdom) each imported from 200,000 to 300,000 tons (0.3 Bcm to 0.4 Bcm) of LNG from Snøhvit during the year (table 5–17). Norway also sells large natural gas volumes by pipeline to the United Kingdom and continental Europe.

Table 5-17. Norway LNG export volumes (MMt)

Destination	2005	2006	2007	2008	2009	2010	2011	2012	2013
France	–	–	0.1	0.1	0.3	0.4	0.4	0.1	0.1
Spain	–	–	–	0.8	1.1	1.2	1.0	1.4	0.9
United Kingdom	–	–	–	–	0.2	0.6	0.3	–	0.2
USA	–	–	–	0.4	0.6	0.5	0.3	0.1	0.1
Other	–	–	–	0.3	0.1	0.3	1.0	1.8	1.6
Total Exports	–	–	0.1	1.6	2.3	3.0	2.9	3.4	2.9

(*Source*: Poten & Partners.)

Russia. Yamal LNG is an integrated project involving upstream production and processing, liquefaction, and export of LNG and condensate from the Yamal Peninsula, in the very north of western Siberia.[45] In December 2013, venture partners led by Russia's Novatek made an FID on the three-train 16.5 MMt/y (22.4 Bcm/y) project. Once Yamal LNG becomes operational, it will break state-owned Gazprom's monopoly on international gas sales from Russia and will give the country its second LNG export project after the 9.6 MMt/y (13.1 Bcm/y) Sakhalin II complex, completed in 2009. Train 1 is scheduled for commercial start-up in 2018. The production is based on the resources of the 907 Bcm (32 Tcf) South Tambey gas condensate field, which will be able to produce up to 1.2 MMt/y of condensate.

Novatek holds a 60% stake in the venture. Foreign partners, France's Total and China National Petroleum Corporation, each have 20% shareholdings. Moscow is supporting Yamal LNG through tax relief and is underwriting the new Arctic port of Sabetta, where construction began in 2012. The seaport is designed to accommo-date ice-class ARC7 LNG carriers. According to Yamal LNG, the project will require as many as 16 ARC7 ships at a cost of more than $300 million each, and 15 conventional LNG carriers that will cost about $200 million each.

Over 70% of Yamal LNG output is committed under long-term sales arrangements with China National Petroleum Corp., Spain's Gas Natural Fenosa, France's Total Gas & Power, and Novatek's trading affiliate. The ARC7 ships will transport cargoes through the Northern Passage directly to Asia during the summer months and to the Zeebrugge LNG terminal for transshipment to conventional tankers all year. Sanctions imposed by the United States and Europe following the invasion of Ukraine threaten the project's access to technology and financing, and in the wake of the oil collapse the Russian state may have to step up its financial support of the project.

Caribbean and South America

Regional exports began in 1999 when the first train at the Atlantic LNG complex in Trinidad and Tobago was commissioned. By 2005, production had climbed to 15.4 MMt (20.9 Bcm) as three new trains were added at the complex, but it dropped to 13.8 MMt (18.8 Bcm/y) in 2013. South America's first LNG export plant, Peru LNG, was commissioned in 2010. This facility produced 4.0 MMt (5.4 Bcm) of LNG in 2013, boosting regional exports to 17.8 MMt (24.3 Bcm) in that year. While Trinidad primarily supplies Atlantic basin markets, Peru LNG, which is located on the country's Pacific coast, primarily exports into Pacific basin outlets. (Peru LNG is discussed in the Pacific basin section of this chapter.) A small bi-directional LNG export and import facility was being developed for the coast of Colombia, but the project has collapsed (fig. 5–18).

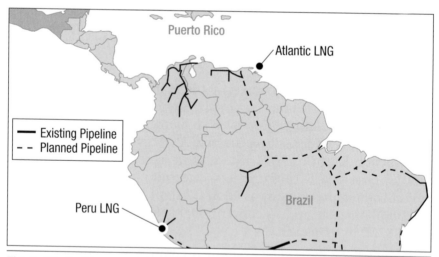

Fig. 5–18. Caribbean and South America LNG export infrastructure and project profiles. (*Source:* Poten & Partners.)

Country	Trinidad	Trinidad	Trinidad
Project	Atlantic LNG T1	Atlantic LNG T2	Atlantic LNG T3
Location	Point Fortin	Point Fortin	Point Fortin
Trains	1	1	1
Capacity (MMt/y)	3.3	3.3	3.3
Status	Operating	Operating	Operating
Start-up	1999	2002	2003
Shareholders	BG, BP, CIC (China), NGC, Shell	BG, BP, Shell	BG, BP, Shell
Liquefaction	Opt. Cascade	Opt. Cascade	Opt. Cascade

Country	Trinidad	Peru	
Project	Atlantic LNG T4	Peru LNG	
Location	Point Fortin	Pisco	
Trains	1	1	
Capacity (MMt/y)	5.2	4.45	
Status	Operating	Operating	
Start-up	2005	2010	
Shareholders	BG, BP, NGC, Shell	Hunt Oil, Marubeni, Shell, SK Corp	
Liquefaction	Opt. Cascade	C3MR	

Fig. 5–18. (Continued)

Trinidad and Tobago. Atlantic LNG, with four trains in operation in Point Fortin, Trinidad, is capable of producing 15.1 MMt/y (20.5 Bcm/y) of LNG and 30,000 bpd of natural gas liquids.[46] When the first cargo was loaded in April 1999, it was the first LNG export project to be commissioned in the Atlantic basin in two decades. The company kept the initial design and construction to a one-train facility geared towards the needs of customers in New England and Spain. By not attempting to create a larger initial project, the schedule was shorter and the unit costs lower than many earlier projects.

Export capacity of Train 1 was designed at 3.0 MMt/y (4.1 Bcm/y); however, a two-step debottlenecking project raised its capacity to 3.3 MMt/y (4.6 Bcm/y). Growing demand for Atlantic Basin LNG supplies, combined with the proving of larger gas reserves, encouraged Atlantic LNG to build Trains 2 and 3, each with 3.3 MMt/y (4.6 Bcm/y) of production capacity. They came on line in 2002 and 2003 respectively. Although Atlantic LNG's original five partners had the right to participate in the expansion venture, shareholders GDF Suez and Trinidad's National Gas Company (NGC) chose not to participate in the Atlantic 2/3 project, owing to the fact that the financing was done through equity contributions rather than through project financing. Moreover, the rate of return on the project was capped at approximately 9% for any downstream partners that participated, limiting NGC and GDF Suez's incentives to do so. A fourth 5.2 MMt/y (7.2 Bcm/y) liquefaction unit shipped its first cargo in January 2006. The shareholding structures of Trains 2, 3, and 4 were different from Train 1—they were tolling facilities. Each partner must secure its own feed gas supply in order to take advantage of this option. An even larger fifth train was proposed and then shelved because of feed gas constraints.

Trinidad LNG production climbed from 3 MMt (4.1 Bcm) in 2000, the first full year of Train 1 operation, to approximately 10 MMt (13.6 Bcm) in 2004 and 2005. Since Train 4 started producing LNG in December 2005, annual production has averaged approximately 14.1 MMt (19.2 Bcm). Through 2007, the US market dominated the Trinidad LNG trade. In many years, more than 90% of Trinidad production was delivered into the United States as buyers rerouted cargoes originally destined for Spain to the Everett terminal, owing to higher prices in the US Northeast. Sales into the United States tumbled to only 1.5 MMt (2.0 Bcm) in 2013, with similar volumes delivered into Spain that year.

Trinidad is strategically located to supply burgeoning South American and Caribbean LNG demand (table 5–18). In 2013, 9.0 MMt (12.2 Bcm) of Trinidad LNG was delivered into these outlets led by Argentina and Chile with 2.4 MMt (3.3 Bcm) and 2.3 MMt (3.1 Bcm) respectively, closely followed by Brazil at 2.0 MMt (2.7 Bcm). In the Caribbean, 1.1 MMt (1.5 Bcm) went to Puerto Rico, and 0.9 MMt (1.2 Bcm) to the Dominican Republic. The balance found its way to Asia-Pacific markets.

Table 5–18. Trinidad LNG export volumes (MMt)

Destination	2005	2006	2007	2008	2009	2010	2011	2012	2013
Argentina	–	–	–	0.3	0.5	1.1	2.1	2.6	2.4
Brazil	–	–	–	–	0.4	0.8	0.1	1.1	2.0
Chile	–	–	–	–	0.1	0.6	1.1	2.3	2.3
Spain	0.4	2.8	1.7	3.4	2.9	2.3	1.8	1.8	1.5
USA	9.1	8.1	9.7	5.5	5.0	4.2	2.7	2.4	1.5
Other	0.5	1.6	2.9	4.3	6.2	6.4	5.3	4.5	4.1
Total Exports	10.1	12.5	14.3	13.6	15.0	15.4	13.2	14.6	13.8

(*Source:* Poten & Partners.)

North America

United States. Growing natural gas resource base, increased production, and low prices have spurred numerous proposals to convert existing LNG import facilities to bidirectional facilities by adding liquefaction trains. The brownfield nature of these projects—building upon the significant infrastructure (storage tanks and marine facilities) that exists at the sites—promises lower construction costs. There are numerous greenfield projects proposed as well, with sponsors betting that low, market-based feed gas costs will make their projects viable, a bet that looks much riskier in a world of $50/barrel oil. As shown in figure 5–19, five projects, four on the US Gulf Coast and one on the Eastern Seaboard, have received requisite regulatory approvals and commenced with construction.

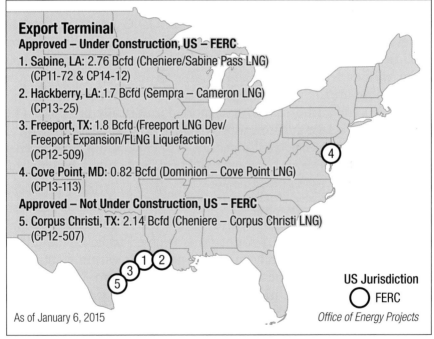

Export Terminal
Approved – Under Construction, US – FERC
1. Sabine, LA: 2.76 Bcfd (Cheniere/Sabine Pass LNG) (CP11-72 & CP14-12)
2. Hackberry, LA: 1.7 Bcfd (Sempra – Cameron LNG) (CP13-25)
3. Freeport, TX: 1.8 Bcfd (Freeport LNG Dev/ Freeport Expansion/FLNG Liquefaction) (CP12-509)
4. Cove Point, MD: 0.82 Bcfd (Dominion – Cove Point LNG) (CP13-113)

Approved – Not Under Construction, US – FERC
5. Corpus Christi, TX: 2.14 Bcfd (Cheniere – Corpus Christi LNG) (CP12-507)

US Jurisdiction
○ FERC

As of January 6, 2015   *Office of Energy Projects*

Fig. 5–19. Potential US LNG export projects. (*Source:* DOE/FE, FERC, and Poten & Partners.)

The US LNG "gold rush" kicked off in August 2010 when Cheniere Energy filed with the US DOE for approval to export domestically produced natural gas as LNG to Free Trade Agreement (FTA) countries from its Sabine Pass terminal in Louisiana. (LNG exports to FTA countries are deemed to be in the national interest unless proven otherwise.) One month later, the project filed for authorization to export LNG to non-FTA countries. (LNG exports to non-FTA countries must be deemed to be in the national interest by the DOE.) This is a key authorization because many large customers are in non-FTA nations, South Korea being a notable exception. This proposal was accepted. Since then, others have followed and more are expected. Owners accounting for all the existing onshore LNG import capacity, with the exception of the Everett terminal in New England, have now proposed or are considering adding liquefaction facilities at their terminals to export LNG.

Regulatory and permitting issues for US LNG export projects. In order to proceed with the construction and operation of liquefaction facilities in the United States, project developers must seek approvals from various federal, state, and local agencies. The highest profile applications are to the Department of Energy's Office of Fossil Energy (DOE/FE) for authorization to export domestically produced gas, and to the FERC for authorization to construct and operate terminal facilities and any associated pipelines. The FERC filing requires that National Environmental Protection Act (NEPA) criteria be satisfied.

An authorization from the DOE to export domestically produced natural gas is required under Section 3(a) of the Natural Gas Act. The United States has entered into FTAs with 18 countries. By law, the export of LNG to FTA countries is deemed to be in the public interest, and DOE generally will grant approvals within a couple of months of submitting the application. Exports to non-FTA countries are also assumed to be in the public interest and, unless the DOE finds otherwise, the DOE will grant approval. The presumptive benefit of the export trade essentially places the burden of demonstrating harm to the public interest on a project's opponents. Authorization to export LNG to non-FTA countries became a focus of political debate, with opponents arguing that it would encourage fracking, which they consider environmentally harmful, and that it would boost domestic gas prices, which would result in higher natural gas bills for US consumers and manufacturers.

The DOE issued Cheniere a conditional order on May 20, 2011, allowing Sabine Pass LNG to export LNG to non-FTA countries.[47] Following the release of this order, politicians and environmentalists and some manufacturing/industrial entities questioned the wisdom of allowing exports of natural gas to any country. In response to filed concerns, the DOE initially noted that opponents of the application had not provided any study or analysis detailing why Sabine Pass LNG exports would not be in the public interest. Responding to mounting criticism, the DOE then authorized two independent studies of this issue.

The first one was prepared by the DOE's Energy Information Administration (EIA) and submitted to the DOE in January 2012. Pursuant to instructions received from the DOE, the EIA considered four export scenarios: 6 Bcf/d (45 MMt/y) with a slow ramp up in volume; 6 Bcf/d (45 MMt/y) with a rapid ramp up; 12 Bcf/d (90 MMt/y) with a slow ramp up; and 12 Bcf/d (90 MMt/y) with a rapid ramp up. The EIA analysis looked at how these LNG export scenarios would impact production, consumption, and prices in the US market (table 5–19). The headline finding from this study was that domestic gas prices would rise to some degree under all four LNG export scenarios. This finding was criticized as resulting from the scenarios proposed and not from rigorous analysis of the overall impact from exports.

A second study was requested by the DOE. This study's purpose was to examine the macroeconomic effects of LNG exports. It was prepared by NERA, an economic consulting group, and issued in December 2012. The key findings from this study, which analyzed 13 scenarios, were that the United States would gain net economic benefits under all scenarios, and that the impact on domestic natural gas prices would be limited. Following receipt of the second report, the DOE requested comments on both reports through February 2013. The DOE received 188,000 initial comments including many bundled by the Sierra Club. Political, environmental, and economic concerns were raised during the comment period, as well as comments in support of exports.

Table 5–19. Proposed US LNG exports from the lower 48 states

Former DOE Non-FTA Order of Precedence	MMt/y	Non-FTA	FERC Progress
1 Sabine Pass, LA	16.0	√	Full Approval
2 Freeport, TX	10.5	√	Full Approval
3 Lake Charles, LA	15.0	√	Awaiting EA
4 Cove Point, MD	5.7	√	Full Approval
5 Freeport, TX	3.0	√	Full Approval
6 Cameron, LA	12.0	√	Full Approval
7 Jordan Cove, OR	6.0	√	Awaiting draft EIS
8 Oregon LNG, OR	9.6	√	Awaiting draft EIS
9 Cheniere Corpus Christi, TX	15.0	UR*	Full Approval
10 Excelerate Liquefaction, TX	10.0	UR	Awaiting draft EIS
11 Carib Energy, TBD	0.3	√	Not required
12 Gulf Coast LNG Export, TX	21.0	UR	Has not filed
13 Southern LNG, GA	4.0	UR	Awaiting draft EIS
14 Gulf LNG Export, MS	11.5	UR	EIS prefiling
15 CE FLNG, LA	8.0	UR	Awaiting draft EIS
16 Golden Pass Products, TX	15.0	UR	Awaiting draft EIS
17 Pangea LNG (North America) Holdings, TX	8.0	UR	Has not filed
18 Trunkline LNG, LA	15.0	UR	Awaiting EA
19 Freeport-McMoran Energy, LA	24.0	UR	Has not filed
20 Sabine Pass Liquefaction, LA	2.3	UR	Final EA scheduled
21 Sabine Pass Liquefaction, LA	1.8	UR	Final EA scheduled
22 Venture Global LNG, LA	5.0	UR	Prefiling requested
23 Eos LNG, TX	12.0	UR	Has not filed
24 Barca LNG, TX	12.0	UR	Has not filed
25 Sabine Pass Liquefaction, LA	6.5	UR	Final EA scheduled
26 Magnolia LNG, LA	8.1	UR	Awaiting draft EIS
27 Delfin LNG, LA	13.0	UR	Has not filed
28 Waller LNG Services, LA	1.5	UR	Has not filed
29 Gasfin Development, LA	1.5	UR	Has not filed
30 Texas LNG, TX	2.0	UR	Has not filed
31 Louisiana LNG Energy, LA	2.0	UR	Prefiling requested
32 Strom Inc., FL	0.02	UR	Has not filed
33 Strom Inc., FL	0.02	UR	Has not filed
34 SCT&E LNG, LA	12.0	UR	Has not filed
35 Venture Global LNG, LA	5.0	UR	Prefiling requested
36 Downeast LNG, ME	3.5	UR	Has not filed
Total	**297.8**		

(*Source:* Poten & Partners.)
*Under review

In December 2012, at the same time the DOE requested comments to the two studies, DOE established an Order of Precedence based on the application receipt date at the DOE, for processing non-FTA export applications. Additional export applications were added in the order received. Proposals totaled approximately 300 MMt/y (400 Bcm/y) as of year-end 2014 (see table above). Finally in 2014, the DOE initiated a rulemaking process and as a result changed the criteria it would use to approve exports to non-FTA countries. Rather than processing applications in the order in which they were received, the DOE is now requiring that the applicants receive their FERC Final Environmental Impact Statement before approving the exports. This approach permits the DOE to meet its obligations under NEPA without carrying out a separate environmental assessment and addresses many of the criticisms raised by environmental groups. However, this elevates the risk for project developers who must go through a full permitting process, which can take three years and $100 million, and risk being denied an export application at the end.

Summaries of the most advanced US proposals. The five most advanced liquefaction proposals represent prospective exports of nearly 60 MMt/y (82 Bcm/y). Two (Cameron and Sabine Pass) are located in Louisiana, two (Freeport and Corpus Christi) in Texas, and the fifth (Cove Point) in Maryland (fig 5–20).

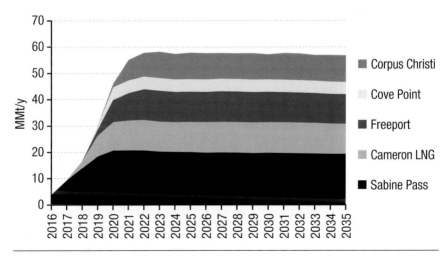

Fig. 5–20. US LNG exports. (*Source:* Poten & Partners.)

Sabine Pass Liquefaction. Cheniere Energy is installing liquefaction trains at its Sabine Pass LNG receiving terminal in Cameron Parish, Louisiana.[48] This will transform Sabine Pass into a bidrectional facility capable of liquefying and exporting natural gas in addition to importing and regasifying imported LNG (table 5–20). The four 4.5 MMt/y (6.1 Bcm/y) liquefaction trains piggyback on existing infrastructure, which includes five storage tanks and two marine berths as well as the 94-mile Creole Trail Pipeline, which is also being reconfigured for bidrectional flows. The site can accommodate two more trains, raising nameplate capacity to 27 MMt/y (36.7 Bcm/y). The first liquefaction train is to be commissioned in 2015, Trains 2 and 3 in 2016, and Train 4 in 2017.

Table 5-20. Sabine Pass FOB sales

Train 1	BG-3.5 MMt/y	Cheniere Marketing-1.0 MMt/y
Train 2	GNF-3.5 MMt/y; BG-0.7 MMt/y	Cheniere Marketing-0.3 MMt/y
Train 3	Kogas-3.5 MMt/y; BG-0.65 MMt/y	Cheniere Marketing-0.35 MMt/y
Train 4	Gail-3.5 MMt/y; BG-0.65 MMt/y	Cheniere Marketing-0.35 MMt/y
Train 5	Total-2.0 MMt/y	Centrica-1.8 MMt/y (Tentative)

(*Source:* Poten & Partners.)

Cheniere has entered into FOB deals with buyers for 16 MMt/y (21.8 Bcm/y) FOB from the first four trains and has tentative deals with Total for 2 MMt/y (2.7 Bcm/y) and Centrica for 1.8 MMt/y (2.5 Bcm/y) from Train 5. Cheniere will have to line up feed gas of approximately 2.5 Bcf/d (25.8 Bcm/y) once all five trains are on line, since the offtake arrangements at Sabine Pass are FOB sales. Unlike traditional projects with dedicated reserves, Cheniere will purchase feed gas on the open market and is responsible for transporting this gas to the project.

Freeport LNG Liquefaction. Freeport LNG plans to add three 4.6 MMt/y (6.0 Bcm/y) liquefaction trains at its import terminal on Quintana Island, Texas.[49] This facility will process approximately 2.0 Bcf/d (20.7 Bcm/y) of feed gas, which will be delivered to the terminal through Freeport LNG's existing 42-inch gas pipeline (table 5–21). At the terminal, it will be liquefied and then stored in two 160,000 m^3 full-containment LNG storage tanks built as part of the LNG import plant. Existing marine facilities are capable of accommodating 267,000 m^3 LNG tankers. Freeport LNG received the final order from the FERC in October 2014 authorizing construction, which began immediately. Under a 48-month construction schedule, Train 1 will be commissioned in late 2018, with the second train following six to nine months later. The facility will still be able to import LNG as market conditions dictate.

In May 2013, Freeport LNG received DOE authorization to export 511 Bcf/y (10.6 MMt/y) of LNG to non-FTA countries. Freeport LNG has signed Heads of Agreement (HOAs) with Japan's Osaka Gas and Chubu Electric for Train 1 and with BP for Train 2. The deals were contingent upon non-FTA authorization. In November 2013, the DOE authorized another 146 Bcf/y (3.0 MMt/y) of exports to non-FTA countries. Freeport has obtained another 170 acres from the Port Authority, leasing this land for 30 years with six 10-year renewal options. This will be used to build the third and a possible fourth train. In September 2013, Toshiba and SK signed 20-year long-term tolling agreements (LTAs) with Freeport LNG, selling out the third train.

Table 5–21. Freeport LNG tolling agreements

Train 1	Osaka Gas–2.2 MMt/y	Chubu Electric–2.2 MMt/y
Train 2	BP–4.4 MMt/y	
Train 3	Toshiba–2.2 MM/y	SK–2.2 MMt/y

(*Source*: Poten & Partners.)

Corpus Christi Liquefaction. Cheniere Energy's Corpus Christi Liquefaction subsidiary is developing an export terminal at one of Cheniere's existing sites that was previously permitted for a regasification terminal (table 5–22). The proposed liquefaction project is designed for three trains with aggregate nominal production capacity of up to 13.5 MMt/y (18.4 Bcm/y). The site is located on the La Quinta Channel on the northeast side of Corpus Christi Bay in San Patricio County, Texas, on over 1,000 acres owned or controlled by Cheniere and is approximately 15 nautical miles from the coast. Corpus Christi Liquefaction has filed for all necessary regulatory permits and has secured the financing. Corpus Christi Liquefaction commenced construction in 2015 with the first train to be commissioned in 2018, with each subsequent LNG train commencing operations approximately six to nine months after the previous train.

Table 5–22. Corpus Christi FOB sales

Train 1 Pertamina-1.52 MMt/y	Endesa-2.25 MMt/y Iberdrola-0.76 MMt/y
Train 2 Gas Natural Fenosa-1.50 MMt/y	Woodside-0.85 MMt/y EDF-0.77 MMt/y
Train 3 Energias de Portugal-0.77 MMt/y	
(*Source*: Poten & Partners.)	

As of early 2015, 8.4 MMt/y (11.5 Bcm/y) of LNG have been contracted to buyers on a long-term FOB basis under sale and purchase agreements (SPAs). Foundation customers include Pertamina, Endesa, Iberdrola, Gas Natural Fenosa, Woodside, EDF, and EDP. Trains 1 and 2 are fully contracted, and Train 3 is partially contracted. Any excess capacity not sold under long-term SPAs would be available for Cheniere Marketing to purchase. Services under the SPAs include procuring the natural gas (up to 1.2 Bcf/d), liquefying the natural gas, and loading LNG onto the customer's vessels.

Cove Point Liquefaction. Dominion Resources started construction in late 2014 on a single 5.25 MMt/y (7.1 Bcm/y) liquefaction train at its Dominion Cove Point LNG Terminal located on the Chesapeake Bay in Lusby, Maryland (table 5–23). Dominion plans to put the LNG train in service in 2017. Dominion has received DOE permission to export 5.7 MMt/y (7.75 Bcm/y) of LNG to FTA and non-FTA countries. The capacity at the liquefaction facility has been fully subscribed by Japan's Sumitomo and India's Gail, who have each contracted for half of the train's capacity for a period of 20 years. Sumitomo in turn has announced agreements to supply Japan's Tokyo Gas Co. and Kansai Electric Power Co., Inc.

Table 5–23. Cove Point liquefaction tolling agreements

Train 1	Sumitomo-2.6 MMt/y	Gail-2.6 MMt/y
(*Source*: Poten & Partners.)		

Cameron LNG Liquefaction. The Cameron LNG liquefaction project includes three liquefaction trains each with nameplate capacity of 4.5 MMt/y (6.1 Bcm/y).[50] In addition, a new 21-mile natural gas pipeline, a compressor station, and modifications to existing pipeline interconnections are proposed to facilitate the delivery of approximately 1.7 Bcf/d (17.6 Bcm/y) of feed gas to run the liquefaction plant at capacity (table 5–24). Construction on the project along the Calcasieu Channel in Hackberry, Louisiana, started in 2014 with the facility expected to start production in the second half of 2017. The liquefaction project will use existing facilities, including two marine berths capable of accommodating Q-Flex-sized LNG ships and three 160,000 m³ LNG storage tanks. Project-lead Sempra Energy has signed 20-year tolling agreements with GDF Suez, Mitsubishi, and Mitsui & Co for the full nameplate capacity of the three-train facility at fixed-tolling charges of about $3/MMBtu. The tolling customers will each acquire 16.6% equity stakes in the regasification terminal and the liquefaction project. Sempra will retain majority 50.2% shareholdings.

Table 5-24. Cameron liquefaction tolling agreements

Trains 1, 2	Mitsubishi-4 MMt/y	Mitsui-4 MMt/y
Train 3	GDF Suez-4 MMt/y	

(*Source:* Poten & Partners.)

Beyond the five projects in advance development, there are scores of other projects proposed for the Gulf Coast, eastern United States, and eastern Canada. A few, such as ExxonMobil's Golden Pass project, could clearly proceed on the basis of the financial, technical, and commercial capabilities of their sponsors, but most are being developed by small independent companies with little or no capacity to fund the permitting costs, let alone the construction costs. Achieving offtake or throughput agreements will likely prove to be a major obstacle given the proliferation of projects in an already glutted LNG market, which is also suffering from falling prices. As a result, most are destined to fail.

Notes

1. Poten & Partners, *Global LNG Outlook*, Proprietary Report.
2. Brunei LNG, "About Us: History & Background," http://www.bruneilng. com/about_history.htm.
3. ExxonMobil, "Indonesia: History," http://www.exxonmobil.com/Indonesia-English/PA/about_history.aspx.
4. Badak LNG, "Plant Operation: Plant Facilities," http://www.badaklng.co.id/ train_facilities.html.
5. BP, "Tangguh LNG," http://www.bp.com/en_id/indonesia/bp-in-indone-sia/tangguh-lng.html.

6. Raras Cahyafitri, "Donggi-Senoro LNG Plant Ready for Operation," *Jakarta Post* (May 12, 2015), http://www.thejakartapost.com/news/2015/05/12/donggi-senoro-lng-plant-ready-operation.html.

7. INPEX, "Changes of Participating Interest in the Abadi LNG Project, the Masela Block, Indonesia" (May 28, 2013), http://www.inpex.co.jp/english/news/pdf/2013/e20130528.pdf?_utma=47397631.32395358.1456431550.1456431550.1456431550.1&_utmb=47397631.2.10.1456431550&_utmc=47397631&_utmx=-&_utmz=47397631.1456431550.1.1.utmcsr=(direct)|utmccn=(direct)|utmcmd=(none)&_utmv=-&_utmk=180156001.

8. Malaysia LNG, "Our History," http://www.mlng.com.my/#/.

9. Petronas, "The Project," https://www.petronasofficial.com/floating-lng/project; and "Petronas Reaches Final Investment Decision on Second Floating LNG Project" (February 13, 2014), http://www.petronas.com.my/media-relations/media-releases/Pages/article/PETRONAS-REACHES-FINAL-INVESTMENT-DECISION-ON.aspx.

10. Woodside Energy, "North West Shelf Project," http://www.woodside.com.au/Our-Business/Producing/Pages/North-West-Shelf.aspx#.Vs9mQpwrKUk.

11. ConocoPhillips, "Australia: Darwin LNG," http://www.conocophillips.com.au/our-business-activities/our-projects/Pages/darwin-lng.aspx.

12. Woodside Energy, "Pluto LNG: Pluto Facilities," http://www.woodside.com.au/Our-Business/Producing/Pages/Pluto.aspx#.Vs4xW5wrKUk.

13. Chevron, "Wheatstone: Chevron and TEPCO Sign Equity and Additional LNG Sales Agreements" (June 18, 2012), http://www.chevronaustralia.com/news/media-statements/wheatstone/2012/06/18/chevron-and-tepco-sign-equity-and-additional-lng-sales-agreements.

14. Chevron, "Gorgon Project," http://www.chevronaustralia.com/our-businesses/gorgon.

15. INPEX, "Investor Briefing—Ichthys LNG Project" (January 24, 2012), http://www.inpex.co.jp/english/ir/library/pdf/presentation/e-Presentation20120124-a.pdf.

16. Shell, "Prelude FLNG," http://www.shell.com/about-us/major-projects/prelude-flng.html.

17. BG Group, "BG Group Sanctions Queensland Curtis LNG Project" (October 31, 2010), http://www.bg-group.com/717/news-&-publications/.

18. Santos GLNG, "The Project," http://www.santosglng.com/the-project.aspx.

19. Australia Pacific LNG, "The Australia Pacific LNG Project," http://www.aplng.com.au/pdf/factsheets/_APLNG012_Fact_Sheet_The_APL_Project_FINAL.PDF.

20. PNG LNG, "About PNG LNG," http://www.pnglng.com/project/about.html.

21. Shell, "About Sakhalin II," http://www.shell.com/global/aboutshell/major-projects-2/sakhalin.html.

22. Gazprom, "Vladivostok LNG Project," http://www.gazprom.com/about/production/projects/lng/vladivostok-lng/.

23. National Energy Board and Alberta Energy Regulator, *Natural Gas Reserves in British Columbia and Alberta* (November 2013).

24. Chevron Canada, "Kitimat LNG," http://www.chevron.ca/our-businesses/kitimat-lng.

25. LNG Canada, "LNG Canada Receives BC Oil and Gas Commission Permit," (January 5, 2016), http://lngcanada.ca/media-items/lng-canada-receives-bc-oil-and-gas-commission-permit.

26. Pacific NorthWest LNG, "The Project," http://www.pacificnorthwestlng.com/the-project/the-project/.

27. Jordan Cove LNG, "Veresen Announces that Jordan Cove Receives License to Allow Export of LNG to U.S. Non-Free Trade Agreement Countries and Provides Commercial Update" (March 24, 2014), http://jordancovelng.com/wp-content/uploads/2014/11/VSN-News-Release-non-FTA-approval-FINAL.pdf; and "Jordan Cove LNG," http://jordancovelng.com/project/.

28. Oregon LNG, "Project Timeline," http://www.oregonlng.com/project-timeline/.

29. ConocoPhillips, "Kenai Liquefied Natural Gas Plant and North Cook Inlet Gas Field, Alaska,"http://alaska.conocophillips.com/Documents/Fact%20Sheet_Kenai%20LNG_CURRENT.pdf.

30. Hydrocarbons.com, "Peru LNG Project, Peru," http://www.hydrocarbons-technology.com/projects/peru-lng/.

31. Qatargas, "Qatargas Company Limited," http://www.qatargas.com/English/AboutUs/Pages/default.aspx.

32. RasGas, "About Us: The Energy to Transform," http://www.rasgas.com/AboutUs.html.

33. ADGAS, "About Us," http://www.adgas.com/En/SitePages/About%20Us/Overview.aspx; and "Chronology," http://www.adgas.com/En/SitePages/About%20Us/Chronology.aspx.

34. Oman LNG, "Oman LNG in Brief," http://omanlng.com/en/TheCompany/Pages/OmanLNGInBrief.aspx.

35. Yemen LNG, "Project Overview," http://www.yemenlng.com/ws/en/go.aspx?c=proj_overview.

36. *LNG World News.* "Anadarko Moves Closer to Mozambique LNG FID" (August 27, 2015), http://www.lngworldnews.com/anadarko-moves-closer-to-mozambique-lng-fid/.

37. Poten & Partners, LNGas Portal.

38. Ibid.

39. UFG, "Liquefaction: UFG Participates in Liquefaction Plants at Damietta and Qalhat," https://www.unionfenosagas.com/en/Negocio/_Licuefaccion.

40. Egyptian LNG, "Welcome to Egyptian LNG," http://www.egyptianlng.com/ELNG/Intro.htm.

41. Shell Nigeria, "Liquefied Natural Gas," http://www.shell.com.ng/about-shell/our-business/bus-nigeria/lng.html.

42. EG LNG, "About EG LNG: Our History," http://www.eglng.com/About_EG_LNG/Our_History/.

43. Angola LNG, "About Us," http://www.angolalng.com/project/aboutLNG.htm.

44. Statoil, "Snøhvit," http://www.statoil.com/en/ouroperations/explorationprod/ncs/snoehvit/pages/default.aspx.

45. Total, "Yamal LNG: The Gas That Came in from the Cold," http://www.total.com/en/energy-expertise/projects/oil-gas/lng/yamal-lng-cold-environment-gas.

46. Atlantic LNG, "About Us: Atlantic at a Glance," http://www.atlanticlng.com/about-us/atlantic-at-a-glance.

47. DOE, *Sabine Pass Liquefaction, LLC: Opinion and Order Conditionally Granting Long-Term Authorization to Export Liquefied Natural Gas from Sabine Pass LNG Terminal to Non-Free Trade Agreement Nations*, FE Docket No. 10-111-LNG; DOE/FEOrder No. 2961 (Washington, DC: Office of Fossil Energy, May 20, 2011), http://www.fossil.energy.gov/programs/gasregulation/authorizations/Orders_Issued_2011/ord2961.pdf.

48. Cheniere Energy, "Sabine Pass LNG Terminal: Trains 1–4," http://www.cheniere.com/terminals/sabine-pass/trains-1-4/.

49. Freeport LNG, "Freeport LNG's Liquefaction and Export Project," http://www.freeportlng.com/The_Project.asp.

50. Cameron LNG, "Liquefaction Helps the Future Look Bright," http://www.cameronlng.com/liquefaction-project.html.

6

LNG Shipping

Introduction

LNG is shipped in specially designed tankers with double hulls and insulated containment systems allowing the cargo to be transported at low temperatures without pressurization. The tankers, possessing unique and specific cargo loading, storage, and discharge capabilities, carry only LNG. A new LNG tanker currently costs about $200 million, making it among the most expensive commercial ships.

Driven by growth in demand and the longer-haul nature of the trades, the LNG fleet grew from approximately 110 vessels in 2000 to about 400 vessels in October 2014. Based on the number of ships on order (about 120 in October 2014), the total fleet will expand by one-third, to more than 500 by 2018. The average age for a currently trading LNG carrier is about 11 years, approximately one-third of the typical vessel life (fig. 6–1).

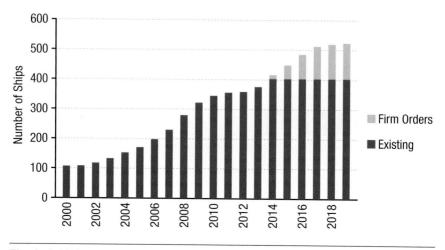

Fig. 6-1. LNG carrier fleet. (*Source:* Poten & Partners Proprietary Service.)

Larger Ship Sizes

LNG carriers have increased significantly in size, further boosting transport capacity. The earliest generation vessels were relatively small and suitable for short-haul, small-scale, intraregional trades such as Algeria to southern Europe. By the 1990s, as the LNG business globalized and trade volumes increased, 125,000 m^3 ships, first introduced in the 1970s, had become the standard, doubling the size of the earlier tankers that primarily traded in the Mediterranean. By the early 2000s, 138,000 m^3 became the new standard, increasing to 145,000 m^3 by 2007. Since then most new vessels have been between 145,000 and 175,000 m^3, with the upper end of this range more prevalent by 2014 (fig. 6–2). The growing number of LNG ships and the larger size of carriers in the fleet have quadrupled aggregate fleet capacity (excluding FSRUs) from 13.2 million m^3 in 2000 to 57.6 million m^3 by October 2014, while the number of ships in the fleet increased by only two and a half times.

Seeking shipping economies of scale to complement their new megatrains, Qatar, through its Nakilat shipping arm, ordered a fleet of 55 giant LNG carriers, including 31 Q-Flex vessels sized between 210,000 and 217,000 m^3 and 14 Q-Max vessels sized between 263,000 and 266,000 m^3. These were originally envisioned for point-to-point trades, and these giant vessels are considered too large for general trading activity. Even though some potential customers have upgraded receiving terminals to accommodate these large ships, many export and import facilities still cannot accommodate them.

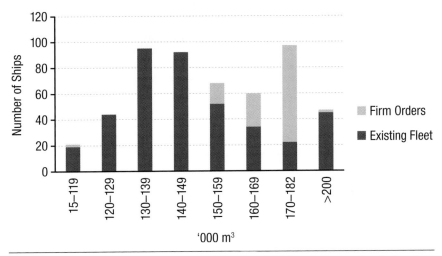

Fig. 6–2. LNG carriers by capacity as of October 2014—existing and orderbook. (*Source:* Poten & Partners.)

Containment Systems

Moss spherical and membrane are the two main cargo containment systems used in the fleet. Membrane tanks have increasingly dominated LNG transportation with 86% of vessels currently on order being built with these containment systems (fig. 6–3). Membrane's popularity stems from its greater cargo capacity relative to hull size. Membrane tanks are built within the inner hull, while the Moss spherical design leads to large void spaces in the hull. Moss vessels account for approximately 27% of the existing LNG carrier fleet. The primary membrane designs are Gas Transport and Technigaz, both owned by GTT of France. IHI of Japan developed a stand-alone membrane tank, which could be constructed away from the ship and lowered into the hull but has seen limited acceptance. Over time, the cargo boil-off rate from these systems has dropped from about 0.25% per day in the earlier days to as little as 0.09% per day today as insulation materials have improved.

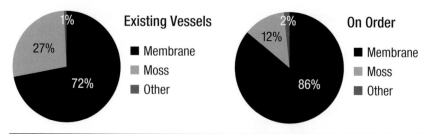

Fig. 6–3. LNG carriers by containment systems—existing and orderbook. (*Source:* Poten & Partners.)

Fleet Ownership

Producers can choose to charter-in their transportation capacity either on a long-term basis, when needed as a project vessel, or under single voyage or short duration agreements for trading or other short-term opportunities. From an exporter's perspective, long-term charters spread the cost of transportation over the life of the project compared to purchasing the shipping outright, adding to the up-front investment, when project development costs are their highest. Generally, it appears that independent shipowners are also willing to accept lower rates of return on capital than companies involved in upstream oil and gas activities, thus further lowering the effective shipping costs. This arrangement has increased opportunities for independent shipowners, who in October 2014 owned approximately half of the LNG fleet and accounted for 73% of newbuildings on order (fig. 6–4). They can ensure a guaranteed rate of return with back-to-back construction contracts and charters for the vessels.

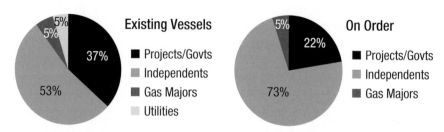

Fig. 6–4. LNG fleet ownership as of October 2014. (*Source:* Poten & Partners.)

Evolution of LNG Shipbuilding

The first LNG ships were built in the United States and Europe, but today, as with almost all commercial shipbuilding, construction is dominated by the Asian yards. Figure 6–5 illustrates the geographic swing in the LNG shipbuilding industry from the Atlantic Basin, where it originated, to fast-paced growth in Asia starting in the early 1980s. Since the main LNG-consuming countries, notably in Japan and South Korea, were also promoting heavy industry, these nations wanted to build ships for their LNG imports—a path China is now following. Other countries such as India, a large importer, and Russia, a potentially large exporter, are considering LNG tanker construction in the future. However, lacking current shipyard capability, large investments in training and facilities would be required for them to compete.

LNG tanker orders and deliveries bottomed in the 1980s, revived in the 1990s, and peaked during the 2007–09 period when the massive Qatari vessels were being delivered. Shipbuilding activity is expected to remain high until the end of the decade.

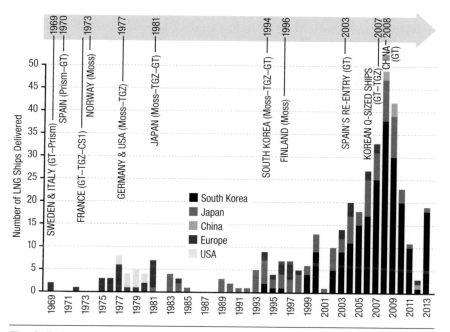

Fig. 6–5. Countries building LNG tankers. (*Source:* Poten & Partners.)

In the early years, European naval architects and LNG tank designers played a key role in developing and building these highly specialized ships. Hence, European yards controlled LNG shipbuilding for some time before containment system licenses and technology transfer arrangements were negotiated with shipbuilders in Asia. Asian shipyards, first Japanese and then South Korean, have come to dominate the LNG shipbuilding industry today. Now China, which is challenging South Korea as the world's top shipbuilding nation, has set its sights on constructing high-value LNG vessels as well and has delivered its first LNG tankers.

As the trade has expanded to accommodate more and more buyers and sellers, so too has the world's LNG ship construction capacity. In 2014, total worldwide capacity was estimated at 70 to 80 ships per year, mostly in the Far East. South Korean yards, which have 71% of the newbuildings on order, also have by far the largest LNG shipbuilding capacity at 40 to 50 ships per year. South Korean yards also built Qatar's Q-Flex and Q-Max ships, the biggest LNG tanker order on record.

Europe

At first, European shipyards received orders from start-up liquefaction projects in both the Atlantic and Pacific markets (table 6–1). The largest order in the early 1970s was placed by Shell Tankers in the United Kingdom on behalf of the Brunei LNG project, which selected three French shipyards to build seven LNG tankers of 75,000 m³ capacity, using two membrane technologies, Gas Transport (GT) and Technigaz (TGZ). These were followed by a series of orders in French shipyards by French and Algerian owners for membrane vessels of 125,000 to 130,000 m³ and by Malaysia International Shipping Corporation, now known as MISC Berhad, which ordered five 130,000 m³ tankers. Meanwhile, in Norway, Moss Rosenberg Verft used its own Moss technology to build two 87,000 m³ tankers, the *Norman Lady* and *LNG Challenger*, the latter renamed *Margaret Hill*. They were soon followed by the 126,000 m³ *Hilli* and *Gimi*, the first large spherical tank vessels.[1] In 1977, Howaldtswerke-Deutsche Werft (HDW) of Germany delivered two LNG ships using the Moss design, *Höegh Gandria and Golar Freeze*, both 125,800 m³.

In the early 1990s, Kvaerner Masa Yard in Turku, Finland, entered LNG shipbuilding with a series of four LNG carriers built for

National Gas Shipping Company in Abu Dhabi, UAE. These 137,000 m³ vessels—*Al Hamra, Mraweh, Mubaraz,* and *Umm Al Ashtan*—were delivered between 1996 and 1997 and still trade between Das Island in Abu Dhabi and Japan. This yard was specially designed to produce the Moss tanks through a special heat-forming treatment process. It was originally owned by Wärtsilä, then Masa Shipyards, and after the LNG vessels were delivered, the yard name was changed once again from Kvaerner Masa to Aker Finnyards. Some 33 years after building the *Laieta* in 1970, Spain reentered the LNG shipbuilding scene in the early 2000s. The major Spanish gas importer, Enagas, placed an order in the summer of 2000 for four LNG newbuildings at the Izar yard.[2] This was the last order placed in Europe, which can no longer compete with its Asian competitors because of much higher labor costs and lower productivity.

Table 6-1. Europe: LNG shipbuilding yards

Shipyard	LNG Shipbuilding Capacity*	Containment Experience
Chantiers de l'Atlantique (France)	3	GT, TGZ, and CS-1
Izar (Spain)	3	GT

(*Source:* Poten & Partners.)
*These are estimated historic capacities. These yards no longer compete in LNG tanker construction.

United States

General Dynamics and Newport News built 13 large LNG ships between 1977 and 1980 using the Moss and Technigaz technologies, respectively. In Quincy, Massachusetts, eight Moss tankers were built for Burmah Gas Transport to serve the Indonesia-to-Japan trade. Newport News built three vessels for the ill-fated El Paso project, which was to bring Algerian LNG to the US East Coast, while General Dynamics constructed two others to trade from Algeria into Lake Charles on the US Gulf Coast. Two of those tankers are now owned by Bonny Gas Transport, a subsidiary of Nigeria LNG, and serve the project's exports; two were acquired by Shell, but have been scrapped; and the fifth, the *Matthew,* is owned by GDF Suez.

Avondale built three LNG tankers, also for the El Paso project, using the Conch independent tank design with a polyurethane foam (PUF) internal insulation system. This design failed its cryogenic trials and, consequently, the ships did not receive the required classification certificate permitting them to load LNG. All three subsequently became the subject of a large insurance claim. One of the vessels was scrapped after grounding and the other two were converted into coal-fired bulk carriers and subsequently scrapped. The United States currently has no viable LNG shipbuilding capacity.

Asia

Japan and later South Korea became world leaders in the shipbuilding industry and the world's foremost LNG carrier builders. Their competitive advantage was achieved through large investments in automation coupled with highly productive labor forces. These advantages were somewhat mitigated by other factors. Construction of the containment systems, not easy to automate, required considerable new investment. Most of the cryogenic equipment and materials required were only available through license from the United States and Europe, thus removing some of the Far East's competitive edge. Although wage costs have escalated considerably, productivity has improved with extensive use of automated processes, thus continually increasing fabrication block sizes and quality of pre-outfitting. The learning curve advantage of South Korean yards may prove hard to reproduce.

Japan. Japanese ship repair yards were able to secure virtually exclusive dry docking and maintenance contracts for the first generation of US- and European-built LNG vessels trading into Japan. These facilities, all associated with the major shipbuilders, allowed these shipbuilders to acquire the knowledge and information that helped their entry into the LNG newbuilding business. In 1981, Japan entered the shipbuilding sector with an order from Gotaas Larsen to construct a Moss spherical containment vessel, *Golar Spirit*, at Kawasaki Heavy Industries.[3]

Spurred on by orders from domestic owners, which were in turn supported by charterers with projects supplying LNG to the domestic market, Japanese yards rapidly increased their market share, focusing almost exclusively on the Moss spherical containment system. Three Japanese shipbuilders—Kawasaki Heavy Industries, Mitsubishi Heavy Industries, and Mitsui Engineering and Shipbuilding—were the dominant builders throughout most of the 1990s. In 2004, the Japanese yards Imabari and Universal Shipbuilding won orders from a Shoei Kisen/K-Line partnership and Sonatrach, respectively. In 2013, Ishikawajima-Harima Heavy Industries (IHI) and Universal Shipbuilding merged to form Japan Marine United Corp. and, in February 2014, secured orders for two LNG ships employing the Self-supporting Prismatic (SPB) design. MI LNG, a joint venture between Mitsubishi Heavy Industries and Imabari Shipbuilding established in April 2013, secured an order for a Moss-type vessel. Japanese yards currently have an annual LNG shipbuilding capacity of about 18 vessels at six shipyards (see table 6–2).

Table 6–2. Japan: Active LNG shipbuilding yards

Shipyard	LNG Shipbuilding Capacity*	Containment Experience
Kawasaki Heavy Industries	4	Moss, GT, and CS-1
Mitsubishi Heavy Industries	2	Moss, GT, and CS-1
Mitsui Engineering & Shipbuilding	2	Moss, GT, and CS-1
Japan Marine United Corp.	3	TGZ and SPB
Imabari Shipbuilding	3	TGZ, GT, and CS-1
MI LNG	4	Moss and TGZ

(*Source:* Poten & Partners.)
* Capacity estimates are based on each yard's maximum delivery capabilities for 2016. Actual capacity may differ from the maximum in any given year because facilities are shared with other vessel construction projects.

South Korea. Japanese shipbuilders were reluctant to share their lucrative shipbuilding market or steam turbine propulsion technology with South Korean yards. Support was not forthcoming from European shipbuilders or containment specialists either. At the same time, foreign shipowners were hesitant to place orders for such highly specialized tankers in shipyards lacking LNG building experience. South Korea faced an uphill task and, like Japan, secured its initial entry through domestic orders. In 1994, Hyundai Heavy Industries delivered the country's first LNG tanker, the *Hyundai Utopia*, to transport Korea Gas's (Kogas) LNG imports.

After an agreement was reached on technology transfer, the 125,000 m³ four-tank Moss vessel, based very closely on a Japanese design and equipment, was constructed in Korea. Kogas had a monopoly for all gas imports into South Korea and was the second largest LNG importer in the world. When Kogas signed new LNG supply contracts in the early 1990s on an FOB basis, it incentivized South Korean shipowners to invest in LNG and collaborate with major South Korean yards. In late 1999, Hyundai Heavy Industries secured its first foreign export order from Nigeria LNG's Bonny Gas Transport, which ordered two 137,300 m³ vessels.

South Korean yards, gaining a reputation for quality construction, soon seized the momentum from Japan by offering lower prices and faster delivery, thus winning the lion's share of newbuilding orders from 1999 onward.[4] In October 2014, 71% of the LNG ship orders were with South Korean yards. Japan had 12% of the order book and China had 17%. The four major South Korean yards have an annual LNG shipbuilding capacity of approximately 50 ships (as summarized in table 6–3).

Table 6–3. South Korea: Active LNG shipbuilding yards

Shipyard	LNG Shipbuilding Capacity*	Containment Experience
Daewoo Shipbuilding & Marine Eng.	15	GT and CS-1
Hanjin Heavy Industries & Construction	2	TGZ CS-1
Hyundai Heavy Industries	18	Moss, TGZ, and CS-1
Samsung Heavy Industries	13	TGZ and CS-1
STX Shipbuilding	3	GT and TGZ

(*Source:* Poten & Partners.)
* Capacity estimates are based on each yard's maximum delivery capabilities for 2016. Actual capacity may differ from the maximum in any given year because facilities are shared with other vessel construction projects.

China. China became an LNG importer in 2006 and quickly adopted the Japanese and South Korean shipbuilding entry strategy by requiring that term LNG sold to Chinese companies be transported by ships built in China (table 6–4). Guangdong LNG Transportation Group[5] ordered three LNG tankers (*Dapeng Sun, Dapeng Moon,* and *Dapeng Star*) at the Hudong-Zhonghua yard to provide transportation for Australian gas deliveries into the Dapeng LNG Terminal in Guangdong Province—the nation's first import facility. Two additional Chinese-built vessels (*Min Lu* and *Min Rong*) are used in lifting LNG from Tangguh in Indonesia to the terminal in Fujian Province. Hudong-Zhonghua has built a sixth vessel (*Shen Hai*) to transport LNG from Malaysia to Shanghai. All these ships use membrane technology, acquired from France through a collaboration agreement. As of October 2014, Hudong-Zhonghua is building 18 carriers for delivery between 2015 and 2019 to support LNG imports from Papua New Guinea LNG and three Australian projects—Gorgon LNG, Queensland Curtis LNG, and Australia Pacific LNG.

While Hudong-Zhonghua is the only active LNG vessel builder, other Chinese shipyards are gearing up to build LNG tankers for the Chinese trade. The Dalian New Shipyard and the Jiangnan Shipbuilding Group are building 28,000 to 30,000 m³ LNG carriers and have acquired licenses for both Moss and membrane containment systems. In addition, Shanghai Waigaogio Shipbuilding has acquired a license to build membrane LNG tankers, while Nantong (NACS) has secured licenses for both membrane and Moss.

Table 6–4. China: Active LNG shipbuilding yards

Shipyard	LNG Shipbuilding Capacity*	Containment Experience
Hudong-Zhonghua Shipbuilding Group	8	GT and CS-1

(*Source:* Poten & Partners.)
*Capacity estimates are based on the yard's maximum delivery capabilities for 2016. Actual capacity may differ from the maximum in any given year because facilities are shared with other vessel construction projects.

LNG Shipbuilding Cost Evolution

LNG carriers are the world's second costliest merchant vessels after large cruise ships. The cost of a newbuilding is determined by the cost of steel, nickel, and aluminum; exchange rates; productivity; and labor costs. Added to these cost factors are a number of both tangible and intangible factors, not least of which is the competition for building slots from other sectors of the shipping industry. In the 1990s, an LNG newbuilding cost more than $200 million. For example, the Australia North West Shelf ships delivered from 1989 through 1994 cost approximately $215 million each. Driven by a strong yen and limited shipbuilding competition, Qatargas ships (135,000 m³) delivered in the mid-1990s were priced at approximately $240 million.

However, competition among shipyards for these lucrative orders also plays a significant role. In order to break into the business, South Korean yards bid aggressively for orders.[6] The decisive entry of South Koreans into the business forced Japanese shipyards to lower their costs. By the early 2000s, the most competitive yards were delivering 138,000 m³ LNG carriers for prices in the range of $160 million.[7]

Since then shipbuilding prices have slowly increased, together with tanker sizes. In the first quarter of 2006, shipyards were pricing conventional 145,000 m³ tankers with steam turbine propulsion in the region of $210 million and 155,000 m³ tankers with dual-fuel diesel electric propulsion for an additional $10 to $15 million. The much larger LNG tankers ordered by the Qataris—specifically the 210,000 to 217,000 m³ ships (Q-Flex) and the 263,000 to 266,000 m³ vessels (Q-Max) carried, at the time of contracting, price tags reported at $250 to $255 million and $280 to $285 million per ship, respectively. In 2015, a modern 170,000 m³ carrier commands a price tag of $200 to $215 million (fig. 6–6).

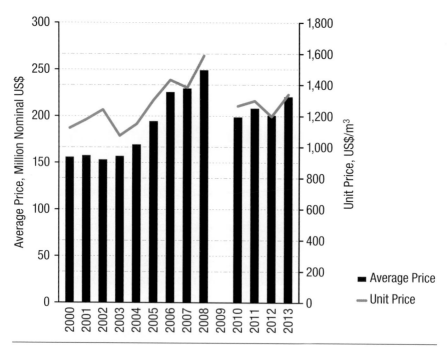

Fig. 6–6. LNG newbuilding costs by order year. (*Source:* Poten & Partners.)

Vessel ownership

Long-term LNG tanker ownership has taken essentially one of three forms:

1. The tanker is owned by an independent shipowner and is chartered to a seller or buyer under long- or short-term agreements.

2. The tanker is owned directly by the LNG seller or indirectly by a separate entity established and owned by the seller (a special purpose vehicle [SPV]).

3. The tanker is owned by the LNG buyers or the buyers' SPV.

It is important to differentiate between tanker ownership, management, and control. In many cases, the owners of tankers are separate shipping entities or consortia[8] servicing their national or home country companies and chartering the ships to the seller or the buyer who takes control of the fleet, depending on the terms of the LNG sale (FOB, DAP, or CIF). Therefore control of the tankers is not through the actual shipowner with whom the ship is registered, but rather through the buyer or seller, known as a disponent or time-chartered owner. Management of the tanker can also vary. Depending on the size of the dedicated fleet and the charterer's in-house ship management resources, fleet management is administered by the actual shipowner, the LNG seller, or an independent ship management company.

To a great extent, these ownership structures still dominate LNG shipping today. However, as the industry has grown and gained a proven track record, ownership and control of LNG tankers have become much more diversified, with arrangements customized to the projects they serve. Furthermore, when the price of newbuildings in 1999 fell by about 40% from a decade earlier, independent shipowners as well as emerging LNG buyers were encouraged to enter the LNG shipping sector.

Large shipowners like Bergesen dy AS, Golar LNG, and Exmar ordered series of tankers without confirmed back-to-back term charters, making these orders somewhat speculative. None of these independent shipping firms have upstream or downstream gas market interests. Even oil and gas majors like Shell and BP and established joint venture projects[9] took the opportunity to order lower-priced LNG ships without firm project dedications.

Many of the so-called speculative orders were subsequently committed under charters upon their delivery, either with export projects or buyers diversifying their portfolio of term purchases. As to other speculative vessels, they typically find employment in growing spot and short-term trades until long-term employment is found. A few months of unemployment for a new tanker can be financially damaging to an independent ship owner, not to mention what could happen if the owner had a fleet of unemployed ships.

LNG shipping trends

LNG shipping will probably retain, at least in part, a traditional structure for most of the long-term fixed trades where the vessels are ordered and built to meet specific requirements. Nevertheless, there has been an increase in speculative ship orders (ships built without secured employment) from fewer than 25 in 2004 to nearly 100 as of 2014. Since LNG SPAs often require tankers to make regular round-trip voyages on a fixed schedule and usually at even intervals, the dedicated tankers have little idle or uncommitted time. Some LNG SPAs require that tankers be dedicated to fulfilling the shipping requirements of a specific contract. Thus, it is difficult to use any operational spare capacity beyond a specific project's transportation requirement, although diversion rights have been negotiated in some SPAs and may only be achievable with shipping spare capacity acquired on a short-term basis.

The advent of spot and short-term trades and the drive to lower transport costs are transforming the industry. As noted above, a surprising number of tankers have been ordered without long-term charter commitments. The key to trading flexibility is the control of shipping capacity and destination flexibility rights to support merchant trading activities. Buyers and sellers alike are increasingly seeking control over ships.

The issue of controlling tankers—and the resulting flexibility that it can provide to the owner or charterer—addresses the issue of simultaneous operational logistics at the LNG export and import terminals. In the traditional LNG project model, tankers were dedicated to specific trades with a regular and predictable schedule of loadings and discharges. The introduction of several buyers who control their own tonnage, perhaps of various specifications and cargo capacities, with plans to discharge at different outlets can present very difficult operational problems for the LNG export venture.

Similar challenges arise at the import terminal owing to multiple suppliers and firm gas throughput obligations. The problem increases with open access terminals serving multiple capacity holders that have multiple supply sources, shipping, and highly variable (and unpredictable) schedules. These problems can become acute as operations at these facilities approach their respective capacities. Alleviation of these logistical problems might require additional investments in facilities to create buffer capacity or, alternatively, restrictions on the very flexibilities that the buyers are seeking.

One solution chosen within the industry is careful scheduling and cooperation among shippers of LNG, especially where the end customer is a utility and perhaps less price sensitive than a trader. However, as more and more trades are added to the global matrix, and complexities increase, this relatively smooth cooperation will come under increasing stress as trading firms see potential (and often fleeting) profit opportunities being aided or eroded by logistics.

Technology developments

Engines and propulsion. The LNG shipping industry has historically been reluctant to adopt new technologies and designs. Steam turbine engines employ a well-proven technology that is highly reliable, has low maintenance costs, and permits the safe and easy disposal of boil-off gas. However, steam propulsion has relatively low fuel efficiency. Moreover, as diesel engines in merchant ships have now become the norm, qualified seagoing steam engineers are in short supply. However, there are still steam turbine vessels on order for delivery as late as 2017, and steam turbine–driven carriers will be in service for 30 to 40 years or more.

Some early LNG carriers were originally powered by gas turbines and subsequently repowered with diesel engines (e.g., *Century, Havfru*). It was not until 2002 that Gaz de France broke away from steam turbine propulsion and ordered a small LNG carrier using new technologies for both containment and propulsion: the new CS-1 membrane tank and a low-pressure dual-fuel diesel electric propulsion system. Gaz de France worked with GTT and Alstom at its Chantiers de l'Atlantique shipyard on the containment system, while Finland's Wärtsilä provided the propulsion technology. Today, several commercially available alternative ship propulsion systems are in operation.

The initial move away from steam turbines was toward dual-fuel diesel electric propulsion but, with the extremely high cost of marine diesel oil, technological advances quickly led to the development of the tri-fuel diesel electric propulsion system capable of running on heavy fuel oil, marine diesel oil, and boil-off gas. The Qatar Q-Flex and Q-Max vessels used slow-speed diesel engines with reliquefaction and were unable to burn boil-off gas. As oil prices moved up sharply in the 2009–14 time frame, the Qatari vessels began an expensive repowering program, swapping their diesel propulsion system for multifuel systems, obviating the need for reliquefaction. Changing the propulsion system of the tanker affects the overall ship design as the different engine types vary structurally in size, shape, and required layout, thus influencing the potential size of cargo capacity. Different methods of dealing with boil-off gas also affect transportation economics.

Boil-off rates. Over the past 40 years, the technology used to build cargo containment systems has changed very little. The main focus has been on reducing boil-off rates to reduce gas wastage concurrent with the shift away from inefficient steam turbine propulsion. This effort has met with considerable success. Boil-off rates have been reduced from as much as 0.25% per day to about 0.09% per day. This has not raised the cost of a newbuilding to the owner, and it has major benefits to the charterer by lowering operating costs. (See chapter 10 for a discussion of cargo containment systems employed in LNG ships.)

Larger LNG ship sizes. Up until the early 1990s, the typical industry standard LNG tanker was in the range of 125,000 m³ to 130,000 m³, the size being dictated by actual or perceived size limitations at existing import terminals in the Tokyo Bay area. During the 1990s, the size of vessels being built increased to around 138,000 m³ but was again restricted by size for access to certain ports such as Boston (where there was an air draft restriction under the Mystic Tobin Bridge) and elsewhere (where displacement limitations existed).

During the 2000s, the Qatar LNG projects designed vessels of 210,000 m³ to 266,000 m³ capacities that were to be used for long-haul bulk cargo movements and that offered the greatest economies of scale. The construction of these ships was also associated with the development of a number of new dedicated import terminals where the marine facilities were sized to accommodate the much larger ships.

For general long-haul shipping of LNG and for wider trading purposes, the industry standard size has, at least for now, settled in a range of 145,000 m³ to 175,000 m³. Vessels up to around 170,000 m³ are being built within the same general length and draft dimensions of their predecessors. Although the beam or breadth of vessels has increased to create the larger capacity, the size increase is mitigated on diesel propulsion ships by removing the steam plant.

In the future, the shape and size of LNG carriers is likely to be governed by the beam restriction of the expanded Panama Canal locks, which will initially accept vessels up to 49-meter beam and possibly up to 51-meter beam after operational experience is gained. This 49-meter beam restriction currently excludes passage for some of the 177,000 m³ to 180,000 m³ Moss vessels coming into service, and for the Qatar vessels, although Q-Flex vessels may be able to pass through the Canal if early operational restrictions are lifted.

Where new LNG berths and/or terminals are built, either for import or export of LNG, consideration of ship size is a key factor; conscious decisions need to be taken about minimum and maximum ship sizes that can be accommodated. This will have an impact on elements of berth design such as moorings, fender loads and spacings, hard arm operating envelope, shore gangway, loading rate, etc. In a broader context, the ship size needs to be compatible with the depth and width of the approach channel, turning and maneuvering area, tug availability, and impact on (and from) other shipping. Dredging may be required for larger ship sizes and this, together with dredged spoil disposal costs, has to be factored into terminal cost calculations.

Notes

1. The first Moss ship was built at Norway's Moss Rosenberg yard in 1973. This ship, the *Havfru*, had a carrying capacity of 29,000 m^3. Her sister ship, the *Century*, was delivered one year later.

2. Spain's leading shipbuilder, Izar, was created after a merger between Bazan and Astilleros Españoles shipyards.

3. Gotaas Larsen was bought by Osprey Maritime in 1997 and subsequently taken over by Jon Frederiksen and renamed Golar LNG in 2001.

4. The Asian financial crisis of the late 1990s was a key influence in the LNG shipbuilding market. The crisis hit Korea's economy hard and the value of the Korean currency, the won, declined by more than 50%. This gave a substantial boost to the competitiveness of Korean shipyards, which saw their labor costs expressed in US dollars decline significantly.

5. The structure of the consortium is as follows: China Ocean Shipping Company (25.5%), China Merchants (25.5%), North West Shelf project participants (30.0%), Energy Transportation Group (3.0%), Shenzhen Marine (8.0%), and Guangdong Yeudian (8.0%).

6. The Asian crisis in 1999 had an adverse impact on the Korean economy, slowing the nation's LNG demand and Kogas's demand for new LNG term supplies. Korea's LNG shipbuilding industry came to a halt after years spent gearing up their yards to accommodate all of South Korea's LNG imports. This hiatus in domestic orders motivated the local shipyards to target the international LNG shipbuilding market.

7. The shipyard price is $160 million. The owner will incur additional costs throughout the building process: supervision during the construction; the cost of finance, which varies according to the financial package and is typically between $10 and $15 million; and, toward the end, deck and outfitting costs in the region of $6 to $7 million.

8. The consortium with the largest LNG fleet in the world in terms of cubic capacity is the MOL (Mitsui OSK Line), NYK (Nippon Yusen Kaisha Line), and K-Line (Kawasaki Kisen Kaisha Line) Consortium, also known as J3 (Japanese 3).

9. The costs of ships ordered for specific projects fall into a different category as they are generally considered as sunk costs. Indeed, LNG vessels are integral in such arrangements where projects are characterized by large capital costs, relatively long lead times, and long-term contracts between buyers and sellers.

7

The Economics
of an LNG Project

Introduction

Unlike oil or most other gas monetization schemes, LNG requires specialized and highly expensive equipment along the entire value chain. LNG is not a commodity like petroleum. Crude oil is fungible; it can, with few restrictions, be loaded onto vessels at any oil export point and delivered to virtually any market. Oil is widely traded across the globe, with transparent pricing and relatively low costs of transportation. Therefore, development of an oil export project does not need to be linked to a particular trade or anchored by a particular buyer with a long-term contract. The assumption of developers is that a market can be found once the oil is flowing (although the price of oil must be taken into account when making an investment decision).

Getting LNG to market, by contrast, requires specialized facilities for liquefaction in the exporting country and for regasification in the importing country, linked by a fleet of specialized ships. Committing to construction of an LNG export facility without having first secured buyer(s) (with import capacity) and adequate transport would represent an enormous risk. Historically, all the pieces of an LNG project have been meticulously laid out from the gas field in the exporting country to the gas distribution grid in the importing country before ground is broken.

The cost of building this LNG delivery chain runs into the billions of dollars with considerable financial risk for investors. Figure 7–1 shows representative capital costs for a generic 4.5 MMt/y greenfield single-train LNG project that would deliver LNG from an export project in East Africa, which is widely viewed by analysts to be among the most cost competitive, to markets in Asia. According to Poten & Partners, these capital costs would range from $9 to $14.5 billion for a one-train project (see fig. 7–1). Just over 80% of expenditures are in the country hosting the liquefaction plant (a little over a third in the upstream and nearly half in the liquefaction complex). It should be noted that a similar project in a high-cost construction country such as Australia would cost significantly more than this, and the share of delivery chain capital expenditures in the host country would be even larger. Capital expenditures for the dedicated shipping fleet to deliver the LNG to the market and the cost of building a regasification facility in the end market add $2–$2.5 billion to the capital costs of this generic delivery chain from East Africa to Asia. For an Australian project that is located closer to the Asian market, there would be shipping cost savings. In this generic example, LNG would be delivered from East Africa to Asia (ex-terminal) for a cost ranging from $7–$10 per MMBtu, which would be competitive against most new LNG projects built elsewhere, with the possible exception of brownfield liquefaction developments at US regasification terminals.

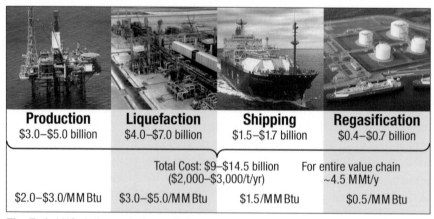

Production	Liquefaction	Shipping	Regasification
$3.0–$5.0 billion	$4.0–$7.0 billion	$1.5–$1.7 billion	$0.4–$0.7 billion
	Total Cost: $9–$14.5 billion ($2,000–$3,000/t/yr)	For entire value chain ~4.5 MMt/y	
$2.0–$3.0/MMBtu	$3.0–$5.0/MMBtu	$1.5/MMBtu	$0.5/MMBtu

Fig. 7–1. LNG delivery chain capital cost—generic example East Africa to Asia. (*Source:* Poten & Partners.)

The large capital costs and the number of participants in an LNG project require that the entire chain be secured through contractual commitments before the project is advanced. Unless a project is financed by a sovereign state (as was the case with Algeria's LNG plants), participation in LNG projects is usually shared between two or more organizations, one of which may be the host government, often through a national oil company. In some instances, such as Indonesia's Bontang project, a national oil company controls the liquefaction facilities while a number of independent oil companies operate the upstream production facilities that supply feed gas to the project. The first two trains of the Oman LNG project involved seven distinct entities, including the Omani government.

Although the involvement of many participants lightens the burden on any single investor, the enormous costs of a grassroots LNG project means that each contributor's commitment is still substantial. This further encourages projects to reduce risks as much as possible by locking up buyers, import capacity, and transport before an FID is made. Usually, specific gas reserves are dedicated to the LNG project, thus further reducing risk or uncertainty. In some cases, the pipeline can be built and owned by a third-party venture, and the liquefaction project agrees to pay a tariff for its use. Yet another variation is apparent in the development of tolling liquefaction plants, sometimes unaffiliated with the upstream owners.

Project Capital Costs

Capital expenditures (CAPEX) for LNG projects are typically categorized as either upstream or liquefaction costs. Upstream costs incurred in the production of project feed gas are similar to the production costs associated with any natural gas project. Liquefaction costs are associated with converting feed gas to LNG for export. Pipeline costs between the production facilities and the liquefaction plant can fall into either category depending on who controls and invests in them or whether they are governed by a production or manufacturing tax regime.

Upstream capital costs

Drilling, completion, and field production costs account for most of the upstream investment, including intrafield flow lines and transmission networks. The characteristics of the field are obviously important: gas fields with high sulfur content or fields in remote areas can incur high upstream expenditures. Offshore gas fields often require expensive platforms, although developers are increasingly using less expensive subsea completions for such developments. These involve wellheads on the seabed that feed the natural gas to a central manifold, which aggregates the individual gas flows and sends them on to a central processing unit located either on a single platform or onshore. Gas fields in deepwater are naturally more expensive to develop than those in shallow water.

The more remote a natural gas field is, the more expensive its development will be. Upstream facilities to supply the Qatargas project's first three trains cost in the range of $450 million per train, because the feed gas comes from Qatar's North Field, a shallow gas reservoir of very good quality that is close to the Qatari mainland and lies in relatively shallow waters. These low upstream costs were a fraction of the investment cost of $25 billion for seven Qatargas trains with 42 MMt/y of liquefaction capacity. This contrasts sharply with Russia's Sakhalin II project, where total project costs escalated to $22 billion for a two-train, 10 MMt/y project. This project was developed in a very remote area off Russia's Pacific coast. Enormous infrastructure investment was required to produce and transport gas from two offshore fields located off northern Sakhalin Island in waters that are ice-covered for part of the year to the ice-free port and liquefaction plant location in the south of the island. Oil production of 400,000 barrels per day (b/d) helps to offset Sakhalin II's upstream investment costs.

In addition to production facilities, upstream costs will also include the supporting utilities needed for feed gas production, housing for both construction labor and field operations personnel, and other temporary and permanent infrastructure. Owners' costs and preproduction expenses are also incurred. In some cases, exploration and delineation/appraisal expenses are considered part of the upstream costs, although initial exploration costs are usually considered sunk by project developers. These types of project formation costs can differ widely from project to project, ranging from almost nothing to $200 million or more.

Upstream costs are largely dependent on downstream gas volume requirements. Producing sufficient feed gas for a two-train, 10 MMt/y (14 Bcm/y) facility will obviously require more investment than for a 3 MMt/y (4.2 Bcm/y) single-train facility. Local labor and materials markets will also have an impact on upstream costs. Upstream projects in high-cost developed nations such as Australia and Norway may be more expensive than comparable undertakings in emerging markets such as Africa or Southeast Asia. While labor productivity in developed nations may be higher, this advantage is usually not sufficient to compensate for the higher wages and benefits that workers in these locations receive. This drives up construction costs.

Not all upstream capital expenditures are incurred prior to start-up. Project participants will usually quantify the reserves available to the project before FID. Often new wells will need to be drilled or new processing facilities added during the life of the gas field as the reservoir is depleted. These future capital expenditures must also be taken into account in the initial budgeting process. They promise to be particularly significant for LNG projects in eastern Australia processing coalbed methane or in Western Canada using shale gas where extensive and continuous ongoing drilling programs will occur after commissioning.

Liquefaction capital costs

While most upstream expenditures are similar to those incurred by any gas monetization scheme, liquefaction capital costs are unique to LNG projects. Liquefaction plant facilities typically represent an LNG chain's most significant capital investment. Until the early 2000s, technological advances and the application of economies of scale led to substantial lowering of the unit capital investment levels for liquefaction. Unit costs for grassroots projects fell to below $200 per ton of production capacity by 2003 from nearly double that in the 1970s.

However, this downward trend has reversed in recent years. Significant increases in raw material costs across key commodities such as carbon and stainless steels, nickel, and cement, combined with increasingly constrained availability of experienced EPC contractors, skilled labor, and equipment suppliers, and the complexities associated with remote and often hostile locations, have caused unit costs for upcoming liquefaction plants to balloon to the $1,000–$2,000 per ton range. Whether these increased costs will slow the LNG industry's high growth rates remains to be seen. These cost increases had been accompanied by a sharp rise in global oil prices, and a resultant rise in long-term LNG prices. Oil prices are experiencing a sharp price reversal with Brent trading below $60/bbl in early 2015, and vigorous debate as to whether this is the "new normal." US liquefaction projects are a notable exception where FOB prices are linked to Henry Hub gas prices; however, these LNG projects now under construction are less expensive on a unit basis, since they are essentially brownfield developments utilizing existing storage and marine facilities. High construction costs and low market prices will prove challenging to greenfield projects almost anywhere in the world.

Project developers have their choice of a number of different liquefaction technologies, including APCI C3-MR and AP-X, Shell Dual Mixed Refrigerant, Phillips Optimized Cascade, Statoil-Linde Mixed Fluid Cascade, Pritchard PRICO technology, and Axens Liquefin (see details later in this book). The choice of technology will usually depend on project participants (for example, Shell and ConocoPhillips would generally choose their own proprietary processes in projects they control). Projects sometimes contract for competing pre-front-end engineering design (pre-FEED) studies prepared by different EPC contractors, conduct competing FEED efforts, and will often place the EPC contract out to bid, either specifying the liquefaction technology or allowing the EPC contractor latitude to select the lowest cost option. Project costs will typically be assessed on a life-cycle cost basis allowing for trade-offs between capital and operating costs.

LNG storage tanks are another aspect of downstream capital costs that are specific to LNG projects. Storage facility costs are highly dependent on the liquefaction plant location, total installed storage (which is usually a function of marine conditions, size of the LNG tankers, and LNG production rates), and the choice of containment system. Storage capacity is intimately linked to ship scheduling. If storage capacity is limited relative to production, the risks that stem from a delay in shipping become greater. Usually, SPAs will allow for the resale of LNG by a producer in the event that the buyer fails to load a cargo (if the cargo is FOB). However, if another ship is not available, or if the sales are ex-ship and storage is not sufficient, a lapse in ship scheduling could lead to storage tanks becoming full (referred to as reaching tank tops), resulting in a production shutdown. Decisions on storage capacity (and capital) are therefore in part dependent on how the developer addresses risk.

In addition to the equipment required for liquefaction and storage, the downstream portion of the LNG project encompasses the marine facilities needed for export. These facilities include the jetty to load ships and another jetty to offload construction and operations material. Other marine costs are heavily dependent on project geography. Plants built on exposed coastlines may require expensive breakwaters to provide a protected area for ship operations. Modern, large LNG ships require water depths exceeding 12 meters, so that in areas with high rates of sedimentation (such as Nigeria's Niger Delta), dredging requirements for both initial construction and channel maintenance can add to costs. In an effort to defray some of these infrastructure costs, including expensive subsea pipelines to shore, some project developers are developing floating LNG production plants.

Although much of the water is usually stripped out of the feed gas stream at or near the producing fields, LNG projects also typically require CO_2, mercury, and H_2S removal, gas dehydration, and additional front-end gas processing capability to strip out heavier hydrocarbons prior to feeding the gas into the liquefaction process. The degree of cleanup is much greater than for gas that is destined for pipeline sales, and is receiving front-end gas treatment on all new US liquefaction plants, even if they are being supplied from the transmission network. Some projects (for example, Nigeria's NLNG expansions) incorporate LPG stripping and export facilities as part of the downstream project. As part of efforts to reduce CO_2 emissions, Norway's Snøhvit project and Australia's proposed Gorgon project will reinject CO_2 into underground reservoirs. Front-end gas processing facilities can account for as much as 20% of liquefaction plant costs.

As with the upstream, construction costs are included in the assessment of liquefaction economics. Again, local wage and productivity factors play a role. Downstream costs will include utilities and support infrastructure. Traditionally, liquefaction plants were "stick built," that is, largely built on-site. However, owners are increasingly turning to modular construction where very large subcomponents of the plant are built in remote locations and shipped to the plant site for final assembly. This method is increasingly employed in Australia where labor is in short supply and costs are among the highest in the world. However, modular construction intrinsically costs more due to substantial additional steel work and carries its own risks, requiring very careful engineering and construction supervision, and a higher degree of schedule coordination to ensure the modules are delivered on time and fit together exactly. Owners' and preproduction costs should also be taken into account.

If reserves are sufficient and enough buyers can be lined up, LNG developers typically plan for more than one train in a grassroots project to capture the economies of scale that come from building additional trains. If the infrastructure for the first train is sufficient to handle the additional feed gas, power demand, and marine traffic for a second train, the owner can double a project's LNG output while increasing capital investment by only about 50% (depending on local factors). Building two trains initially can therefore greatly lower a project's unit costs and save the cost of demobilizing and then mobilizing contractor personnel if the trains were built sequentially.

These economy-of-scale advantages also apply to later expansions of existing facilities. Again, if the existing infrastructure is sufficient to handle the increased production, new trains can be built for much less than the original trains. The second and third trains at Trinidad's ALNG project, for example, cost approximately $1.1 billion combined, only marginally more than the first train's $950 million capital cost. The cost of the third train in Oman is estimated at less than $750 million, compared to the $2 billion required to establish the first two trains. Unlike the first two trains, the third train did not require spending on marine or storage infrastructure.

Pipeline capital costs

Pipelines deliver feed gas from the producing fields to the LNG plant. Their costs are highly location specific. A pipeline built through a flat, empty desert in a developing country where labor is low cost will be substantially less expensive than an offshore pipeline or a line built across a mountain range. The number and capacity of compression stations needed along the pipe's length also affect total costs. Generally, pipeline costs are between $1 million and $3 million per mile, although more expensive projects have been built in challenging environments. For example, multibillion dollar pipelines will be required to deliver feed gas from shale gas fields in eastern British Columbia in Canada across two mountain ranges to proposed LNG plants on the coast. Projects with associated liquids production will often build separate condensate pipelines, further increasing total capital investment, though the growing use of multiphase pipelines may provide mitigation.

Operating Costs

Generally, annual operating expenses for both the upstream and the liquefaction portions of an LNG project are approximately 2%–5% of the project's total capital cost, not including fuel costs. Grassroots projects located in harsh or remote environments are at the higher end of this range, and expansions at the lower end.

Like CAPEX, operating expenditures (OPEX) are highly dependent on location. In developed countries such as Norway or Australia, labor costs are substantially higher than in developing regions. Operations are more difficult in remote environments, and the costs to transport materials to the site are also higher. Overall unit OPEX is lower for expansion projects, since many operations are shared among trains.

LNG plant economics are fundamentally driven by plant availability (i.e., the days of actual production achieved from the plant each year). Typically, an LNG plant targets availability of around 93% once in stable production. In the first year or so, there will be minor issues taking the plant down and modifications that are needed to achieve reliable operations. Even the most conservative of plant designs usually has some issues to resolve. Major factors impacting plant availability are:

- Upstream gas availability—even dedicated upstream facilities cannot be expected to deliver 100% of design production at all times.

- Plant downtime due to planned maintenance—gas turbine driver servicing is a major source of this and the reason the industry has started to use aero-derivative units that can be swapped out for service rather than serviced in situ.

- Plant downtime due to unplanned outages—even a minor problem may take a day for the plant to stabilize operations after an upset. Estimates of this are usually developed during the design process in a Reliability, Availability, Maintainability, and Safety (RAMS) study.

- Reaching tank tops due to inability to load LNG—marine facilities may need to be designed with breakwaters to protect the loading berth and reduce weather downtime.

These factors contribute to a very conservative approach in the industry, and innovative developments are carefully evaluated and contingencies allowed in the availability prediction until the design is proven.

Liquids Production

A project's condensate and/or LPG production can provide a separate revenue stream alongside LNG production and have a very positive effect on a project's economics. Liquids production and exports do add capital costs to a project for extraction, pipelines, storage, and ship-loading infrastructure. Nonetheless, for many projects (including Indonesia's Arun and Bontang plants, Australia's NWS and Ichthys LNG projects, Qatar's LNG projects, and Abu Dhabi's ADGAS LNG), the relative ease of production and handling of liquids compared with LNG make them a valuable second income source. Liquids production can add substantial credit to a project, depending on the gas quality of the upstream resource. In some cases, LPG or condensate exports can generate extra revenues equivalent to a credit of $0.50/MMBtu or more to the LNG price. LPG production can also benefit the project, as propane can be used as a refrigerant gas in the LNG production cycle as well as an export product with lower shipping costs.

In the preceding examples, liquids production is an adjunct to the LNG chain. In other projects which focus on oil, condensate, or LPG, LNG is almost an afterthought, a way of monetizing the associated gas. An example is Australia's Darwin LNG project, which is supplied by the Bayu Undan field controlled jointly by Australia and East Timor. The field's gas liquids reserves were developed via an offshore production platform and a floating storage and offloading (FSO) facility. A gas pipeline to shore delivers feed gas to the LNG plant.

Since the liquids portion of Bayu Undan's development is a stand-alone project, the LNG project can be viewed as a separate entity. When the developers analyzed the economics of the project over its lifetime, they did not credit the liquids produced upstream to the LNG project. However, the liquids project derives economic benefits from LNG production, since piping the gas to shore, instead of rein-jecting it, allows faster extraction of the liquids and eliminates the capital and operating costs associated with reinjection. The upstream development of the Bayu Undan field, including construction and well completion, is part of the gas liquids project. As a result, while Darwin LNG does not receive any credit for the associated liquids, it incurs very little of the upstream capital costs.

When developers analyze the effect of liquids production on a potential LNG project, they need to be careful as to what they deem to be within the bounds of the project. For example, one of the main drivers of LNG development in West Africa has been the desire of national governments to eliminate gas flaring. Much of the feed gas for LNG projects in Nigeria and Angola is, therefore, associated gas generated as a by-product of oil production. In Australia at the Darwin LNG project, crude oil is being produced from the same wells that supply the LNG project with gas, but this crude is not a credit to the LNG project. At the same time, the cost of upstream production platforms is not borne by the LNG venture.

An integrated project's development can be hampered by an absence of liquids. Delays in the launch of the Greater Gorgon LNG plant in Western Australia were partially due to the dryness of the project's gas fields. A stream of condensate revenue would have considerably improved the overall project economics and would have also facilitated the FID decision. The trains at Indonesia's Bontang LNG plant and Qatar's LNG projects, on the other hand, enjoy substantial liquids credits. In contrast, not all LPG production is value added. At high, oil-linked LNG prices, it may be better to leave the LPG (especially ethane) in the LNG and avoid the CAPEX to build LPG export facilities. While LPG traditionally has traded at prices closely linked to oil prices, the surge of shale oil and gas production in the United States has been accompanied by surging US LPG production and export, placing downward pressure on global LPG prices, and threatening to break the traditional linkage to oil prices.

Government Terms

Most countries have established tax, royalty, and other obligations for hydrocarbons that cover oil exploration and production. The difficulty in finding markets for gas meant that many petroleum producers ignored, flared, or reinjected gas, so that no specific terms for gas were developed. Now that global gas production is increasing and countries have a wider range of gas monetization choices, governments are spelling out explicit terms for upstream gas and, where applicable, LNG facilities. The investment required to commercialize gas is usually substantially higher than for an equivalent volume of oil, so that typically the total profits are lower for gas projects, even though the total impact on the local economy may be greater.[1]

The regime under which a project is structured relies on the sophistication of the host government, the openness of the local economy, the government's desire to spur further gas production, and the need to promote economic development in the region, among other considerations. There is no set template for LNG terms. The most fundamental term that will apply is whether the upstream development is characterized as a production sharing or license agreement. Under the production sharing structure, the resource is developed by an independent company with the government retaining the right to participate, either at no cost, or after the independent has paid the initial exploration and development costs. Generally, the costs incurred by the independent company can be expensed against any income received until they are fully amortized. No royalties are generally collected under this approach. In the case of a license agreement, the government's take is limited to royalties and taxes, with the investment being deductible against taxable income. Other obligations, such as reserves set aside for the domestic market, can also be applied.

Upstream taxes within countries are usually higher than taxes on manufacturing or services to capture the economic rent generated by a depleting resource, which is often viewed by the host country as the property of the state. Downstream LNG fiscal terms are usually simpler. Often the terms applied to a liquefaction venture are the same as those for typical manufacturing or refining. They usually include a fixed corporate income tax rate, depreciation schedule, and possibly a duty on imported equipment. All of these terms are likely to be negotiated with the host country.

Under the license structure, governments typically get a cut of a project's gross revenue through royalty payments. High royalties are onerous to an LNG project, since they involve a payment to the sovereign made directly out of the sales revenues before the project can make any deductions for depreciation or operating expenditures. These royalties can be on oil and/or gas; in the case of an LNG project, any liquids produced are subject to the same royalties as oil revenue. Royalties are usually expressed as a percentage of wellhead revenues. For current LNG projects, they are in the range of 5%–20%. Less commonly, they can be calculated as a fixed unit cost ($/MMBtu).

After royalties are paid, the developer calculates expenses that are subject to cost recovery. Expenditures that can be recovered usually include operating costs (which are expensed) and capital costs (which are recovered according to some specified depreciation schedule). These schedules can be straight-line, in which a fixed percentage of the total capital spent on the project to that point can be deducted, or front-loaded, in which a higher percentage of the capital costs can be deducted in the first years of the project and lower percentages in subsequent years. Malaysia and Brunei both utilize a front-loaded schedule. For other projects, straight-line schedules are employed, generally ranging between 10% and 25% a year. Norway's Snøhvit project was granted a special tax regime in 2002 by the Oslo government that allows 33.3% per year depreciation.

The amount of the costs that an LNG project can recover varies greatly from country to country. A number of governments provide the upstream segments of LNG projects with incentives in the form of uplifts, which are additional percentages that can be added to the total recoverable costs. The majority of LNG fiscal terms around the world allow developers to recover 100% of the costs; countries that have implemented uplifts (such as Nigeria, Norway, and Indonesia) allow, instead, projects to recover between 105% and 120% of costs. Uplifts are only one potential fiscal incentive that governments are willing to offer. Nigeria LNG, for example, has been given multiyear tax holidays as an inducement for foreign companies to invest.

Other governments, however, actually cap the costs that a project can recover. Malaysia caps the total recoverable costs at 60% for gas and 50% for condensate. In Egypt, the caps vary between about 30% and 50%, depending on the nature of the upstream fields. Trinidad's cost recovery scheme is adjusted according to total production. It is important to note that in both the uplift and the cost recovery scenarios, the total expenses and depreciable capital costs that the project has incurred are eventually recovered; the different schemes primarily affect timing.

In the cash waterfall, which allocates cash inflow and outflow, once applicable royalties are levied and costs (including interest paid on any nonrecourse debt) are recovered, the project revenue is subject to some sort of profit-sharing plan. There are a number of different ways that profit sharing can apply in the upstream, depending on the agreed tax structure between the host government and the project sponsors. Commonly, governments levy a corporate income tax (CIT) similar in structure to the one imposed on all concerns in the country, although usually the CIT for oil and gas projects have a higher rate. For LNG projects, those countries that do impose a CIT on profits take between about 30% and 50%. The Snøhvit project in Norway is subject to a 78% tax for both the upstream and liquefaction segments, but allows the write-off of new developments against existing revenue. The average for current and prospective LNG-exporting countries is about 37%.

A profit-sharing tax can be a straight percentage or linked to production. In the Malaysian fiscal regime, the government's share of profits increases once a set cumulative volume of gas has been produced by the project. In Trinidad, the upstream fiscal take increases progressively with daily gas production.

Another flexible method for governments to get revenue from gas extraction is the use of a resource rent tax. Such taxes allow the sovereign government to share in the upside if a project is very profitable, but it does not front-load the project with a high tax burden if profitability is weak. Resource rent taxes link the project's tax burden to the economic rent it receives, which is defined as any revenue over and above costs and some specified rate of return approved by the government. There are two main types of resource rent taxes: R-factor based and rate-of-return (RoR) systems.

R-factor terms are based on the ratio of the project's total cumulative receipts over total cumulative costs (the R-factor). When this ratio exceeds one, the tax is levied. Algeria is an example of an exporting country that adopted such a system.

RoR applies after some target rate of return has been realized from the upstream investment. This target rate is usually the return on a risk-free investment (that is, US bonds) plus some country-specific risk premium. In some countries, the tax is increased if higher returns are achieved.

Australia uses a Petroleum Resource Rent Tax (PRRT) that applies a set tax on income over a certain RoR plus the long-term bond rate. The RoR threshold depends on the nature of the income: 5% for development, 15% for exploration. For example, a development project would be liable for tax if its RoR rose above 9% in a period when the long-term bond rate was 4%.

One of the drawbacks of a resource rent tax is that, taken on its own, it has a tendency to delay the government's income until later in the development's life. Some countries therefore use both a rent tax, which does not penalize a project with low profitability and therefore discourage investment, and a royalty, which front-loads at least some of the tax revenue.

Indonesia has an innovative system for receiving up-front income. Upstream projects in Indonesia are subject to a first tranche petroleum (FTP) obligation, under which the government receives a set percentage of the initial oil and gas production. In addition, projects have a domestic market obligation (DMO), which compels them to sell a certain proportion of the oil and/or gas into the domestic market at a set (less than global market) rate. Both of these mechanisms are in effect royalties. Indonesia has a number of different vintage fiscal terms packages it has applied to upstream developments since the 1970s. Jakarta has encouraged development in times of lower investment by adjusting fiscal packages.

As an alternative approach, governments can assure themselves returns from oil and gas developments by becoming participants. Production-sharing contracts (PSCs) are agreements under which a foreign company develops a project jointly with a national oil company. The foreign entity assumes all pre-production costs and risks and recovers these costs from the production. Many upstream Indonesian contracts are PSCs, as is the agreement between Shell, Mitsubishi, Mitsui, and the Russian government for the Sakhalin II project. However, these structures present other issues. Under some PSCs, the foreign entity is permitted to recover its entire investment through revenues before it owes any taxes or other government take. This can result in the host government absorbing far more of the project risk than it intended if there are cost overruns, and any government take is deferred until all the costs are recovered.

Governments can also take direct equity in a project. They can either buy project equity outright (either on market or, more likely, concessionary terms) or can be granted a carried interest (in which the government's share is paid out of production proceeds or by some kind of noncash incentive [possibly tax concessions or infrastructure]). This noncash incentive arrangement can, however, become an economic burden for developing countries. Equity positions can also lead to conflicts of interest within a government.

The majority of these complex terms apply to upstream exploration and production. In most cases, the liquefaction segment of the LNG chain is subject to the same tax regime as any other manufacturing concern in the host country. In some cases, the liquefaction and pipeline segments are incorporated within a larger special tax package along with production facilities (Australia's NWS, Norway's Snøhvit), but when liquefaction is separate from the upstream, the host's normal fiscal terms may apply. Given the scale of LNG projects for most countries, it is usual for the investors and the government to negotiate a package of depreciation incentives, tax holidays, and other project-specific terms to stabilize the project's obligations to the government and help facilitate the financing.

Business Structure

As briefly discussed in chapter 2, a variety of commercial structures are available to participants for setting up an LNG venture. The choice of business structure can have a profound effect on the project's total economics, since it affects the project's total tax burden, division of ownership, and distribution of risk. The main difference between business structures is the division between upstream and liquefaction assets, and the transfer pricing of gas between the segments.

Integrated joint venture

The simplest structure in theory is the integrated joint venture, in which the ownership of and the investment in the upstream, pipeline, and liquefaction segments of the LNG plant are shared equally among all project participants. There is therefore no transfer price between the upstream and liquefaction, so that all project segments are financed and taxed as a single entity.

Risks—price, production, and political—are shared equally among all the project participants. Setting up an integrated joint venture, therefore, requires full agreement by the partners about the project's prospects. It is imperative to reduce uncertainty to the extent possible. Norway's Snøhvit project is an example of a true integrated joint venture, since both the upstream and liquefaction are owned by five project partners, and the entire project is subject to a single fiscal structure. Snøhvit has been placed in a Norwegian upstream fiscal regime, which subjects it to high taxation but gives it a faster depreciation schedule.

Downstream tolling facilities

Increasing numbers of LNG projects are treating the liquefaction plant as a tolling facility, which operates on essentially a stand-alone basis, processing the feed gas into LNG. This structure generally provides a steady income for liquefaction and pushes economic rent into the upstream or, in the United States, into the LNG marketing. If operated on a pure tolling basis, the allowable RoR for the trains is fixed (usually somewhere between 8% and 12% of the actual capital cost) so that liquefaction receives a predictable revenue stream "off the top." The feed gas price is based on a netback from the LNG sales price (deducting tolling fees from FOB sales and tolling and shipping costs from DAP sales). Again the US tolling model differs: the tolling fee is added to the domestic gas price (Henry Hub) to derive an FOB LNG price. The liquefaction segment therefore recovers its costs plus a return. The equity return can be enhanced through leveraged financing, taking advantage of the predictable tolling fees to raise more debt than might be the case for an integrated project.

Within the tolling structure, the owners and operators of the liquefaction venture are less exposed to price and production risk. Their share is essentially guaranteed as long as the project has cash flow equal to or greater than the tolling fees. However, there is often little or no upside for the liquefaction segment. Any economic rent over and above the tolling fees usually goes to the upstream, though liquefaction revenues can benefit based upon actual performance—the more volume produced, the higher the tolling revenues with few or no additional costs. Most of the risk on the project is therefore borne by the production segment or by the capacity holders in the US projects.

Pushing rents upstream is a good way to remunerate the upstream participants for the risks they have taken in exploration and development of the gas fields. However, this system exposes more revenue to upstream taxation, which is typically higher than for liquefaction. The tolling facility model is best applied either in situations where the liquefaction segment is controlled by a government entity or national oil company that is not involved in the producing fields or in cases where gas is sourced from a number of different fields with different owners, some of whom do not participate in liquefaction (or where the owners of the liquefaction plant may not all participate in the upstream).

As is often the case, the exceptions may prove more common than the rule. Trinidad's Atlantic LNG is a good example of a downstream tolling structure. Ownership of ALNG's first train is split among five participants, including the national gas company, while only three partners participate in trains two and three. In each case, the gas is sourced from different fields operating under different license agreements or PSCs. The liquefaction segment operates on a cost-plus-return system, while the feed gas price is based on a netback from the FOB sales. However, the liquefaction plant can participate in the upside, since tolling fees are also collected on excess production, creating an incentive for the liquefaction plant owners to maximize production through better operating practices and capital investment to increase throughput. In the case of ALNG Train 1, the liquefaction margin is a function of the LNG sales price, another possible model of risk and return sharing. These approaches also have the advantage of aligning the economic interests of the owners and the capacity holders. A similar structure is used for the LNG facility in Idku, Egypt, where the partners have built two trains, each liquefying gas from different Egyptian sources on a tolling basis.

Indonesia uses a unique business structure for its older LNG facilities. Pertamina, the national oil company, owns the plants at Bontang and Arun. Feed gas is sourced from a number of different PSCs in distinct packages, each with different ownerships. The liquefaction facility is a separate venture, paid on a cost-recovery basis, and operated by a company that includes representatives from Pertamina, the PSC contractors, and the lenders. Revenues from the project are paid to a trustee, who pays for shipping, debt service, and LNG plant operating expense. There is a revenue allocation to pay for the LNG administration office in Jakarta, which handles contract negotiation, invoicing, etc. The trustee then pays remaining proceeds to the upstream ventures where it is distributed under the PSC.

An exception is the tolling structure that has been adopted for US projects where there is no dedicated upstream and the companies holding tolling rights plan to benefit by LNG resale and trading arbitrage. In some instances, the owners of the liquefaction have even more upside. In the case of Cheniere's Sabine Pass, the tolling revenue pays for the entire costs of and returns on the liquefaction project. But Cheniere has retained the right to use excess capacity for itself, enabling it to access a stream of LNG for which it will only pay the feed gas cost and the marginal cost of liquefaction.

Formula-price feed gas

Some projects specify a certain feed gas price formula, which is a function of the LNG price. This formula may set the feed gas price to a fixed percentage of the FOB or the ex-ship LNG price, in proportion to the project's annual revenue. Alternately, the formula may set a per-unit payment for feed gas that increases over time. The defining characteristic of this type of project structure is that, as opposed to cost-recovery-based projects, the upstream shoulders very little of the risk and receives none (or very little) of the upside if LNG prices jump.

In the cases where the upstream receives a fixed fraction of the LNG price, be it FOB or ex-ship, the upstream percentage is usually less than 25%. This limits the risk on production; it would also limit the fiscal burden on the upstream, except that in many projects with this type of structure (for example, Oman's first two trains or Abu Dhabi's ADGAS project), the government owns the gas reserves and sells feed gas to the LNG project.

In Oman, the government owns the feed gas and the upstream project, and it appointed Petroleum Development Oman, which is owned by the government of Oman, and three shareholders of Oman LNG to operate the upstream. Feed gas is sold to the project at a set per-unit price that is indexed to LNG prices received, and is also stepped-up over the life of the project. By comparison, the national oil company in Abu Dhabi, ADNOC, controls the producing fields and operates them in consortia with commercial oil companies. The feed gas price is a function of the LNG project's annual revenue.

Some projects do away with the upstream segment altogether. Union Fenosa's LNG facility in Damietta, Egypt, buys most of its feed gas from Egyptian General Petroleum Company (EGPC) directly from the natural gas transmission grid. In the proposed US liquefaction projects, the companies leasing tolling capacity will buy feed gas from the transmission grid. While there are no up-front upstream investments, there is a cost over the life of the project embedded in the feed gas purchase price, which can be higher than for a dedicated upstream development.

Shipping Costs

Shipping costs, until the 1990s, were the most opaque component of the LNG chain. Some observers went as far as to describe them as black boxes, into which revenues disappeared with no clear understanding of how or where they were subsequently distributed. Many LNG tankers were owned by the shareholders in the liquefaction plant (North West Shelf, Brunei), chartered from independent owners (Bontang and Arun), or owned by affiliated state companies

(Sonatrach). For CIF or DAP sales, the specific commercial terms of shipping arrangements were not normally of great interest to the customers. However, over time, the shipping segment has become the most transparent element of the chain. This transformation has been driven by open competition between a growing number of shipyards for the construction of LNG tankers and the entrance of many new shipowners willing to bid to supply shipping to LNG projects. The project sponsors themselves have been more inclined to view shipping as less of a core activity, preferring to contract it out to unaffiliated shipowners. The competitive tendering process has driven down shipping rates to a level where it does not make sense for the project sponsors to own the ships since the implied return on capital embedded in the current charter rates is well below the return on invested capital typically required by companies engaged in the exploration and production business.

A typical LNG ship delivered in the 1980s was 125,000 m³ capacity. In the 1990s, the size grew to 130,000 m³. Ships delivered since 2010 have an average capacity of 164,000 m³, and LNG ships on order are up to 174,000 m³. The shipyard price of LNG ships delivered over the past five years averages about $210 million, to which should be added the cost of finance during construction and owner's costs to give a delivered cost between $225 and $230 million. Term charter hire rates for ships delivered in 2014 appear to be in the range of $70,000 per day depending on specific ship characteristics and length of the charter. These rates imply financial returns on the high single-digit range, but are often augmented by highly leveraged borrowing or tax-advantaged lease arrangements, secured by bankable long-term charters. Most LNG shipowners enjoy favorable tax treatment, as many of them are located in tax-advantaged countries.

A further component of value is the residual value of the LNG tanker following the end of the primary term of the charter party. Even after 25 years, a well-maintained LNG tanker is still fit for service. Charterers have become conscious of this value and have in some cases negotiated rights to share in it at the end of the charter term. The charter hire payments include the costs of crewing and maintaining the vessel, which can run about $4 to $6 million per year, with another $3 million annually going to insurance and other operating costs on the owner's account. Charter hire will often be adjusted to allow the owners to pass along these costs to the charterer, although in recent years some owners have been bidding these components on fixed escalation factors.

Other operating costs, such as bunker fuel, port charges, and canal tolls (for ships coming from the Middle East to the Mediterranean and Atlantic) are passed through to the charterers, and along with boil-off, which is a charterer's cost as well, can account for a considerable portion of the transportation expenses. Representative marine transportation costs for a new LNG project can run from around $0.25 per MMBtu (from Algeria to France or Spain) to as much as $1.60 per MMBtu (from West Africa to Japan/South Korea).

LNG ships under long-term charters generally have charter hire rates based on the cost of newbuildings delivered from the shipyard and the period of the charter. Depending on trading conditions (LNG supply availability and pricing in various markets, as well as shipping availability), LNG tankers have traded under spot charter arrangements for as little as $20,000 per day and for as much as $150,000 per day, with a corresponding impact on the unit transaction cost.

Import Terminals

The last leg of the chain is the import terminal. Traditionally, import terminals were owned by utilities in the importing countries and operated as integrated components of the utilities' networks. As such, their commercial terms were generally not ascertainable on a stand-alone basis. However, as a part of the utility, one could assume that the terminal was included within the rate base and that its costs and return were included within the rates charged by the utility to its customers. On this basis, the implied cost of terminaling is on the order of $0.50 to $1.00 per MMBtu, with the more expensive terminals being in Japan where high land costs and the need to maintain large storage capacity tend to drive up the required investments to construct the terminal.

More recently, with the advent of open access and the development of terminals on a proprietary but stand-alone basis, it is possible to assess these costs more accurately. For example, the terminals developed by Cheniere Energy in the Gulf of Mexico have a terminaling cost to third parties of around $0.35 per MMBtu, plus the cost of vaporizer fuel (about 1.75% of throughput). Pipeline costs to tie into the transmission grid are usually add-ons, on the order of $0.05 per MMBtu. Older US terminals with their depreciated rate bases, now operating on a third-party access basis, enjoy lower costs. Where terminaling services are available in Europe, their costs are comparable to those in the United States.

Note

1. Thomas Baunsgaard, "Primer on Mineral Taxation," working paper WP/01/139 (Washington, DC: International Monetary Fund, September 2001).

8

Upstream Natural Gas

Introduction

Natural gas has historically been the poor relative of oil. In areas with ready access to domestic natural gas markets, this has not been the case. Natural gas exploration and development has long been a target of producers in North America and western Europe. Increases in natural gas consumption in other regions in recent years, coupled with the increasing difficulty and expense of accessing oil reserves, have led companies to develop known natural gas deposits and to target exploration at new natural gas fields.

Exploration for and production of natural gas is broadly similar to that of oil, and often the two hydrocarbons are found together. There are some substantial differences, however. While the physical characteristics of natural gas can make transportation difficult (hence the need for LNG technology), its gaseous nature makes its production much easier than that of oil, which usually requires some sort of assisted recovery technique, especially as oil fields mature.

The Physical and Chemical Properties of Natural Gas

Hydrocarbons are chains of molecules composed of hydrogen and carbon atoms. They are generally in the chemical form C_nH_{2n+2}. The higher the n is, the heavier the molecule will be. The lighter hydrocarbons are gases at normal temperature and pressure, hence the term "natural gas." Natural gas is primarily methane, with varying proportions of heavier molecules depending on its source. A typical hydrocarbon composition of natural gas from the mid-continent United States is shown below.[1]

- Methane (CH_4): 91%
- Ethane (C_2H_6): 5%
- Propane (C_3H_8): 2%
- Butane (C_4H_{10}): 1%

- Nitrogen (N_2): 1%

In addition to these constituents, natural gas can contain water, carbon dioxide (acid gas), nitrogen, and, less commonly, hydrogen sulfide. Trace elements of other constituents can also be present. Methane is extremely volatile, with a boiling temperature of $-161\,°C$ ($-258\,°F$) and a specific gravity of 0.6 (compared to a value for air of 1.0). The heavier constituents—ethane, propane, and butane—can be liquefied at higher temperatures and lower pressures than methane, and are referred to as liquefied petroleum gas (LPG).

The composition of natural gas varies depending upon its source. Heating values (the amount of heat that is produced by the combustion of a given quantity of fuel) for gas range from 900 to over 1,200 Btu/ft³, with higher heat content associated with greater proportions of heavier hydrocarbons.[2] Natural gas is colorless and odorless.

Natural gas in a reservoir can occur as either associated gas, which is found in the same geological strata as crude oil, or nonassociated gas. Associated natural gas can exist either in solution within the oil or as a separate natural gas cap distinct from the liquid phase. Because natural gas is lighter than oil, and both are lighter than water, natural gas not in solution will occupy the highest sections of a reservoir. Nonassociated gas can be further characterized as "wet" (with a higher proportion of heavier hydrocarbons) or "dry" (mostly methane with few heavier components).

A field developer has a number of choices when encountering associated gas. Flaring is one option, usually the cheapest and easiest way to dispose of a product that can be a nuisance. Given concerns over global warming and general environmental quality, flaring is viewed as an increasingly unacceptable solution for dealing with associated gas. In fact, the development of Nigeria's first (and thus far only) LNG project, Nigeria LNG, was in part driven by the government's desire to end associated gas flaring. Many oil field operators reinject gas into the oil and gas reservoir as it is produced to maintain the formation pressure that drives the oil into the well, with the Prudhoe Bay field on Alaska's North Slope being a prime example of this process. Operators can also choose to monetize the gas in its own right, independent of oil production. However, not all reservoirs are susceptible to reinjection of the gas, nor can all associated gas be economically captured for sale or reuse, so flaring continues.

Nonassociated gas occurs in a reservoir without significant oil or other heavy hydrocarbon liquids. Gas-condensate fields are gas accumulations in which the ambient pressure is high enough that the heavier hydrocarbons remain in a gas phase, only to fall out of the gas stream in liquid form when brought to the surface during production. These liquids, known as condensate, generally contain the C_5H_{12}, C_6H_{14}, and C_7H_{16} molecules, and, along with butane, propane, and ethane, are often referred to as natural gas liquids after being removed from the natural gas stream during processing.[3] Many of the stranded-gas reserves worldwide, which are candidates to supply LNG projects, are in nonassociated gas fields.

Natural Gas Reserves

The term "reserves" has a variety of meanings and classifications. Generally, reserves refer to the quantity of technically and commercially recoverable hydrocarbons in an oil and/or natural gas reservoir (as opposed to the total quantity of oil or natural gas in place, much of which may not be recoverable with the aid of current technology or under current market conditions). To further confuse matters, the term "resources" is used to describe oil and gas that is likely to be present in the reservoir, but for which there are no specific data supporting the estimate. Resource estimates are always higher than reserve estimates, because resource estimates ignore technical and economic considerations that apply to reserve estimates.

There are three major classifications of reserves: proven, probable, and possible. Generally, proven reserves are estimates with a high degree of certainty, supported by the geological or geophysical delineation of the reservoir, the drilling of exploration and appraisal wells, and testing of these wells. The wells' major characteristics (pressure, flow, gas composition, etc.) are analyzed, along with rock cores and drilling logs to establish the porosity and permeability of the reservoir rocks. Information from producing reservoirs in the vicinity of the field being appraised may also be used in conducting the analysis. The designation of reserves as probable or possible is more of a statistical exercise based on a wide variety of geological, physical, and economic criteria, and may have much less information derived from actual drilling results. Usually, reserves estimated to have a greater than 50% likelihood of being developed are labeled probable; those less likely to be developed are called possible.

The question of whether identified resources can be technically produced applies much less to natural gas than to oil. The development of oil reserves often requires expensive, enhanced techniques in order to recover much of the oil in place, whereas gas reserves are more easily produced because the essential mechanism that brings the natural gas to the surface is its own pressure. However, the commerciality test of proven reserves is more certain for oil than for natural gas since it is much easier to demonstrate a market and price for potential future oil production than for natural gas production.

Under US accounting guidelines, companies are required to report their year-end reserves based on the average price of the oil and natural gas that they sold for the prior 12 months, which can lead to some interesting reserve fluctuations if energy prices change dramatically. US companies cannot book gas reserves until they have demonstrated that they are marketable, for example, by demonstrating they can be connected to the existing gas grid (normally the case in the domestic market) or by producing a sale and purchase agreement (SPA) or binding letter of intent with a buyer for gas in remote locations. Non-US companies generally have greater latitude, unless they are listed on US stock markets, in which case they must meet the US criteria. This can result in anomalous disclosures with a US company classifying reserves as probable or possible while a foreign company might classify the same reserves as proven.

Reserve estimates are constantly updated over the life of a reservoir, both to reflect the production of the oil and/or natural gas and continually reassess the remaining hydrocarbons in place and the likelihood of their recovery. As more wells are drilled and more data are collected, the nature of a reservoir becomes increasingly clear. This will often lead to a reclassification of hydrocarbons from the probable or possible categories to proven. Thus, even as oil and natural gas are produced, the reservoir's remaining proven reserves may often increase.

Worldwide, the largest deposits of estimated natural gas reserves (table 8–1) are found in Iran with 1,192.9 Tcf (33.8 Tcm), which, recent estimates suggest, are slightly larger than its nearest rival, Russia, with 1,103.6 Tcf (31.3 Tcm). Qatar is currently estimated to have the world's third largest gas reserves of 871.5 Tcf (24.7 Tcm), followed by Turkmenistan with 617.3 Tcf (17.5 Tcm). Together, these four nations contain approximately half of the world's proven gas reserves. These estimates illustrate the difficulty of distinguishing between reserves and resources. For example, while the United States has reserves of 330.0 Tcf (9.3 Tcm), its resources including shale gas are far larger.

These reserves rankings, however, do not reflect production. In 2013, the United States at 688 Bcm (66.5 Bcf/d) was the largest producer, followed by Russia at 605 Bcm (58.5 Bcf/d). Iran, the largest reserves holder, produced just 167 Bcm (16.1 Bcf/d) with Qatar accounting for 159 Bcm (15.3 Bcf/d) and Turkmenistan 62 Bcm. (6.0 Bcf/d). In terms of market, the United States is the largest, consuming about 737 Bcm (71.3 Bcf/d) of natural gas in 2013. Russia was second at 414 Bcm (40.0 Bcf/d). Iran at 162 Bcm (15.7 Bcf/d) and China at 162 Bcm (15.6 Bcf/d) were third and fourth. Of the rest of the world, only Japan (117 Bcm/11.3 Bcf/d), Canada (104 Bcm/10.0 Bcf/d), and Saudi Arabia (103 Bcm/10.0 Bcf/d) consumed more than 100 Bcm (9.7 Bcf/d) in 2013.

Table 8–1. Estimated proved reserves for top gas reserve holders (January 1, 2014)

Country	Reserves (Tcf)
Iran	1,192.9
Russia	1,103.6
Qatar	871.5
Turkmenistan	617.3
Saudi Arabia	290.8
United States	330.0
UAE	215.1
Venezuela	196.8
Nigeria	179.4
Algeria	159.1
Australia	129.9
Iraq	126.7
China	115.6
Indonesia	103.3
Norway	72.4
Canada	71.4
Egypt	65.2
Kazakhstan	63.9
Kuwait	63.0
Libya	54.7
India	47.8
Malaysia	38.5
Uzbekistan	38.3
Oman	33.5
Azerbaijan	31.0
Netherlands	30.1
Ukraine	22.7
Pakistan	22.7
Vietnam	21.8

(*Source:* BP Energy Statistics 2014.)

The geology of natural gas

A hydrocarbon system (a term describing the elements that make up an oil and gas reservoir) is made up of four essential parts: a source rock, a migration route, a reservoir rock, and a trap (including a cap formation). The source rock is the organic-rich layer created from prehistoric plant and animal remains. Over time, the organic material is subject to great heat and pressure as it is buried under successive layers of new rock. This combination of heat and pressure essentially cooks this carbon-rich material until it forms hydrocarbon molecules.

Often this source layer takes the form of shale, a dark, carbon-rich rock comprising grains of sand or silt interspersed with organic material. Carbonate rocks (rocks that are rich in calcium carbonate formed almost entirely from the remains of living organisms) and sandstones can also be source rocks. Shales, sandstones, and carbonates are all examples of sedimentary rocks, composed of material that has been transported to and deposited in a geological basin. Sedimentary is one of the three major rock classifications, the others being igneous (resulting from the cooling of liquid magma) and metamorphic (rocks, whether originally sedimentary or igneous, that have been altered by heat and/or pressure).

The organic material within a source rock can be of either plant or animal origin, which can influence whether the hydrocarbon created is more likely to be oil or natural gas. A large amount of woody plant material will produce a dry natural gas (that is, low in liquids content), while high proportions of algae in a source rock will produce more oil (all else being equal). The depth of burial also plays a role. As the source rocks are buried deeper, they are subject to greater temperatures and pressures, which "crack" the hydrocarbon chains into smaller and smaller molecules. Eventually, these hydrocarbons can become methane (CH_4). Geologists refer to "windows"—below certain depths, source rocks are in the "natural gas window," where they are capable of forming only natural gas. Above this depth is the "oil window." Generally, the oil window occurs between depths of 7,000 to 18,000 feet, while the natural gas window is deeper than 18,000 feet.[4] These windows are a general rule of thumb, as their actual depths are dependent on a number of other geological factors.

Natural gas can also be formed close to the surface via bacterial action or buried organic material. This form is referred to as biogenic gas (swamp gas being a good example). Most biogenic natural gas is not trapped and dissipates directly into the atmosphere. Nearly all commercial natural gas is thermogenic, meaning that it is formed by the burial of organic material as described above.

Like all fluids, hydrocarbons will move from areas of higher pressure to areas of lower pressure. They seek migration routes to move upwards, away from the high-pressure zones where they were created. Often this movement is aided by the migration of water, which is squeezed out of the source rock (and even out of individual minerals within the source rock) as pressure increases. As they are created, hydrocarbons can move through the source rock either via fractures and bedding planes, or via pores between the grains making up the source rock. The ratio of a rock's pore space to total volume is known as porosity; as oil and gas are formed, the organic material from which they are formed shrinks, increasing the rock's porosity. The ability of fluids to move through a rock is a measure of the rock's permeability: the higher the porosity and permeability, the better the quality of the reservoir.

Hydrocarbons move from source rocks into reservoir rocks (often passing through carrier formations, which allow the passage of the fluids without actually trapping them). Reservoirs can be made up of a number of different rock types, the most common being sandstones. As the name suggests, sandstones are made up of compacted grains of sand that have been eroded from terrestrial rocks, deposited (in deltas, beaches, channels, etc.), and, over time, buried. Sandstones often have high porosity and permeability (owing to the spaces among individual grains), making them ideal for both transferring and trapping hydrocarbons. Carbonates can also make good reservoir rocks, depending on how fractured or porous they are. A large proportion of Middle Eastern oil is trapped in carbonate reservoirs.[5]

Given the opportunity, hydrocarbons would continue to move through these sandstones or carbonates, as long as they were driven by a pressure differential, until they reached the Earth's surface. In order to accumulate, oil and gas must migrate into a trap. Traps come in a variety of forms; commonly, hydrocarbon-bearing sandstone is overlaid by an impermeable cap rock that prevents the fluids from migrating farther. If the layers of sandstone and cap rock are folded in any way as a result of movement of the earth's crust, the oil and/or natural gas will accumulate in the highest sections of the reservoir rock and remain there (fig. 8–1). Faults can also form excellent traps (fig. 8–2), as can impermeable salt domes (fig. 8–3). Salt domes form a large proportion of oil and gas traps in the Gulf of Mexico. Traps can even be formed by gradual diminishing of the porosity and/or permeability of the reservoir rock.

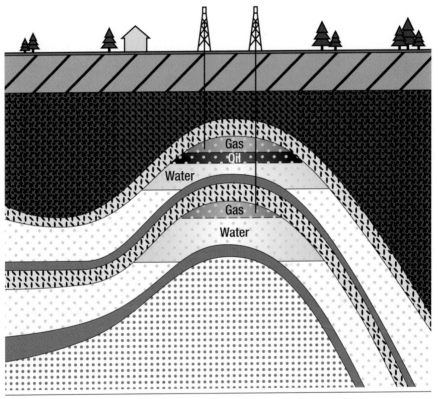

Fig. 8–1. Gas trap created by folding. (*Source:* Poten & Partners.)

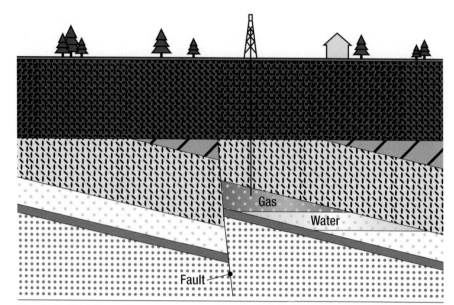

Fig. 8–2. Fault-bounded gas trap. (*Source:* Poten & Partners.)

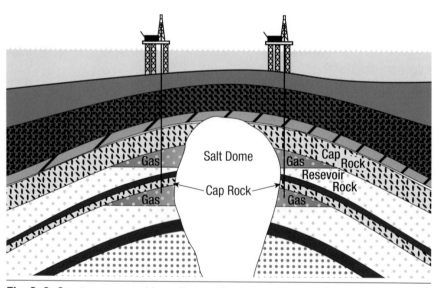

Fig. 8–3. Gas traps created by salt dome intrusion. (*Source:* Poten & Partners.)

Natural gas exploration

At least initially, an explorationist or prospector usually looks for geological conditions favorable to the presence of hydrocarbons. The oldest and least certain of these methods is direct geological investigation. Geologists use both contour maps of the Earth's surface, showing highs and lows of the topography, and the composition of outcrops (surficial exposures of underlying rock layers) to make informed guesses as to the likelihood of finding hydrocarbon traps in an area. Geologists also consider the prehistoric conditions of a region, such as the presence of an ancient river delta where organic material would have been swept into concentrated accumulations and subsequently buried. Somewhat more directly, geochemical analyses, which examine the chemistry of plants and soils for the presence or effects of hydrocarbons seeping to the surface, can give an indication of buried oil or gas. The most effective method for hydrocarbon prospecting, though, is geophysics.

Geophysicists have three main types of tools at their disposal. Measurements of gravity and the Earth's magnetic field can indicate anomalies, which give insight into the nature of the underlying rocks. Gravity is a function of a rock's density, and abrupt changes in density can indicate significant variations in rock types or the presence of salt domes that may point to potential hydrocarbon traps.

Magnetic surveys analyze the variations in different rock layers' magnetic fields, which are largely controlled by the presence or absence of the magnetic mineral magnetite, a common constituent of basement rocks (those rocks, usually igneous, that underlie the sedimentary layers that produce and contain oil and gas). Knowledge of the basement's depth allows a geophysicist to quantify the thickness of a sedimentary layer within a basin, which aids in estimating whether an area was conducive to the generation of hydrocarbons. Both gravity and magnetic measurements are also useful in identifying salt deposits that may act as traps.

The primary tool of geophysicists in the search for oil and gas, however, is seismology, which measures the paths of sound waves propagating and reflecting through the Earth's crust to draw a picture of the underlying geology and expose stratigraphic traps that may contain hydrocarbons. These sound waves are generated from a source that imparts a shock wave to the subsurface rocks. (On land, the source is often explosives or heavy weights dropped on the ground, while at sea air guns or electrical sparkers are used.) These waves travel through the Earth's crust, and as they pass through a boundary between different layers of rock, some of their energy is reflected back to the surface. These reflected waves are detected by carefully spaced arrays of detectors referred to on land as geophones, and at sea as hydrophones (fig. 8–4).

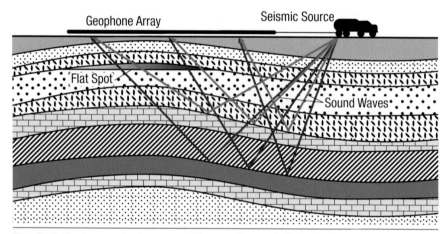

Fig. 8–4. Onshore seismic survey. (*Source:* Poten & Partners.)

Analysis of the energy and travel time of these reflected waves gives enormous insight into the nature of the subsurface, allowing geoscientists to paint a picture of the different layers of rock in a basin and the contacts between these layers. The amount of time it takes for a wave to travel to a boundary layer and back to a detector gives an indication as to the depth of that layer. Strong differences in the amplitude and frequency of wave reflections can indicate abrupt changes in density among rock layers, which are often signs of gas-saturated reservoir rocks (gas is not dense, so sound waves travel at a much lower velocity through a sandstone rich in gas). Oil and gas can also show as "bright spots," which are highly reflective layers within traps—the bright reflector being often not a rock contact at all, but a contact between oil or gas or water. The analysis requires massive amounts of computing power to analyze the seismic data and map the subsurface formations. The oil industry is the largest user of supercomputers in the world.

By systematically setting a series of explosive charges or other energy sources along a line, geophysicists can map a cross section of the subsurface. Using arrays of geophones along a series of lines parallel to one another and a set distance apart allows an investigator to map a "block" of a basin, a process known as 3-D seismic. A relatively new technique is 4-D seismic, which entails taking a series of seismic readings in the same area over time, allowing one to track the evolution of a hydrocarbon field as it is produced. These advances in exploration techniques have dramatically improved subsurface mapping, allowing a more accurate prediction of reserves.

Seismic data have proved invaluable in the identification of potential prospects, which are areas that are shown to have all of the components necessary for an oil or gas field—a reservoir rock, a cap rock, and a trap that has a reasonable probability of containing commercial reserves of oil or gas. Independent seismic contractors often gather seismic data and sell them to exploration and production companies (E&P companies) with the capital and resources to fund further exploration. Improved computer techniques also allow for the reprocessing of vintage seismic data, which can be used to identify formations that could not be seen through earlier analysis.

Notwithstanding all of the above, the only way to establish conclusively whether a prospect contains oil or gas is by drilling. Operators begin with a wildcat or exploratory well, which will confirm (or refute) the presence of hydrocarbons estimated through geology and geophysics. As the well is drilled, data known as well logs are taken from the wellbore. These measurements are taken via wireline, in which a recording device is lowered into a well, taking measurements of the rock formation through which it passes. These logging devices can establish a formation's thickness, porosity, permeability, pressure, and electrical resistivity, all valuable factors in determining the potential productivity of a given formation. Wireline logs can also directly establish the presence of oil or gas. Successful wildcat wells are usually followed by the drilling of appraisal wells in the same area.

Modern drilling is almost entirely done by rotary rigs (figures 8–5 and 8–6 show onshore and offshore rigs). The drilling is achieved with a long, jointed steel pipe (drill string), at the end of which is a drill bit. The drill string is extended or shortened by adding or removing joints of pipe. The drilling rig rotates the length of pipe, which in turn rotates the bit and causes it to dig through the rock creating a wellbore (the actual hole). Drill bits can be in a variety of configurations; the most common modern bit being a tricone one, which comprises three separate toothed cones on the end of the bit that spin independently of one another, essentially chewing through the rock. Very hard rocks are often drilled using a diamond bit, a static bit covered in industrial diamonds. Bits are subject to considerable pressures (generally between 3,000 and 10,000 psi, depending partially on the depth of the hole and the weight of the drill string) and are usually changed every 40 to 60 hours of drilling.

Fig. 8–5. Onshore drilling rig. (*Source:* img.tfd.com.)

Fig. 8–6. Offshore drilling rig. (*Source:* mits.doi.gov.)

As wells are drilled, a heavy, viscous mixture of clay and water and/or oil, known as drilling mud, is injected down hole through the drill string and the bit. This mud performs several functions. As it is injected through and forced to the surface, it brings rock cuttings. Analysis of material brought to the surface by drilling mud, known as mud logging, can give valuable insight into formation conditions. The mud also serves to lubricate the drill string. The weight of the mud regulates the borehole pressure and reduces the chances that pressurized hydrocarbons may blow out of the wellbore. As the well is drilled, successive lengths of steel tubing, known as casing, are inserted into the wellbore and cemented to the side, preventing the formation from collapsing in on the drill string.

Actual drilling is undertaken by a drilling contractor, who owns and operates the drilling rig. The cost of drilling is largely determined by the location—onshore drilling, using a simple rotary rig, is substantially less expensive than offshore drilling, the main difference being the type of rig used. For offshore fields, a number of different configurations are available, depending largely on the water depth.

In shallow water, rigs can be placed on drilling barges, which are especially useful in river delta regions such as those in Nigeria or the Mississippi Delta. For somewhat deeper water, jack-up rigs provide a stable platform that can be moved if needed. Jack-ups consist of two distinct platforms: a hollow lower "mat" connected by three or more vertical legs to an upper platform upon which sits the drilling equipment. The rig is towed to the drilling site, the mat is flooded and sinks to the bottom, and the upper platform is then jacked up the legs (in a manner similar to the one used by a common rack-and-pinion car jack) until it is well above the waterline.

For deeper water, semisubmersible platforms are towed to the drilling site and anchored into place. These platforms have flotation pontoons below the waterline, which keep them afloat while maintaining stability in rough seas. Semisubmersibles are relatively easy to move once drilling is complete. Drill ships are used for very deep seafloor drilling. These ships have a hole cut in the center of their hull (the moon pool) through which they can drill the seabed. This type of drilling relies heavily on global positioning system (GPS) technology—the ship's position relative to the wellbore is constantly monitored, while its position is maintained by a system of thrusters.

Like advances in seismic techniques, innovations in drilling technology have led to improvements in exploration technology. Wells can now be drilled deeper into formations, more than 30,000 feet below the surface. Offshore rigs can operate in deeper and deeper water, at depths exceeding 10,000 feet. Slimmer wellbores and improvements in muds, drill bits, and drill strings allow for faster and more accurate drilling. Even so, a deepwater exploration well can take months to drill and evaluate, often at costs exceeding $100 million. Horizontal drilling, a procedure in which the drill string is deflected from the vertical to drill horizontally, allows drillers to penetrate much more of a reservoir layer than was traditionally possible. This also allows greater overall oil or gas production from a given well. Horizontal drilling has been a key component (along with advances in hydraulic fracturing) in the development of oil and gas shales.

Natural gas production

Once exploration and appraisal drilling has been completed, and it has been established that a prospect will produce commercial hydrocarbons, development drilling can begin. Reservoir management plays a key role in well placement and operation. All wells, normally including the exploration and appraisal wells, are completed. This entails setting casing strings composed of connected lengths of casing pipe. These pipes, usually 25 to 35 feet long and up to 30 inches in diameter, are joined together and run down a well, providing a steel liner which prevents the adjacent rock from collapsing on the wellbore and allows the drill string to reach ever greater depths. Casing diameters generally decrease with depth, with successive casing strings set within one another going deeper and deeper, so that a casing string looks somewhat like a telescope, with the narrowest portion in the producing formation (fig. 8–7). The space between the casing string and the wall of the drilled borehole or the previous casing string (the annulus) has cement pumped down the inside and forced up the annulus; the cement makeup depends on the conditions of the wellbore.

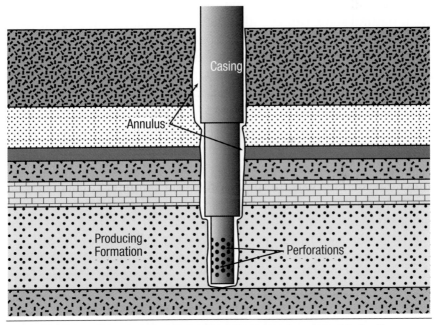

Fig. 8–7. Well casing, showing diminishing diameter of casing with depth (exaggerated). (*Source:* Poten & Partners.)

Once the cement surrounding the casing is set, hydrocarbons can flow into the well in one of two ways. The simplest is to set the casing only to the top of the producing formation, then continue drilling, so that the borehole goes past the bottom of the casing directly into the producing formation, allowing oil or gas to flow directly from the rock into the well. However, this allows very little control over the production process and runs the risk of subsequent collapse, thus keeping it reserved for the simplest and most stable formations. In most wells, the casing is set through the producing formations, and the casing string is perforated through the same depths as in the producing formations. These perforations are created by puncturing the casing string at the depth of the producing formation, often by shooting holes through the casing.

Unlike oil, which often requires the application of various recovery techniques to bring it to the surface, gas is driven by its own pressure and therefore flows into the well and out of the ground of its own accord (groundwater pressure can also drive production). Formations with especially low permeability (and therefore low gas flow), or fields that are reaching the end of viable pressure drive in their natural state, can be "stimulated" to enhance their production in a number of ways. The introduction of large volumes of water- or nitrogen-based foam at very high pressure into a well can cause the formation to fracture, opening up pathways in the rock in a process known as fracking. The injection water is mixed with proppants (materials such as sand or ceramics that hold the cracks open once they have formed) and a small quantity of additives (on the order of 0.5% of the total injection volume).

These additives are mainly bactericides, gelling agents, and surfactants. The composition of the additive package depends primarily on the well conditions: pressure, temperature, and proppant quantity. In addition to sterilizing and preventing bacterial contamination of the reservoir, the additives serve to improve the efficiency of the process. After the fracturing job is completed, the fluid is allowed to flow back to the surface, leaving the sand behind to prevent the fractures from closing. With limestone (and some sandstone) reservoir rocks, the injection of acid through the well casing can also open up gas paths (limestone is especially vulnerable to acid). Gas reservoirs that need these initial forms of treatment, often "tight" sands or shales, were rarely considered candidates for supplying LNG projects because the costs of gas development and production are so high as to render the upstream uneconomical compared to other more productive formations. However, advances in combining horizontal drilling with fracking have turned the conventional wisdom on its head.

Once the wellbore has been completed, a string of hollow pipe called production tubing is run into the wellbore, and the oil or gas flows to the surface through this tubing. In wells with multiple formations, there may be more than one string of production tubing.

The surface completion of a gas well is usually a wellhead fitting with a series of valves, known as a Christmas tree (fig. 8–8). The Christmas tree is attached to the head of the casing string and production tubing, with valves permitting flow to be controlled within each string (multiple strings are often used when more than one formation is tapped through the same wellbore). The tree connects to a flow line or gathering line (a pipeline connecting the tree to field treatment facilities) to move the gas into a pipeline for transportation. The valves and the choke regulate the flow of gas. Trees also incorporate pressure gauges to monitor the flow rate and pressure. Control of reservoir pressure is crucial in gas field development, not only in order to regulate the flow and maintain the pressure necessary to drive production, but also to prevent heavier hydrocarbons from liquefying and dropping out of the gas stream while still in the reservoir, thus making recovery of these liquids much more difficult.

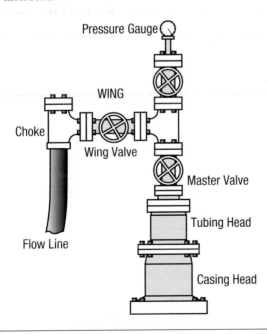

Fig. 8–8. Christmas tree. (*Source:* Poten & Partners.)

Large gas fields usually comprise a number of wellheads and associated Christmas trees. These are integrated by a field-gathering system, essentially a series of flow lines that feed gas from each well into a central processing unit where some of the initial liquids and other contaminants can be removed.

Traditional onshore fields are easier to develop, as the wells can be spaced throughout the field, minimizing the amount of drilling that must be carried out to tap a given reservoir. Offshore is a different matter. With an offshore field, the production phase must be managed from a specially built platform, which carries all the equipment that would be found in the onshore environment (fig. 8–9). These platforms are either fixed to the seabed or are floating and anchored to the seabed. The development of an extensive gas field may require the installation of multiple platforms. As operators move into deeper and deeper water to access new gas fields, subsea completions that place most of the wellhead equipment on the seafloor allow development with more limited platform requirements. Flow lines from each completion feed into a subsea central manifold, which then feeds the gas into a pipeline leading either to a central platform or to shore. A number of gas developments (including the giant Ormen Lange field in Norway and the Chevron-led Greater Gorgon LNG project in Australia) use subsea completion configurations.

The combination of improved seismology and enhanced drilling techniques has brought down exploration and field development costs over the past decade, by as much as one-third, with resulting improvement in field economics. Seismic technology alone has raised the success rates of exploration drilling from around one well in ten in the 1970s to around one well in three today.

Fig. 8–9. Offshore gas production platform. (*Source:* CSIRO.)

Unconventional gas

In terms of their chemical composition (primarily methane), unconventional gas resources are identical to conventional natural gas (fig. 8–10). They are called unconventional because of their atypical geological locations. Unconventional gas is found in highly compact rock or coal beds and requires a specific set of production techniques. Unconventional gas covers three main types of natural gas resources: shale gas, tight gas, and coalbed methane (also known as coal seam gas).

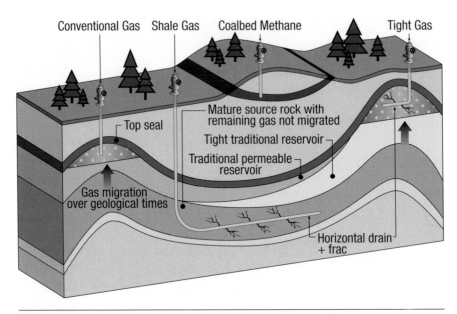

Fig. 8–10. Unconventional natural gas resource structures. (*Source:* Total.)

Shale gas and tight gas (mainly natural gas, but some oil as well) are trapped in subsurface reservoir rock formations (see fig. 8–10) just as in conventional gas reservoirs. As described above, the quality of a reservoir rock is determined by its porosity and its permeability. A common feature of shale gas and tight gas is that both are trapped in very low-permeability rock—ultra-compacted structures that prevent or sharply limit the migration of the gas.

Tight gas is trapped in reservoirs characterized by both very low porosity and permeability. The rock pores that contain the gas are minuscule, and their interconnections are so limited that the gas can only migrate through them with great difficulty. The unit of permeability measurement is the darcy. A good-quality hydrocarbon trap will have permeability of 1 darcy or more, while tight gas reservoirs, more compact than brick, may have permeability of only a few dozen microdarcys.

Shale gas is extracted from a geological layer known as the source rock rather than from a conventional petroleum reservoir structure. This clay-rich sedimentary rock has naturally low permeability. The gas it contains is either adsorbed (that is, closely "inserted" into the organic matter) or remains in a free state in the void spaces (pores) of the rock. The permeability values of gas shales are as little as one one-thousandth of the permeability of tight gas formations. The unit of measurement here is the nanodarcy.

Development drilling to exploit unconventional reservoirs has advanced rapidly, mainly by the adoption of horizontal drilling coupled with hydraulic fracturing. Horizontal drilling begins in the same way as conventional drilling, but, shortly before the drill bit penetrates the target formation, the drill string is pulled out of the hole. Then, a hydraulic motor is inserted between the drill string and the drill bit. Powered by the circulating drilling mud, this down-hole motor can be used to steer the bit into a horizontal direction, then follow the formation for several thousand feet. This drilling technique is much more expensive than vertical drilling, but by using this technique, two benefits are realized. First, much more of the field can be covered from a single drilling point, keeping costs lower, especially in the offshore environment. Second, by steering the bit within the producing formation rock strata for extended distances, the wellbore will intersect a much greater proportion of the formation and result in enhanced productivity. Cement and casings are perforated only in the horizontal section of the wellbore (several thousand feet underground) to access the productive layer. Hydraulic fracturing (fracking) is the final step before the well is brought on stream. Increasingly the industry employs a technique of fracturing the rock in stages, opening numerous holes in the horizontal portion of the well, then staging sequential fracturing operations to allow the maximum energy to be focused on the rock, in a technique known as multistage fracturing. On long horizontal wells, over 30 fracture stages might be deployed. Wells in an unconventional formation drain a smaller volume of rock than wells in a conventional gas reservoir. To limit the physical footprint of the operations, wellheads are grouped together in clusters and horizontal wells radiate out from the central location in a star-like pattern.

Coalbed methane, as its name suggests, is trapped in coal deposits. It is also known as coal seam gas. Most of the gas is adsorbed on the surface of the coal. The recovery of coalbed methane is less technology-intensive than producing other types of unconventional gas, but it does require a specific approach—gas is trapped in the coal seams with water, and the water pressure must be reduced in order to free the gas. This is achieved by pumping the water out of the coal bed's natural fracture system. Coalbed methane can be produced via conventional vertical wells. However, depending on the geology of the deposit (depth and thickness of the coal bed), it may be necessary to drill horizontal wells radially from a centralized location to optimize the cost of surface developments. Wells must be equipped with bottomhole pumps to lift the water from the coal seams to the surface. During the initial production phases, the water volume exceeds the volume of gas. Peak productivity is not reached until one to five years after starting up a well. To increase the wellbore contact with the reservoir—and thereby maximize gas recovery—localized hydraulic fracturing may prove necessary, particularly in the deepest layers of the deposit. This process helps propagate the natural fractures in deeper coal seams that are less developed than in layers closer to the surface.

Natural gas field processing

Natural gas straight out of the well, whether conventional or unconventional, can contain various impurities that may need to be removed prior to injecting the gas into a pipeline. Quality specifications are dependent on the feed gas supply contract. For LNG production, further processing often takes place immediately prior to liquefaction. Flow lines are the pipes that make up an oil and/or gas field's gathering system, and are of much smaller diameter than transmission pipelines. The pressure inherent in gas wells is usually sufficient to push gas through flow lines to the field's central processing facility, although gathering system compressors may be installed in cases of lower-pressure wells or later in the life of a field as natural pressures decline.

Hydrocarbons heavier than methane, whether liquid or gas, can be removed from the feed gas stream near the production field, depending on the configuration of the pipeline, a process known as stripping. Condensates are the easiest to remove; they could otherwise cause problems if left in the pipeline, where they can drop out of the gas stream and block pipeline flows. Other natural gas liquids can also be removed in the production area. Chilling the feed gas stream in a propane-cooled vessel can cause natural gas liquids to drop out. Percolating the feed gas through a series of trays containing heavier liquid hydrocarbons can also remove natural gas liquids.

Gas needs to be dehydrated prior to being introduced into a pipeline since water can form hydrates, methane molecules surrounded by ice. These solid hydrates can form at temperatures up to 20°C, depending on pressure, and can block pipelines. Water is removed from the feed gas by passing the stream through a glycol bath, which draws the water out of the feed gas.

Most pipeline contracts require the removal of carbon dioxide (referred to as acid gas) and hydrogen sulfide before transport, since both have a corrosive effect on pipeline steel. Hydrogen sulfide is also toxic as a gas, and combustion of the compound creates sulfur dioxide. Gas with high carbon dioxide and/or hydrogen sulfide content is referred to as sour gas, as opposed to sweet gas. Both contaminants are removed by a sweetening process, which entails passing the feed gas through liquid amines. Shell's patented Sulfinol process, which adds the use of a physical solvent to the amines, is commonly used to extract acid gases. CO_2 is increasingly reinjected underground ("sequestered") after its removal to prevent its introduction into the atmosphere.

Natural gas pipelines

Once gas has been processed to pipeline quality, it is sent via transmission pipeline to the LNG plant. Onshore gas transmission pipelines are made of sections of steel pipe, welded together, and buried in trenches (fig. 8–11). Offshore pipelines can be positioned by large pipe-laying ships or lay barges, which carry lengths of continuous pipeline (up to about 26 inches in diameter) on enormous spools that lay out as the ship moves along a mapped route, or where the pipe is welded and dropped to the seabed over the stern of the ship or barge (fig. 8–12). Offshore lines are often laid directly on the seabed. However, if the seafloor is soft enough, these lines may be trenched and buried using underwater ploughs.

Fig. 8–11. Onshore gas pipeline construction. (*Source:* ZGG GmbH.)

Fig. 8–12. Offshore gas pipeline construction. (*Source:* Offshore Technology.)

Downstream of the processing facilities, larger pipeline diameters are used for gas transmission and may require higher pressures. Diameters for the larger pipelines transporting gas from the Gulf of Mexico to markets in Chicago and the northeast United States range between 24 and 36 inches. In order to maintain pressure, compressor stations may be built along the pipeline's route, giving the gas stream periodic pressure boosts. The number and power of these compressors is a function of the pipeline's total length, diameter, and the volume being transported. For transmission lines linking gas fields to LNG plants, the objective is to keep the line as short as practicable, and use as little compression as possible. Larger diameter pipelines need less compression, but the initial installation is more expensive. In Trinidad, a 56-inch transmission line was laid across the island to link gas fields off the east coast with the liquefaction plant on the west coast, moving large volumes of gas while minimizing the need for compression.

Where it is impractical or too expensive to strip out all the liquid hydrocarbons in the gas stream in the field, the field developer may utilize a multiphase pipeline, in which the gas stream is specially handled to ensure that the liquids (oil, natural gas liquids, and/or water) move with the gas until the line reaches a treatment area. There are a number of different flow regimes that describe how the liquids and gas move through the pipeline—as a mist, as separate stratified layers, as gas and "slugs" of liquids, or as a liquid flow impregnated with gas bubbles. This type of line is most commonly used to move gas from offshore fields where the cost of removing and exporting the liquids in situ may be prohibitive.

The environment in which a pipeline will operate is a major determinant of its design. Pipelines designed for Arctic environments need to be able to operate at temperatures well below freezing, and in many cases are not buried at all, since burial in permafrost or in soil that is continuously freezing and thawing can have adverse effects on the pipe (the 1,300-kilometer Trans-Alaska pipeline is the best known example). Onshore pipelines that cross rivers may require complex bridges or trestles, or special burial processes (the two 800-kilometer Sakhalin Island pipelines, one for oil and the other for natural gas, include 126 kilometers of swamp crossings and 110 kilometers over mountainous routes). In rain forest environments, special attention must be paid to construction and restoration techniques to avoid permanent damage to the sensitive ecosystem and to prevent future erosion which could rupture the pipeline.

The stress of laying pipeline offshore is much greater than that for burying onshore pipeline, and the steel strength required will be chosen accordingly. Undersea pipelines are also much more susceptible to corrosion. Pipeline routes that cross over coral reefs near shore can require the use of horizontal directional drilling (HDD) to permit laying pipe directly under sensitive habitats without disturbing them. Globally, pipeline construction is attracting much more attention than ever, given the potential for serious environmental damage—Peru, Sakhalin, and Florida are among the regions where environmental concerns have complicated pipeline construction. Once the gas has been produced, treated, and transported, it is ready to enter the liquefaction plant.

Notes

1. Gas Processors Association, *The Gas Processing Industry: Its Function and Role in Energy Supplies* (Tulsa, OK: Gas Processors Association, 1998).
2. "Whitepaper on Liquid Hydrocarbon Drop Out in Natural Gas Infrastructure" (NGC+ Liquid Hydrocarbon Drop Out Task Group, Dec. 17, 2004).
3. Hyne, *Dictionary of Petroleum Exploration*.
4. Ibid.
5. Malcolm K. Jenyon, *Oil and Gas Traps: Aspects of Their Seismostratigraphy, Morphology and Development* (New York: John Wiley & Sons, 1990).

9

The Liquefaction Plant

Introduction

The liquefaction process transforms natural gas into liquefied natural gas (LNG) by cooling it to –163°C, after which it is stored in the plant's storage tanks until it can be loaded on board LNG tankers (fig. 9–1). There are variations in liquefaction plant designs, but liquefaction processes are very similar and, setting aside the scale, quite straightforward. Liquefaction is nothing more than a giant refrigeration process, involving no application of chemical processes to the natural gas other than pretreatment for impurities.

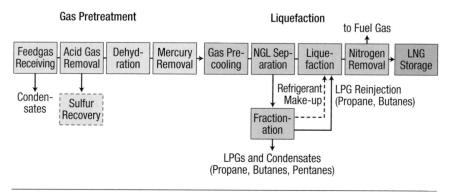

Fig. 9–1. Natural gas liquefaction into LNG, flow diagram. (*Source:* Poten & Partners.)

Siting Considerations

The LNG plant is a large and complex industrial undertaking. It typically includes gas processing and purification, liquefaction, product storage and marine loading systems, utilities, control rooms, material and equipment warehousing, maintenance shops, and infrastructure facilities, which often include housing for the operating staff.

Because the feed gas pipeline to the plant can be costly, sponsors generally seek to minimize the distance between the gas fields and the liquefaction plant. For example, a two-train 6.6 MMt/y (9.2 Bcm/y) LNG project requires more than 1 Bcf/d (10.3 Bcm/y) of feed gas. A 40-inch pipeline to transport this quantity of gas daily could cost from $2 to $5 million per mile, depending on a variety of factors including whether the gas fields are onshore or offshore, the topography and complexity of the terrain between the fields and the plant (tropical jungle being more challenging than desert), environmental considerations, the remoteness of the facilities, and the availability of pipe-laying equipment. On the other hand, the cost to construct liquid-loading lines and harbor infrastructure is so high that there are clear advantages in locating the plant close to deepwater. These are examples of the trade-offs that play a major role in site selection.

Other issues relate to the environmental and physical aspects of the site. Factors such as water depths, tides, currents, and wave activity directly impact the marine facilities. For plants located in far northern latitudes, the potential for ice buildup in the harbor is a consideration. Onshore soil and seismic conditions and site topography also influence the design of storage and other structures.

The need to build roads, airstrips, and marine docks for construction material and equipment delivery, utilities, and housing for employees can raise the overall cost of liquefaction projects substantially. In remote or challenging areas, the cost of labor also increases, as employees receive higher compensation for harsher work conditions; the costs associated with transportation and employee care are also much higher. In this respect, the Qatar LNG projects gained advantages because of their location close to the gas field and access to infrastructure developed for earlier projects.

Australia's North West Shelf (NWS) and other Western Australian projects, for example, encountered high costs as a result of the remote location, extreme heat, and cost of dealing with difficult Australian labor unions. The Gorgon LNG project suffered from the challenge of building the large liquefaction plant on the remote and environmentally sensitive Barrow Island, coupled with a CO_2 content of 14% for the feed gas. The island, located 50 kilometers off the coast of northern Western Australia in the Indian Ocean, is a Class-A nature reserve, which complicated the construction process. To minimize the impact on the environment, the partners and the EPC team decided to build fully modularized LNG trains and the world's largest CO_2 sequestration project. The CO_2 injection facility reduces the project's overall greenhouse gas emissions by between 3.4 and 4.1 MMt/y at a cost of approximately $2 billion. The EPC contractor also implemented a comprehensive quarantine management plan for Barrow Island to protect against the introduction of nonendemic species, adding costs and delays to the construction process.

At the other environmental extreme, Russia's Sakhalin Island, with its below-freezing temperatures, tundra-like conditions, and ice-prone waters, represents a remote and tough project environment, as well as one of the costliest at the time.[1] The two-train Sakhalin II LNG facility entailed the construction of offshore oil and gas production facilities, construction in extremely cold weather, a pair of 500-mile oil and gas pipelines to bring the production to a nearly ice-free port, and substantial support infrastructure. Theoretically, these disadvantages are balanced by Sakhalin's proximity and lower shipping costs to its target markets of Japan and South Korea.

The Yamal LNG project has an even more challenging Arctic environment. While the Yamal Peninsula holds very large gas reserves, it has some of the toughest working conditions in the world. The land is permafrost for most of the year and boggy for a couple of months during the summer when temperatures are above freezing, making drilling operations a challenge. In the winter, there are only short periods of daylight during which the majority of drilling operations and complex construction need to be carried out. The area is very remote, and there is no local workforce. Moreover, the nearest potential market is located 3,000 kilometers away, and during winter months ice-breaking LNG tankers will have to pass through thick ice entering and exiting the Port of Sabetta, which will service the project. However, the high-cost environment is being cushioned by support from the Russian government, including tax relief, construction of port facilities, and the provision of icebreakers.

Political risk considerations also have to be considered before choosing a potential location. Some locations are more vulnerable to disruptions including sabotage, kidnappings, and threats of terrorism. Yemen LNG, for example, had dealt with sabotage of the feed gas pipeline and shipping through pirate-infested waters, before finally succumbing to closure after the nation slipped into all-out civil war in early 2015.

Liquefaction Process—Technology Overview

Feed gas purification and pretreatment

The natural gas entering a liquefaction facility will often contain heavier hydrocarbons and contaminants that must first be reduced to acceptable levels to ensure satisfactory liquefaction plant performance or meet LNG sales specifications, or both. The level of each depends largely on the characteristics of the production field and the extent of upstream gas treatment.

First, any condensates still present in the incoming feed gas stream are separated and recovered for sale or use as a plant fuel. Acid gas removal comes next, reducing carbon dioxide levels to around 50 parts per million (ppm) in order to prevent freezing in the main cryogenic exchanger. Hydrogen sulfide levels are similarly reduced to below 3 ppm to meet normal sales gas specifications.[2] LNG facilities typically use either Sulfinol or one of the many available amine-type systems to meet these specifications. If the feed gas contains significant amounts of hydrogen sulfide, it may be necessary to convert this compound to sulfur by running the separated acid gas effluent stream through a sulfur recovery unit.

Feed gas leaving the acid gas removal system is saturated with water vapor that must be removed to prevent freezing in the main liquefaction exchanger. This dehydration is achieved in two steps. First, the feed gas is cooled using air or water and a precooling refrigerant to condense much of the water out of the feed gas stream. Then it is passed through a molecular sieve to reduce residual water vapor to very low levels (below 0.1 ppm).

The feed gas may contain mercury. This contaminant, which is always assumed to be present, must be removed down to the smallest detectable amount, about 0.1 microgram per normal cubic meter (Nm^3), because even trace quantities of mercury can cause corrosion of aluminum heat exchanger equipment.[3] Mercury removal is typically achieved by passing the feed gas through a sulfur-impregnated carbon bed, where mercury reacts with the sulfur to form mercuric sulfide.

Finally, any remaining heavier hydrocarbon components, such as pentanes, hexanes, and aromatics (such as benzene), must be separated to prevent freezing in the main cryogenic exchanger. These heavy hydrocarbons are removed by condensing the liquids using a precooling refrigeration process ahead of the main liquefaction process. The separated liquids are typically fed into a fractionation system comprising a succession of distillation columns, typically a de-ethanizer, de-propanizer, and de-butanizer in sequence, with each column stripping out one component from the liquid mix. Any recovered lighter hydrocarbons from this fractionation process, such as ethane, propane, or butane, are used as refrigerant makeup, reinjected into the liquefaction feed gas (up to the gas specification limits), used for plant fuel, or sold as separate products.

Liquefaction processes

Several proprietary processes are marketed today for large-scale baseload natural gas liquefaction plants. These processes fall into the following broad categories:

- Pure refrigerant Cascade process
- Propane precooled mixed refrigerant processes
- Propane precooled mixed refrigerant, with back-end nitrogen expander cycle
- Mixed refrigerant processes
- Nitrogen expander-based processes

All liquefaction process designs are proprietary. The choice of an initial design is critical since future trains will almost certainly utilize the same design. Expansion licenses are generally negotiated at the front end of each EPC contract for a new plant development. Table 9–1 shows the processes used in liquefaction facilities.

Table 9-1. Liquefaction processes at LNG export plants worldwide

Country, Project	Start-up	Process
Alaska, Kenai	1969	ConocoPhilips Opt. Cascade
Libya, Marsa el Brega	1970	APCI APM SMR
Brunei, Brunei LNG	1972	APCI C3-MR
Algeria, Skikda T10	1973	Technip Teal
Abu Dhabi, ADGAS	1977	APCI C3-MR
Indonesia, Bontang A–H	1978–1996	APCI C3-MR
Indonesia, Arun NGL, Phases 1 &2	1977–1986	APCI C3-MR
Algeria, Bethioua GL1Z, GL2Z	1978, 1981	APCI C3-MR
Algeria, Skikda T5P, 6P	1981	Pritchard PRICO SMR
Malaysia, MLNG Satu, Dua, Tiga	1983,1994, 2003	APCI C3-MR
Australia, NWS T1–3	1989	APCI C3-MR
Australia, NWS T4, T5	2004, 2008	Shell PMR
Nigeria LNGT1– T6	1999–2008	APCI C3-MR
Qatar, Qatargas T1–3	1997	APCI C3-MR
Qatar, Qatargas T4–T7	2009–2011	APCI AP-X
Qatar, RasGas T1–T5	1999–2007	APCI C3-MR
Qatar, RasGas T6,7	2009	APCI AP-X
Trinidad, Atlantic LNG T1–T4	1999–2005	ConocoPhillips Opt. Cascade
Oman, Oman LNG, Qalhat LNG	2000, 2005	APCI C3-MR
Egypt, SEGAS LNG (Damietta)	2005	APCI C3-MR
Norway, Snøhvit	2007	Linde Mixed-Fluid Cascade
Egypt, Egyptian LNG (Idku)	2005	ConocoPhillips Opt. Cascade
Australia, Darwin LNG	2006	ConocoPhillips Opt. Cascade
Russia, Sakhalin II	2009	Shell DMR
Equatorial Guinea LNG	2007	ConocoPhillips Opt. Cascade
Indonesia, Tangguh LNG T1,2	2009	APCI C3-MR
Yemen LNG	2009	APCI C3-MR
Peru LNG	2010	APCI C3-MR
Australia, Pluto T1	2012	Shell PMR
Angola LNG	2013	ConocoPhillips Opt. Cascade
Algeria, Skikda Replacement	2013	APCI C3-MR
Papua New Guinea, PNG LNG	2014	APCI C3-MR

(*Source:* Poten & Partners.)

Pure refrigerant Cascade process. The Cascade process is currently used or specified for around 20% of LNG projects worldwide: Kenai in Alaska, the Atlantic LNG facility in Trinidad and Tobago, Egyptian LNG in Idku, Australia's Darwin LNG and Wheatstone, Equatorial Guinea LNG, Angola LNG, the three coalbed-methane-to-LNG projects (Gladstone, AP LNG, and QCLNG) in Queensland, Australia and Sabine Pass on the US Gulf Coast. Starting with Idku, recent plants use an improved and more efficient version of the basic Cascade process, developed by Phillips, called the Optimized Cascade™ process. ConocoPhillips and Bechtel have an alliance agreement for design and construction of liquefaction plants incorporating this process.

The process consists of three separate pure-component refrigerant cycles that provide cooling at progressively lower temperatures in order to liquefy natural gas. As shown in figure 9–2, each refrigerant cycle consists of a compressor, a condenser, an expansion valve, and an evaporator.

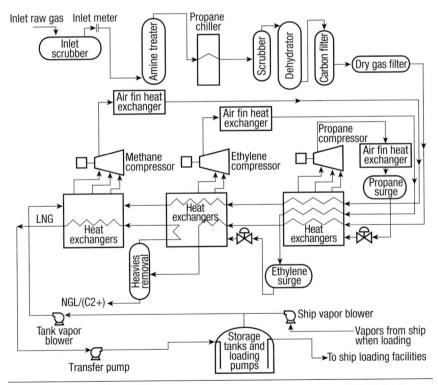

Fig. 9–2. ConocoPhillips Optimized Cascade™ process—Atlantic LNG Train 4. (*Source:* ConocoPhillips.)

After feed gas pretreatment, high-pressure feed gas is first cooled to a temperature of around −30°C using a propane refrigeration cycle. In this cycle, the propane refrigerant is condensed at high pressure, using either air- or water-cooling. The condensed liquid, later expanded to lower pressures and temperatures through a throttling valve, is completely vaporized to cool feed gas and methane refrigerant flows, as well as to condense the ethylene refrigerant used in the subsequent refrigeration cycle. The vaporized propane is then compressed to its initial high pressure to complete the refrigeration cycle.

The cooled feed gas subsequently goes through the remaining two cycles in much the same way. In the second cycle, the feed gas is further cooled to around −100°C through an ethylene refrigerant cycle that is also used to condense the methane refrigerant used in the third refrigeration cycle. Finally, expansion of high-pressure methane refrigerant in the third cycle cools the gas to the liquefaction temperature, around −163°C.

The major advantage of the Cascade process is its simplicity of operation and control due to its use of pure component refrigeration systems. The Optimized Cascade process can also lower capital costs by doubling the methane refrigerant compressor as a fuel gas compressor, using nonproprietary brazed aluminum plate-fin heat exchangers for liquefaction service, and importing makeup refrigerant rather than producing and blending these on site.

However, pure-component refrigeration processes have somewhat lower thermodynamic efficiencies than precooled mixed-refrigerant processes, increasing the compression power and fuel gas consumption needed for natural gas liquefaction. Process efficiencies can be improved by using additional refrigerant steps, but this increases the complexity of the equipment and piping needed, increasing plant investment costs. The unequal distribution of horsepower load among the three refrigeration services (propane, ethylene, and methane) also complicates compressor/driver selection and maintenance requirements. Finally, importing the refrigerant makeup—particularly ethylene—increases operating costs.

Propane-precooled mixed-refrigerant (C3-MR™) cycle. This process—the workhorse of today's LNG industry—is used in over 60% of the world's completed trains, including projects in Algeria, Abu

Dhabi, Brunei, Indonesia, Malaysia, Qatar, Australia, Nigeria, Egypt (Damietta), Yemen, Peru, Russia, Papua New Guinea, and Oman, as well as prospective projects in the United States. Each C3-MR project has unique features depending on the technology available at the time of design and construction and local environmental characteristics, but the fundamental concept does not vary. New projects being constructed or developed also use variants of this process.

The C3-MR system uses a multicomponent refrigerant—usually nitrogen, methane, ethane, propane, butane, and pentane—to condense and evaporate natural gas in one cycle over a wide range of temperatures. The mixed refrigerant used is Air Products & Chemicals Inc. (APCI) proprietary multicomponent refrigerant (MCR).

Dry, treated gas is first precooled to around –30°C using propane refrigerant. The precooled (–30°C) feed gas is then sent to the main cryogenic heat exchanger (MCHE), where it is condensed and then subcooled at elevated pressures. The subcooled LNG leaving the MCHE is then flashed to near-storage tank pressure, cooling the LNG to approximately –161°C and releasing a nitrogen-rich stream that can be used for fuel gas.

The APCI MCHE consists of thousands of small-diameter, spiral-wound tubes extending the entire length of the heat exchanger. The precooled natural gas and MCR flow upward through these tubes, in which they are cooled and condensed. As the MCR emerges at two levels, it goes through pressure-reducing throttling valves and passes into the shell of the MCHE, where it flows downward over the outside of the tubes, vaporizing to provide refrigeration to the tube-side fluids. The low-pressure MCR vapor leaves the bottom of the MCHE at about –30°C and is then recompressed to about 650 psi in a multistage compressor, with water or air cooling provided between each compression stage. The MCR flow is then further precooled using the propane-precooling refrigerant before reentering the tube side of the MCHE (fig. 9–3).

The main advantages of the C3-MR process are its proven technology and high efficiency. C3-MR liquefaction systems have been widely used for nearly 40 years in a wide variety of process and environmental settings. Start-up and operation of these facilities have generally been free of major problems, and plants have generally exceeded design throughputs on a continuous basis.[4] The C3-MR process also achieves high efficiencies by adding the propane precooling stage for both feed gas and the mixed-refrigerant loop, allowing the MCR vaporization temperature curve to closely match the natural gas liquefaction curve. This, of course, comes at an extra equipment cost, but the operational gains have proved this additional investment to be justified.

A disadvantage of the MCR process is limited flexibility to shift refrigeration load between the propane and mixed-refrigerant cooling circuits, although some of this problem has been alleviated in recent designs by having propane and MCR compression stages on a single turbine driver to better balance the available turbine power.

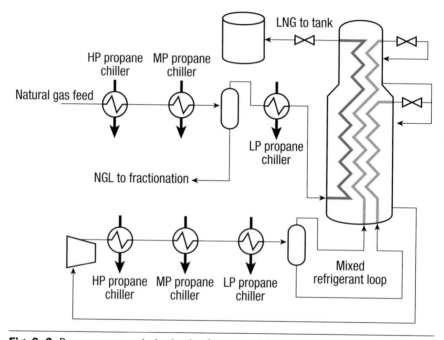

Fig. 9–3. Propane-precooled mixed-refrigerant (C3-MR) process. (*Source:* Air Products.)

The APCI AP-X™ process. The drive toward increased liquefaction economies of scale led APCI to develop a variant of their C3-MR process that increases the liquefaction train capacity from 5 MMt/y to 8 MMt/y (7 Bcm/y to 11.2 Bcm/y) while still utilizing established MCHE designs and compressor configurations. This new AP-X process, introduced in 2001, adds a third nitrogen expander (N_2)

refrigeration cycle to the back end of the C3-MR process's propane (C3) mixed refrigerant (MCR) cycles (fig. 9–4). This N_2 cycle takes the LNG subcooling duty off the MCR cycle, increasing the natural gas capacity and reducing the refrigeration loads on the first two cycles. While nitrogen expander cycles are new to baseload LNG, they have been widely applied in air separation applications and peak shaving LNG facilities. AP-X is employed in the six 7.8 MMt/y Qatari megatrains: Qatargas II (two trains), RasGas III (two trains), Qatargas III (one train), and Qatargas IV (one train). While this offers economies of scale in the liquefaction unit, there are offsets, which may defeat these apparent economies, such as the need to custom-fabricate many of the components at a scale that is not available "off the shelf," or which limits the number of vendors. Since the Qatari megatrains, no projects have been developed using this scale of liquefaction process.

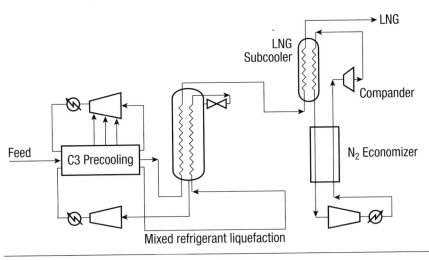

Fig. 9–4. APCI AP-XTM process. (*Source:* APCI.)

Other mixed-refrigerant processes. The efficiency and reliability of the C3-MR process have led to the development of several other mixed-refrigerant processes. These variations aim to eke out additional process efficiencies by using mixed refrigerants rather than propane for the precooling stage, and also introduce a wider variety of heat exchanger equipment, compressor, and driver configurations to reduce equipment costs, increase vendor competition, and in some cases, reduce carbon dioxide emissions.

Shell Dual Mixed Refrigerant (DMR) process. Shell has introduced both propane-precooled and mixed-refrigerant precooled variations on the APCI C3-MR process that are not dependent on APCI's proprietary spiral-wound MCHE exchanger technology. The propane precooled design was used for Trains 4 and 5 at Australia's NWS LNG project and Pluto, while the Dual Mixed Refrigerant (DMR) process is used at Russia's Sakhalin II project and Australia's Prelude FLNG venture.

Axens Liquefin™ DMR process. The Axens process liquefies natural gas in plate-fin heat exchangers (PFHEs) using a two-stage DMR process (fig. 9–5). Using PFHEs rather than spiral-wound exchangers makes the process more expandable, increases the number of available exchanger suppliers, and eliminates the need to separate refrigerant phases in the exchanger, thereby reducing equipment costs. The DMR design enables refrigeration loads to be readily balanced across the two similar refrigerant compressors, reducing overall power requirements, increasing the applicable range for the more-robust centrifugal compressor designs, and reducing spares costs. No baseload Liquefin facilities have been built to date, although FEEDs incorporating this design have been performed for several potential projects.

Statoil-Linde mixed Fluid cascade (MFC) process. The Statoil-Linde process liquefies LNG using three mixed-refrigerant cycles (fig. 9–6). The first precooling cycle uses a mixture of ethane and propane refrigerants in a PFHE, while the intermediate liquefaction and final subcooling stages take place in Linde's spiral-wound aluminum heat exchangers. Refrigeration power can come from three turbines or from large electric motors. The MFC liquefaction technology is employed at Norway's single-train 4.3 MMt/y (6 Bcm/y) Snøhvit project.

Nitrogen expander processes. The APCI AP-X™ process is not the only design incorporating a nitrogen expander-based refrigeration process. Several nitrogen expander-based designs are available from process vendors such as Black & Veatch and ABB Lummus that perform the entire liquefaction process. Although these processes are not as scalable as the Cascade and mixed-refrigerant alternatives, they are appropriate for small baseload, offshore, and peak shaving applications because of their simple, robust, and compact designs.

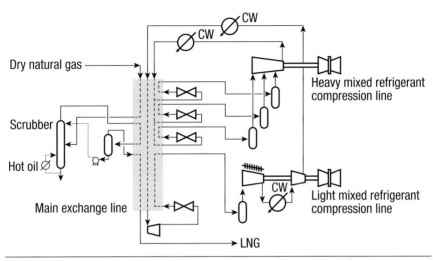

Fig. 9–5. Axens Liquefin™ DMR process. (*Source:* Axens, BP.)

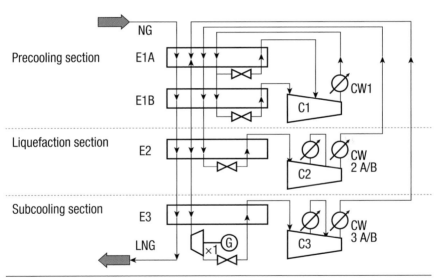

Fig. 9–6. Statoil-Linde Mixed Fluid Cascade (MFC) process. (*Source:* Linde.)

High-pressure nitrogen vapor from a nitrogen compressor is cooled against water and then further cooled and expanded in a series of expanders and heat exchangers to provide refrigerant flow at the required temperatures and pressures. After cooling and liquefying the natural gas flow, the low-pressure nitrogen is partially compressed using energy from the expanders, so as to reduce the power required to return the circulating gas to high pressure in the main nitrogen compressor.

Driver technology

The choice of drivers for the refrigeration compressors has been a source of ongoing debate for several years and seems likely to continue. The very earliest plants used steam turbines to drive the compressors, with the notable exception of the Kenai project, which used gas turbines (more efficient in Alaska's colder ambient temperatures). Steam turbines offered the advantage of reliability, though the need for vast amounts of treated cooling water was a major drawback. By the late 1980s, as industrial gas turbines became more broadly accepted in a variety of applications, especially in power generation, these became the drivers of choice. However, there are several differing approaches—Atlantic LNG uses multiple, smaller gas turbines in a very flexible arrangement, while the Nigerian and Qatar projects use much larger turbines coupled to multiple compressors. A further variance is the use of electric drive compressors with the electric power provided from gas turbines operating as power generators. Air cooling has generally replaced water cooling. While air cooling has a much larger footprint, the technology appears more reliable and avoids the need for complex, maintenance-intensive water-cooling systems. In the United States and Canada, environmental considerations make it essentially impossible to achieve approval for once-through, water-cooled plants. Due to site limitations and environmental considerations, the Freeport LNG project is planning to use all-electric drivers with power supplied from the local grid. This reduces the plant footprint and the emissions associated with running multiple gas turbines.

Utilities

As liquefaction plants are often located in remote regions, the provision of utilities is often a critical feature of the plant design. The plants usually have their own power generation facilities, with 100% redundancy to minimize operational interruptions. Water supply may also be a factor, especially if water-cooled equipment is used in the process.

LNG storage tanks

Although not an integral part of the liquefaction process, LNG storage plays a critical role in project cost and scheduling. LNG tanks are large and expensive, with their costs potentially representing up to 10% of the total installed plant value. Tank construction is often on the project's critical path for construction, but technological advances and decades of experience with established tank designs have helped shorten LNG storage tank construction times.

The required storage capacity for a particular facility is largely governed by planned LNG tanker size, while additional capacity is also required to allow LNG tanker scheduling flexibility and to handle planned and unplanned plant outages. In addition, LNG tanks must provide for minor allowances to maintain minimum liquid suction head requirements for the in-tank LNG pumps. The result typically is the installation of storage capacity equivalent to twice the cargo capacity of the largest LNG tanker expected for the initial trains of the liquefaction plant; for instance, a project that uses 138,000 m^3 LNG tankers would typically include approximately 276,000 m^3 of storage volume. Individual storage tank sizes have increased over the years, from 36,000 m^3 at Alaska's Kenai facility to 200,000 m^3, with 240,000 m^3 tanks in the design and permitting stage. As expansion trains are added, LNG storage capacity tends to be added at about half the level provided for the initial trains.

There are several types of LNG storage tanks. The choice of storage tank design is largely determined by safety and operational considerations, which are based on plant location, layout constraints, engineering design standards, and code requirements. The two broadest classifications are aboveground and inground storage. In this section, we focus entirely on the aboveground tank design because it is more commonly used in liquefaction plants. Inground storage tanks have only been used at some import terminals, particularly in populous settings such as Tokyo Bay, for both safety and aesthetic reasons. LNG tanks at liquefaction plants and import terminals share common designs.

All LNG storage tanks have a double-walled design, with insulation inserted between the two walls to maintain the cryogenic temperature of the LNG. Tank designs are typically classified as single-containment, double-containment, or full-containment.

Single-containment tanks. Single-containment refers to a freestanding, open-top inner tank made of 9% nickel steel with a carbon steel outer tank (fig. 9–7). A layer of several feet of perlite insulation is sandwiched between the two tanks. The base of the inner tank rests on rigid foam blocks for insulation, then on the foundation. The choice of foundation, dictated by site soil conditions, can have a ring wall, pile, or stone column design. The tank base usually includes a heater system to maintain the ground at a constant temperature and prevent tank disturbance due to frost heaves. The tank has a steel roof designed to contain gas vapor and support a suspended ceiling that insulates the top surface of the inner tank. The outer steel tank wall will not contain any LNG in the event of a breach in the inner tank. This type of tank, therefore, must have external secondary containment, usually a berm area with sufficient capacity to contain the entire contents of the tank plus a margin of safety. This tank design is the least expensive. Single-containment tanks have operated worldwide without major incident for more than 30 years in locations that have sufficient plot area for the secondary containment berms and to allow for significant separation between the tank and adjacent process facilities and other tanks.

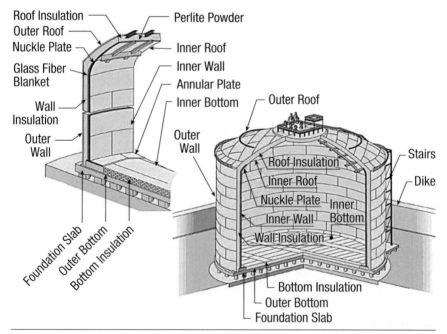

Fig. 9–7. Single-containment LNG storage tank. (*Source: PCI Journal*, Fall 2013.)

Double-containment tanks. Double-containment tanks are similar to single-containment systems, except that the outer tank is capable of containing liquid spills in the event of a breach in the inner tank wall (fig. 9–8). This tank design has a freestanding 9% nickel steel inner tank and an outer tank made of either a prestressed reinforced concrete or poured-in-place reinforced concrete strengthened by an earthen or rock embankment. Double-containment tanks require less plot area than single-containment designs because of their concrete outer wall. However, the roof is still constructed of steel and will not contain vapor produced by an inner tank failure. The approximate cost of a double-containment tank is around 40% higher than for a single-containment design. Steel and nickel cost increases have driven prices for this tank design far nearer to the cost of the all-concrete full-containment design, as discussed below.

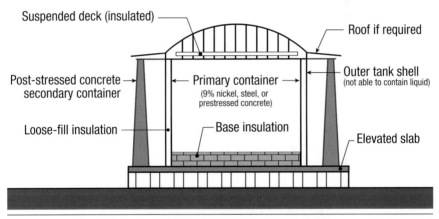

Fig. 9–8. Double-containment LNG storage tank. (*Source: PCI Journal*, Fall 2013.)

Full-containment tanks. Full-containment tank designs add a concrete roof to the double-containment tank's concrete outer walls (fig. 9–9). This roof can withstand higher pressures and lower temperatures than the steel tank design, providing containment for any vapor produced by a breach in the inner tank. Full-containment tanks provide the greatest design integrity and allow the closest spacing between tanks and process equipment when land area is constrained. However, this is the most expensive aboveground tank design, with a cost about 50% higher than for a single-containment design.

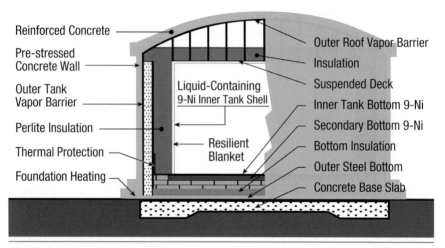

Reinforced Concrete

Pre-stressed Concrete Wall

Outer Tank Vapor Barrier

Perlite Insulation

Thermal Protection

Foundation Heating

Liquid-Containing 9-Ni Inner Tank Shell

Resilient Blanket

Outer Roof Vapor Barrier

Insulation

Suspended Deck

Inner Tank Bottom 9-Ni

Secondary Bottom 9-Ni

Bottom Insulation

Outer Steel Bottom

Concrete Base Slab

Fig. 9–9. Full-containment LNG storage tank. (*Source: PCI Journal*, Fall 2013.)

Membrane tank. Membrane tanks are similar to the designs found in membrane containment LNG tankers. Primary LNG containment is provided by a flexible stainless steel membrane, supported by a layer of insulation mounted on an outer prestressed concrete wall (fig. 9–10). This outer concrete wall and roof can contain both the liquid and vapor from a leak in the primary membrane. Membrane tanks have similar or slightly higher costs than free-standing full-containment designs. Very few membrane tanks are in service, as they are perceived to be less durable than other tank designs.

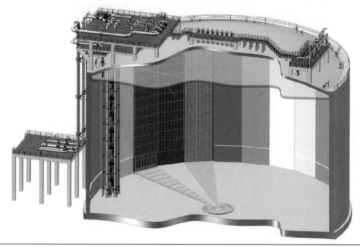

Fig. 9–10. Membrane LNG storage tank. (*Source:* BOUYGUES Travaux Publics.)

All-concrete tank. Developed in the 1970s, these tanks comprise an inner and outer prestressed, post-tensioned concrete wall with insulation between. The LNG comes in direct contact with the inner wall, which is fabricated from special concrete. The design fell out of favor in the 1980s, though there has been discussion of reintroducing it in projects in high-cost environments, where it can reduce both the material and, more importantly, on-site labor by the use of prefabricated concrete panels.

LNG Loading Jetty

LNG jetties are designed to berth and load the LNG tankers. Tanker mooring is typically accomplished with the assistance of three or four tugboats. Once it is moored and the loading and vapor return arms on the jetty are connected, loading pumps in the facility's storage tanks transfer LNG into the LNG tanker's cargo tanks through a cryogenic piping system supported by the loading jetty. Product loading usually takes between 12 and 13 hours for a 138,000 m³ LNG tanker. However, 12 additional hours are usually required for berthing, loading arm connection/disconnection, cargo measurements and customs clearance. LNG tankers that are entering service or have been waiting for extended periods to load cargoes may arrive with LNG cargo tanks at temperatures higher than the LNG to be loaded. To load the LNG tanker, the cargo tanks must first be cooled down, a process that must be performed at very low loading rates taking up to 12 hours and generating large quantities of boil-off gas. Boil-off gas or vapor generated during loading is returned to the shore facilities via a vapor return arm and pipeline and reliquefied. The practical limit for jetty utilization at a loading port is around 60%, accounting for scheduling variances, weather, and regulatory restrictions. This results in a limit of about 210–220 loadings a year per jetty, the equivalent of about 13 MMt/y (18.7 Bcm/y) of throughput if larger LNG tankers are used.

Project Development and Execution

LNG liquefaction projects require substantial capital investment, a high degree of technical expertise, and significant project management skills. Complicating the endeavor is the need for coordination between the development of the upstream production facilities, the liquefaction facility, the shipping, and where necessary, the regasification terminal. This makes the construction of a baseload liquefaction facility the domain of major engineering/construction firms equipped to perform such complex projects in remote locations worldwide. A phased approach is typically employed in developing the facility (fig. 9–11).

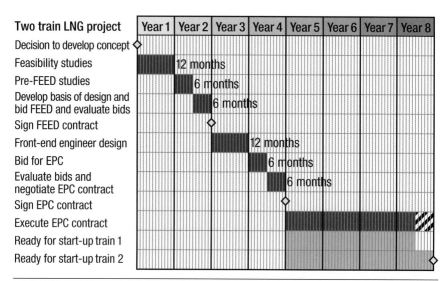

Two train LNG project	Year 1	Year 2	Year 3	Year 4	Year 5	Year 6	Year 7	Year 8
Decision to develop concept	◇							
Feasibility studies	12 months							
Pre-FEED studies		6 months						
Develop basis of design and bid FEED and evaluate bids			6 months					
Sign FEED contract			◇					
Front-end engineer design				12 months				
Bid for EPC					6 months			
Evaluate bids and negotiate EPC contract					6 months			
Sign EPC contract					◇			
Execute EPC contract								
Ready for start-up train 1								
Ready for start-up train 2								◇

Fig. 9–11. LNG export facility development phases. (*Source:* Poten & Partners.)

The development of a project starts after the identification and confirmation of sufficient gas reserves. A broad-based feasibility study follows in order to determine the conceptual design and identify potential plant locations. This study typically includes a preliminary economic analysis to guide sponsors in deciding whether to pursue development. It also addresses adequacy of gas reserves, environmental issues, and potential obstacles. The steps involved in project development are as follows.

Basis of design (BOD)

BOD follows the feasibility study. At this stage, the conceptual design is finalized, and the fundamental engineering parameters are identified. A more rigorous analysis is performed on expected feed gas quality, LNG supply parameters, storage, marine facilities, utilities, modules spacing, and safety. Contractors may be asked to use their own engineering design standards or adopt the standards of the sponsors. The BOD is a necessary step either done before issuing a formal FEED request to potential contractors or as phase 1 of the FEED. The BOD is also described as pre-FEED.

Front-end engineering design (FEED)

The FEED will further develop the design, including detailed specifications and layout, in order to prepare a more accurate cost estimate and the basis for issuing tenders for competitive EPC bids. The FEED identifies equipment and provides a detailed design of the facility. Depending on the contracting philosophy, critical items may be preordered at this stage to accelerate the schedule. The FEED contractor usually prepares engineering documents and bid packages to a sufficient level of detail to be used in the EPC bidding.

The FEED deliverable can greatly influence the overall cost of the project, as the liquefaction process and storage capacity are determined during this phase. Historical data suggest that the largest cost-saving opportunities are captured during the FEED phase. FEEDs are funded by the project sponsors. There is evidence that funding multiple FEEDs may result in more competitive bidding and improved plant design, but this approach places additional demands (and costs) on the project management team and extends the schedule.

Engineering, procurement, and construction (EPC)

EPC contractor selection generally takes several months and culminates with the selection of the contractor through either open bids or direct negotiations. Typical EPC contracts are structured as: lump-sum turnkey (LSTK), where the EPC contractor bids a lump sum to deliver a plant built to the FEED design by a target date; on an open-book basis, where the contractor provides the owner with detailed insights into the various cost components of the project, and a negotiated overhead and profit margin; or as a fully reimbursable project with the owner controlling all aspects and the EPC contractor(s) acting as subcontractor(s) to the owner under a negotiated cost overhead and profit reimbursement. The LSTK approach transfers most of the risk to the contractor, but has become harder to implement as project costs have mounted. The open-book model shares more of the risk between the contractor and the owner. A further variation is an EPC contract that starts as an open book and converts to an LSTK during the construction process once the risks can be clearly defined and allocated. Due to unique cost issues in Western Australia, the owners have been forced to take almost all of the risk, and projects have been executed as fully reimbursable.

The EPC contractor is responsible for bringing a project to completion and start-up of the facility according to the sponsors' design and specifications. The EPC award is followed by the process of detailed engineering and procurement. This colossal effort requires substantial experience and can take more than two years. Procurement of proprietary heat exchangers, turbine drivers, and process compressors are particularly important, as vendors for these specialty items may require lead times in excess of two years, often creating scheduling and cost problems.

Shell, a company involved in numerous LNG projects, has pushed for standardization in an effort to substitute large proprietary equipment items with smaller standard ones to promote competition and shorten the overall schedule (see the section on standardization later in this chapter). The EPC contractor oversees construction by supervising a number of subcontractors and suppliers, and manages a construction workforce that could be as large as 10,000 workers.[5] Knowledge of local resources, regulations, and local culture can help immensely in navigating a potentially difficult construction period.

Project risk allocation

In addition to the EPC risk allocation described above, final transfer of the project risk from EPC contractor to the project sponsors occurs at the final-acceptance stage. Delays can have a detrimental effect on the overall LNG chain, since the plant construction is designed to coordinate with upstream development, LNG tanker construction, regasification plant commissioning, and commencement of contractual obligations under the SPAs. Delays in one part of the chain can wreak havoc on the entire development plan by delaying shipment of LNG and the flow of revenues. Furthermore, the transfer from finish-construction to start-operation takes on greater significance in a project finance environment, where it signals transition from recourse to nonrecourse financing for the sponsors.

Liquidated damages (LDs)

The importance of a timely finish has induced project sponsors to incorporate liquidated damages (LDs) as well as early completion and performance bonus provisions into construction contracts. LDs are compensation the contractor has to pay to the sponsors in case of an unexcused failure to meet critical milestones, primarily the in-service date, or the failure of the plant to meet its design output and operating performance. The LDs are usually designed to cover interest, loss of revenues, and overhead expenses incurred by the sponsors during the delay period or during periods of below design performance. LDs do not cover delays due to events of force majeure, which stem from circumstances beyond the contractor's control. Force majeure is best mitigated through insurance. The design and placing of appropriate insurances covering project construction also require detailed and expert review and negotiation.

Finish-construction milestone

Reflecting the importance of the finish-construction milestone, the EPC contract incorporates detailed completion and testing procedures. Typically, mechanical completion occurs upon completion of construction. It is followed by operational completion, which integrates punch-list items and commissioning activities. Final acceptance follows a successful completion of all test runs, and signals final transfer of responsibility to the project sponsors. Warranties start on achieving this milestone.

Debottlenecking

After the plant is operating, maintenance and debottlenecking activities have a great impact on future production. Plant debottlenecking typically adds 5%–10% capacity by upgrading or modifying critical equipment items that limit production. Debottlenecking efforts are typically easy to justify, as the costs are relatively small (on the order of $20 million to $50 million) compared to the revenues from the incremental production.[6] Earlier plants, with their more conservative designs, offered much greater opportunities for debottlenecking, but as designs have improved, many of these gains have been built into the initial trains.

Expansion trains

Sponsors can also take advantage of economies of scale by leveraging existing infrastructure through expansions. The cost to expand a liquefaction plant is based on the original project development plan. If surplus capacity is built into the initial design and construction of major components (storage tanks, marine facilities, gas supply pipelines, and utility systems) then these facilities may not need to be replicated for an expansion train, providing significant economies. Before expansions can be fully considered, additional proven gas reserves must be available to the project; however, it is not unusual for the expansion process to be started before the initial trains are operational, thereby avoiding the cost of demobilizing and remobilizing the EPC contractor, the sponsors' project management team, and the construction workforce.

Recent Trends

Larger trains

As a result of bigger individual trains and larger venture production capacity, economies of scale have enhanced the competitiveness of LNG in international energy markets. In the 1980s and the early 1990s, the largest LNG production train was about 2.5 MMt/y (3.5 Bcm/y). By the mid-to-late 1990s, 3 MMt/y (4.2 Bcm/y) became the benchmark, rising to 4 to 5 MMt/y (5.6 to7 Bcm/y) by the turn of the century. The LNG megatrains in Qatar, commissioned during the first decade of the 21st century, have nameplate capacities of 7.8 MMt/y (10.9 Bcm/y). These so-called megatrains still represent the largest size benchmark. This progression in liquefaction train size is illustrated in figure 9–12.

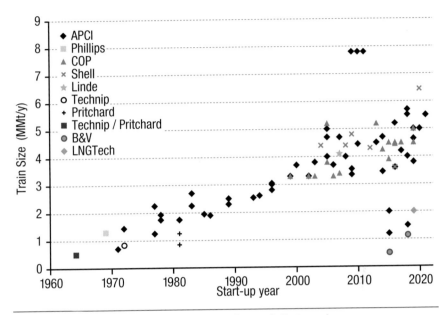

Fig. 9–12. LNG train size growth. (*Source:* Poten & Partners.)

One limitation to train size has been the feasibility of manufacturing larger heat exchangers and transporting them from the manufacturing site to the field. For the spiral-wound heat exchangers used by APCI, the transportation restrictions from the company's manufacturing location to the sea have constrained the feasible exchanger diameters to 18 to 20 feet. However, APCI overcame this constraint with the AP-X megatrain process, not by increasing the length and diameter of the exchanger, but by adding an additional refrigeration cycle in a separate process unit that reduced the refrigeration duty of the existing exchanger design. As a result, the AP-X process can produce 7.8 MMt/y (10.9 Bcm/y) of LNG by using spiral-wound exchangers similar to those found in their 5.0 MMt/y (7 Bcm/y) C3-MR train designs.

LNG plants have progressively employed ever-larger gas turbine drivers to power refrigerant compressors as train sizes increased. Earlier gas-turbine-driven LNG plants used smaller, GE Frame 5 (or equivalent) dual-shaft gas turbines as compressor drivers. LNG train capacities were then increased to 3.3 MMt/y (4.6 Bcm/y) or more for the C3-MR liquefaction process by stepping up to GE Frame 6 gas turbine drivers for the propane-precooling refrigeration cycle compressor and GE Frame 7 gas turbine drivers for the mixed-refrigerant compressor. This was followed by the use of GE Frame 7 gas turbine drivers for both the propane and mixed-refrigerant cycles, taking individual train capacity up to 4 to 5 MMt/y (5.6 to 7 Bcm/y). The Qatar megatrains have taken this trend further, using GE Frame 9 gas turbine drivers for the propane, MCR, and nitrogen cycles. Large electric motors are used as compression drivers on Norway's Snøhvit project.

However, there is a downside to increasing plant size. Beyond certain sizes, equipment such as gas turbines, compressors, cryogenic valves, and other components of the plant may not be readily available, except through custom specification and ordering. In addition, the number of vendors may be more limited, and regular maintenance cycles take longer on the bigger machines, losing valuable production. These considerations have contributed to a swing back to somewhat smaller trains in the 4–5 MMt/y size following the completion of the Qatari projects.

While megatrains are technically feasible, establishing their commercial viability in specific locations is a far more complex issue. They require a larger natural gas reserve base than typical projects, and place more capital at risk. Finding sales outlets for megatrains also constitutes a greater challenge than for traditional projects, as volumes are substantially larger. Only a few buyers may be willing to make large purchase commitments from single trains, as LNG buyers typically seek multiple smaller LNG contracts to diversify their supply sources and match the requirements of newly liberalizing and competitive LNG markets. Thus, the trend with the Qatari megatrains has been to have the sponsors emerge as the buyers for the entire train production and assume the risk of placing the LNG in the market after project completion.

Cost trends

The industry has grown rapidly in recent decades. By 2014, there were 53 operating plants and 19 projects under construction, compared with only 11 plants operating in 1995. One major reason for the industry growth rate has been the substantial drop in liquefaction unit costs and the resulting increase in cost competitiveness for LNG relative to other fuels and pipeline gas through 2003. According to the International Energy Agency (IEA), the average unit investment for a liquefaction plant in 2003 was under $200 per ton per annum (tpa), compared with $350/tpa in the 1970s and 1980s (fig. 9–13).[7] Since then, liquefaction plant construction costs have increased dramatically, with some Australian projects reaching $2,500/tpa. However, higher energy prices, with crude oil at $100/bbl or more, supported the financial viability of new LNG projects.

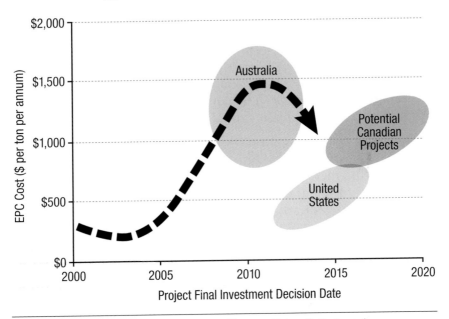

Fig. 9–13. Liquefaction plant unit costs. (*Source:* Poten & Partners.)

The cost reductions observed up until 2003 were due to many factors. An early push away from steam turbines (used, for example, in Algeria, Indonesia, and Brunei) toward gas turbines (in Australia's NWS and Malaysia's LNG Dua and all subsequent projects) and increased train capacities helped reduce costs, as gas turbines are generally cheaper than steam turbines, which typically require boilers, boiler feed water makeup facilities, and steam condensers. The transition from water-cooling (Brunei) to air-cooling (at NWS) also allowed further cost reductions. In the LNG storage sector, the use of larger-sized storage tanks helped bring down costs. Early tank designs were typically smaller than 40,000 m³. In the 1990s, tank volume rose to 150,000 m³, and today, tanks with capacities of 200,000 m³ have been built. Further cost reductions resulted from increased engineering experience, reductions in design margins, and a competitive FEED and EPC bidding environment. Increased application of larger turbine drivers to reap economies of scale allowed less conservative designs to emerge and costs to fall further.

However, recent cost increases across several key raw materials—such as carbon and stainless steel, nickel, and concrete—combined with strong demand for qualified engineering contractors, project personnel, and skilled labor (in addition to labor unions) have reversed this downward cost trend. Currently, EPC costs greatly vary by location. For example, brownfield projects on the US Gulf Coast could be as low as $500 per tpa, benefiting from existing LNG import terminal infrastructure and reaching as high as $2,500 per tpa for Australian projects, where host country labor policies are having a major impact on LNG plant capital costs.

The effect that rising costs will have on LNG project feasibility remains to be seen. While LNG facility costs have risen sharply, the accompanying increase in global oil price and a seller's market for LNG helped to maintain LNG production expansion momentum. The fall in oil prices to the $50–$60 per barrel range in early 2015, if prolonged, combined with constrained equipment and construction resources, could bring this capacity surge to a halt and call into question the economics of some of the more costly projects under construction. It could also put downward pressure on EPC costs.

Standardization and use of off-site fabricated modules

The critical path for construction of an LNG plant is usually driven by the delivery of long-lead items, such as heat exchangers, compressors, large turbines, and LNG storage tanks. Manufacturing delays thus can have dire consequences for cash flow and revenue collection. Substitution of large pieces of proprietary equipment by smaller, more standardized ones can benefit both schedule and cost by introducing more competition between suppliers. When the vendor base is larger and the equipment is less specialized and complex, the costs are generally lower and the delivery time shorter. This rationale was one of the drivers for Shell's DMR process, and Shell continues to drive standardization initiatives further.[8]

This concept is almost opposite to the megatrain concept where economies of scale are achieved through pushing critical components to their design limits. Standardization allows for repeat design and ease of operation, as equipment is not pushed to its limits but operates well within proven technical parameters. Design is based on selected standard equipment, rather than engineering overoptimized equipment specifically to accommodate a particular design strategy.

Standardization has already left a positive mark on the LNG construction industry. LNG storage tanks, where design and construction have been largely standardized, have experienced some reduced costs. Shell is pushing this concept further by prefabricating pipe and preassembling skid-mounted units for compressors and pumps. Some of these modular fabrication techniques were applied to Train 5 at Australia's NWS project. Statoil took this concept even further at Norway's Snøhvit liquefaction plant, building many of the process units on barges and towing the completed units to the site. However, much of the motivation in this case was to minimize the effect of Arctic conditions on the construction schedule rather than to provide a broadly applicable construction approach. Ironically, part of the delay in the Snøhvit project came about as a result of delays in shipyard fabrication of the liquefaction modules, illustrating the risks associated with innovative techniques.

Modularization is being widely adopted for recent Australian projects such as the Pluto LNG project completed in 2012. Project sponsor Woodside and the EPC consortium opted to build the world's first complete modular LNG processing facility in order to offset Australia's labor shortage. The train, fabricated in modular form in Thailand, was shipped to the site in 264 modules. The strategy suffered the inevitable glitches associated with a first-of-its-kind, innovative strategy, which contributed to delays and cost overruns. Moreover, lower-than-expected productivity in assembling the modules and problems with the modules contributed to project delays and rising costs. Modular construction is also being used at the three CBM-to-LNG projects in Queensland, Australia, and at the Gorgon LNG plant on Barrow Island off Western Australia, where this construction scheme is far less intrusive on the environmentally sensitive island. However, it is unclear based on these examples whether modularization will in fact offer any meaningful savings, other than in very constrained labor environments such as Australia and perhaps British Columbia.

One drawback to extensive off-site modularization is that most LNG-exporting countries would rather see in-place construction techniques that tend to generate more local construction jobs. Certain components of the plant, the tanks being the most obvious, cannot be modularized, which also limits the value of this approach. Modularization also results in more materials and labor hours than stick-built construction due to the steel structures needed and interfaces created. However, there are cost savings if low-cost module yard labor is substituted for higher-cost site labor.

Offshore liquefaction facilities

The increased demand for LNG has led to the conceptual development of several new fixed or floating technologies aimed at exploiting stranded offshore gas reserves either too small or too remote for conventional land-based liquefaction facilities (fig. 9–14). The gravity-based structure (GBS) concept is intended for shallow waters and provides an artificial offshore island for LNG production and storage. The concept involves prefabricating massive concrete structures at an onshore location and then floating and towing the finished facilities to the installation location, where they are flooded and sunk to rest on the seafloor. GBS designs allow for height above water level to provide for sufficient freeboard for expected sea conditions at the facility location.

Fig. 9–14. Offshore liquefaction: Prelude FLNG. (*Source:* Shell.)

GBS technology for building offshore oil and gas production facilities has been proven and applied to a multitude of projects worldwide, particularly in the North Sea. The challenge remains to apply these technologies cost effectively to the LNG liquefaction process. Compared to oil and gas GBS production structures, LNG facilities will require much larger footprints to be feasible in a much narrower range of water depths. As a result, no GBS liquefaction projects have been taken to a planning and permitting stage, let alone deployed.

While GBS technology is stalled, floating LNG (FLNG) has gained traction. As of 2014, there are five floating liquefaction units under construction and another dozen or more announced projects under study. Those units now under construction are Shell's Prelude LNG in Australia, Petronas's PFLNG-1 and PFLNG-2 in Malaysia, EXMAR's Caribbean FLNG, a floating liquefaction, storage, and regasification unit (FLSRU), which was planned for deployment on the coast of Colombia, and Golar's conversion of the *Hilli* from an LNG tanker to an FLNG unit. In the designs for FLNG, the LNG storage tanks are enclosed within the structure. The liquefaction modules that rest on top of the structure must be adequately spaced to ensure safe operations.

The FLNG units are being customized to suit the individualized needs of each project. For example, the two Petronas units, which will be used to process gas at smaller fields, will have liquefaction capacities of 1.2 MMt/y and 1.5 MMt/y. Shell's Prelude LNG, which will be used off the coast of Australia, will have a liquefaction capacity of 3.6 MMt/y with the gas processing plant and liquefaction facilities mounted on the deck of the vessel and LNG and LPG storage in the hull. The Prelude LNG facility will be the largest floating structure ever built, the length of four soccer fields and the weight of six aircraft carriers. By contrast, Caribbean FLNG will have just 0.5 MMt/y of liquefaction capacity, but the unit's regasification capabilities will allow for importation of LNG during periods when alternative power sources are in low supply (for example, during summer months when low rainfall curbs hydropower). The Colombian project also featured a floating storage unit (FSU) stationed alongside the barge. However, due to a failure to secure sufficient upstream reserves, the Columbian project has been canceled, and the future of the barge is uncertain.

When comparing FLNG to conventional onshore LNG lique-faction plants of similar size and scope operating or being built throughout the world, a major difference is determined by the cost and productivity of labor used to build the plant. In Australia, for example, labor is in short supply, and restrictions prevent the temporary importation of labor. Promoters of FLNG suggest that, since their units can be built in a shipyard outside of the country with the gas resource, the labor cost savings associated with FLNG units can result in a lower overall cost relative to onshore facilities. In addition, FLNG units do not require a pipeline connection to shore or a port to load LNG carriers, both of which can be major cost components for onshore LNG export projects.

Over time, repeatable construction of FLNG units in a shipyard environment could eventually lower costs as experience is gained. However, except for Prelude, these projects generally deploy liquefac-tion technologies based on nitrogen expansion liquefaction, which has not been proven in baseload service.

The feasibility of regular ship-to-ship cargo transfers at the production field is a key issue and may limit the areas where such FLNG vessels can be located. Handling complex gas streams and liquids can also complicate design and operational considerations. As of this writing, Excelerate had just announced that it was suspend-ing development of its FLNG project in Port Lavaca, Texas, in part due to the low oil price environment and the difficulty of gaining offtaker interest, which could be coupled to perceived risks with the technology. Additionally, since the project relies on once-through water-cooling (as do all FLNG facilities), it struggled to meet US environmental standards. FLNG technology still appears to have some way to go before it achieves the success of its FSRU cousin.

Notes

1. In July 2005, the Sakhalin Energy Investment Company (SEIC) revised its capital cost estimates for Phase II of the Sakhalin II project to $20 billion, roughly double the original cost estimate.
2. Malcolm W. H. Peebles, Natural Gas Fundamentals (London: Shell International Gas Ltd., 1992).
3. Poten & Partners and Merlin Associates, LNG Cost and Competition (New York: Poten & Partners and Merlin Associates, 2004).
4. Ibid.

5. Ibid.
6. Ibid.
7. Sylvie Cornot-Gandolphe and Ralf Dickel, *Flexibility in Natural Gas Supply and Demand* (Paris: Organisation for Economic Co-operation and Development, International Energy Agency, 2002).
8. Cas Groothuis, Dave Fletcher, and Rob Klein Nagelvoort, "Changing the LNG Game" (paper presented at the 13th International Conference & Exhibition on Liquefied Natural Gas, Seoul, South Korea, May 14–17, 2001).

10

LNG Tankers

Introduction

The idea of transporting liquid methane, or LNG, was conceived by Godfrey L. Cabot, who received a patent in 1915 for transporting liquid gas by river barge. It was not until 1951 that William Wood Prince, chairman of the Union Stock Yards of Chicago, put the concept to use. Prince had the idea of liquefying and storing gas produced in Louisiana and then transporting it by barge up the Mississippi River to Chicago. There it would be unloaded, regasified, and consumed by Union Stock Yards. Willard S. Morrison, a cryogenic consulting engineer, led the project.

This first effort failed to materialize because the American Bureau of Shipping (ABS), a marine classification society, refused to class the barge, claiming that the project had not taken sufficient security measures. Later, a number of gas companies and shipowners joined together to further the concept, with the aim of economically transporting the gas over long distances. These pioneers of ocean-going methane transportation included Continental Oil Company (Conoco), Union Stock Yards, and Shell. The commercialization of the shipping concept was largely led by the British Gas Board and Gaz de France. Their interest came from converting customers from town gas to natural gas. They were also breaking new ground in developing insulated containment systems for maritime gas transportation. Competition was fierce as research and development was taking place on both sides of the Atlantic, with little collaboration or sharing of technology.

The Prototype LNG Ships

The first oceangoing LNG vessel was created by converting a surplus wartime cargo ship, the *Normati*, into a 5,000 m³ LNG carrier named *Methane Pioneer*. This prototype was a joint effort by a company called Constock International Methane, a partnership of Conoco and Union Stock Yards, and was funded by the British Gas Council. The *Normati* was converted to an LNG tanker at the Alabama Dry Dock shipyard in Mobile and delivered in October 1958. The cargo containment consisted of five aluminum, self-supporting, prismatic cargo tanks, with balsa insulation panels and a secondary barrier for protection in the event of an LNG leak. In 1959, ownership transferred to Conch International Methane Limited, a partnership of Conoco (40%), Union Stock Yards (20%), and Shell (40%).

The *Methane Pioneer* started her maiden voyage with a full cargo of LNG on January 25, 1959, from Constock's LNG storage facility near Lake Charles, Louisiana. After a 27-day transatlantic voyage, she reached Canvey Island, on the River Thames, in England. In all, the *Methane Pioneer* carried eight LNG cargoes between Lake Charles and Canvey Island, the last being in March 1960. She was then converted to carry LPG and renamed *Aristotle*, acting as floating LPG storage at Recife, Brazil. The vessel was scrapped in 1972.

A second prototype was the *Beauvais*. It began its life as a wartime Liberty vessel, the *John Lawson*, owned by the Compagnie Générale Maritime. The conversion was undertaken by the Methane Transport Company, a group led by Gaz de France working closely with French shipyards, which at the time were studying LNG as a new business activity. This vessel had a capacity of 640 m³ contained in three different tank and insulation designs. Each was built by a different yard, testing slightly different technologies.

The *Pythagore*, a third experimental ship with 610 m³ capacity built in Le Havre in 1965, used the Technigaz membrane cargo containment system. This ship was later converted into a fish carrier.

The First Purpose-Built Commercial Ships

The *Methane Princess* and *Methane Progress* were sister ships (the term for ships constructed with the same specifications) built in 1964 by British shipyards—Vickers Armstrong in Northwest England and Harland & Wolff in Northern Ireland, respectively. These tankers were owned by Conch International Methane. Each had nine prismatic cargo tanks of the Conch design, giving a total capacity of 27,400 m³ per tanker. They served the Algeria-to-Canvey Island LNG trade for British Gas. The *Methane Progress* had a service life of 22 years and the *Methane Princess* operated for 28 years.

The *Jules Verne*, a 25,840 m³ tanker with a cylindrical cargo containment system, was built by Ateliers et Chantiers de la Seine Maritime and owned by the Société Gaz-Marine. She began operating in January 1965, servicing the Algeria-to-France trade, and later operated between Algeria and Spain until 2004, under different management and with a new name, the *Cinderella*. The ship was sold to Taiwan Maritime Transport Company in 2004, and used for training purposes. *Cinderella* was scrapped in 2008.

After 1970, LNG tanker technology and scale developed rapidly, and an increasing number of shipyards began building LNG tankers according to specific project requirements. In 1971, Gazocean ordered a 50,000 m³ LNG tanker from Chantiers de l'Atlantique that used the Technigaz membrane concept. This tanker, the *Descartes*, operated in the Atlantic and Mediterranean basins until 2007, when it too was sold to Taiwan Maritime Transport and renamed *Prince Charming* (to go with *Cinderella*). This vessel, subsequently renamed *Charming Junior*, was scrapped in 2008.

By January 1980, there were 54 LNG vessels in operation worldwide and a further 17 vessels, each of 125,000 m³ or more, were on order (fig. 10–1). By 2014, the fleet had grown considerably, spurred by the construction of 45 Q-Flex and Q-Max vessels, which were delivered between 2007 and 2010 for the Qatari projects. As of October 2014, the fleet stands at 404 trading vessels with a further 126 on order for delivery up to 2017. The most popular LNG vessels being ordered in 2014 are approximately 170,000 m³ capacity. Owing to the recent expansion of the fleet, the average age is approximately 11 years, representing only about one-third of the average expected vessel life. Growing LNG production from Australia and the anticipation of US LNG exports are driving current new ship orders. The long-haul nature of many of these trades is making the larger carriers more attractive.

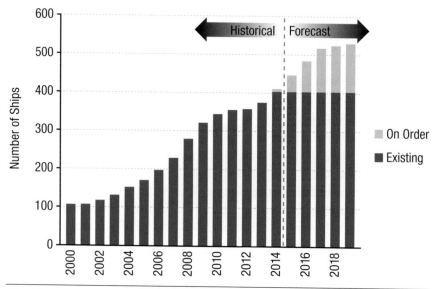

Fig. 10–1. LNG carrier fleet—operating and on order (excluding FSRUs and FLNGs). (*Source:* Poten & Partners.)

LNG Tanker Cargo Containment and Handling Systems

The cargo containment system has to serve several important functions. It has to seal the LNG in a gastight compartment to ensure it does not leak into the ship's hull spaces, insulate the LNG from the influx of heat and thus minimize boil-off, and protect the vessel's hull from the very low cargo temperature which would cause steel brittleness. In the early years, many experimental containment designs were developed, including variations of membrane, prismatic, and spherical tanks. Most received patents. The industry quickly rationalized and chose two containment systems, self-supporting independent and membrane, both of which have been extensively developed. Each design can be broken into subcategories (see table 10–1).

Table 10-1. Main LNG ship containment tank designs

Category	Self-Supporting Independent Tank Designs	Membrane Tank Designs
Particularity	Cargo tanks are constructed independent of the hull structure.	The ship's inner hull provides the structural strength of the cargo tank.
Containments	Kvaerner Moss Spherical Tanks IHI-SPB* Prismatic Containment Others: Conch and Esso (were utilized on small ships built in the 1960s)	Technigaz Tanks (TGZ Mark I & III) Gaz Transport Design (GT-96) CS-1** Containment

(*Source:* Poten & Partners.)
* Ishikawajima-Harima Heavy Industries (IHI) developed this design. The SPB acronym is derived from the description of the tank: Self-supporting, prismatic, independent type B tank. This tank design, a direct descendant of the earlier Conch system, was used in two 89,900 m³ ships (the *Arctic Sun* and the *Polar Eagle*) completed in 1993 by Japan's IHI for Phillips-Marathon's Alaska LNG export project to Japan. Although successful, no follow-on ships have been constructed with this tank design.
** The Combined System 1 (CS-1), as its name indicates, is a hybrid of the Mark III and the No. 96 containments. CS-1 combines the advantages from both systems, using existing technology and proven materials. The first LNG ship to install this containment system was delivered in 2004 by Chantiers de l'Atlantique for Gaz de France.

Self-supporting, independent tanks are heavy, rigid structures built to support the weight of the LNG. They are expensive to build since they require extensive use of cryogenic materials.[1] Membrane tanks are box-shaped and made of light, flexible metals that require a rigid, load-bearing insulation system around the entire tank to allow the transfer of loads from the tank to the tanker's hull. This insulation system must provide rigid support over the whole membrane surface while allowing it to contract and expand as necessary.[2]

Each containment system has its own technological and commercial advantages and drawbacks. Commercial and operational issues are summarized in figure 10-2.

LNG Ship Containment Systems—Overview

Containment Sections and Profiles	Commercial and Operational Implications
Section 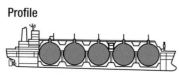	***Kvaerner Moss***

Section

- Tank Cover
- Insulation
- Tank Shell
- 9% Ni Steel or Aluminum Alloy
- Insulated Drip Tray
- Water Ballast

Kvaerner Moss

Advantages

Most proven of all second-generation containments

Excellent operating history

No tank-filling restrictions

No slosh damage potential

95% of welding is automatic (reducing defect risk during construction)

Disadvantages

Larger-dimensional ships (for the same carrying capacity as membrane ships); higher air draft

Less maneuverability (high wind area)

More affected by weather and poor navigation visibility

Higher canal charges (40% higher gross tonnage than for membrane ships)

Slightly higher fuel consumption

More difficult deck access and maintenance

Profile

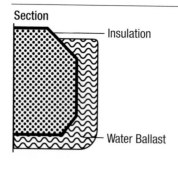

Section

- Insulation
- Water Ballast

Membrane Tanks (GT-96 and TGZ Mark III)

Advantages

Lower fuel consumption than for Moss

Lower canal charges (smaller gross tonnage)

Maximum cargo volume for given hull size

Primary barrier has first-class history

Unrestricted navigation visibility (flat continuous deck)

Good maneuverability

Low wheelhouse and cargo control room give lower air drafts

Disadvantages

Potential slosh damage to cargo tanks in partial fill

Membrane fatigue life is difficult to measure

Difficult to access containment system

Labor intensive construction—increased probability of defects

Profile

Fig. 10–2. Overview of LNG ship containment systems. (*Source:* Adapted from Peebles, *Natural Gas Fundamentals.*)

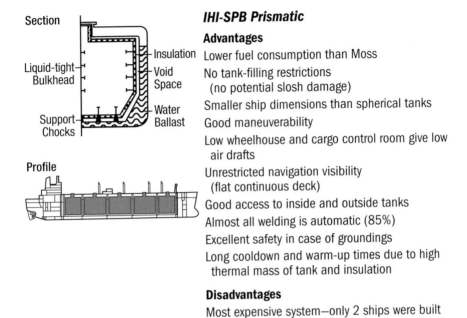

Section

Liquid-tight—
Bulkhead

Support—
Chocks

Profile

Insulation
Void
Space

Water
Ballast

IHI-SPB Prismatic

Advantages

Lower fuel consumption than Moss

No tank-filling restrictions
(no potential slosh damage)

Smaller ship dimensions than spherical tanks

Good maneuverability

Low wheelhouse and cargo control room give low
air drafts

Unrestricted navigation visibility
(flat continuous deck)

Good access to inside and outside tanks

Almost all welding is automatic (85%)

Excellent safety in case of groundings

Long cooldown and warm-up times due to high
thermal mass of tank and insulation

Disadvantages

Most expensive system—only 2 ships were built
using the IHI-SPB technology

Fig. 10–2. (Continued)

Over the past 40 years, the technology used to build LNG cargo containment systems has changed very little. Indeed, many of the technologies in use today are similar to those designed in the 1970s. Today, there are just three vendors offering designs for cargo systems: Kvaerner Moss, which inherited the original Moss Rosenberg design, GTT, and IHI. The developers of containment technologies patented their work to ensure the exclusive right to build and sell their technologies. At least 500 patents have been registered for marine transportation containment. However, as the market rationalized and the two primary designs became dominant, the licensing process for the patents has been focused on these systems.

Membrane containment technology

Until the mid-1980s, Gaz Transport and Technigaz each received many contracts for LNG ships ordered and constructed in France. After that time, Gaz Transport began to struggle financially as shipbuilding moved to Japan, which had primarily adopted the Moss system. Technigaz, on the other hand, was more secure, as it also sold technology for onshore LNG storage tanks.

The two groups had been slow to market their membrane licenses to foreign yards in fear of losing business at French yards. In 1994, Technigaz merged with Gaz Transport to form Gaz Transport & Technigaz (GTT). Initially owned by Gaz de France (40%), Total (30%), and Bouygues Offshore (30%), the company is now publicly traded, with GdF Suez owning 40% and Temasek 10%. GTT was more successful in Korea, where membrane technology established itself as the market leader.

The French shipyard Chantiers de l'Atlantique recognized that marketing its LNG know-how would be more profitable than trying to compete for contracts against Asian shipyards. When China was seeking assistance to build its first series of LNG ships, GTT and Chantiers de l'Atlantique agreed to provide the necessary technical support. A similar strategy was adopted when Spain's Izar received an LNG tanker order from Enagás in 2000.

During the 2000s, GTT developed a new membrane containment system, CS-1, which combined elements of GTT's two principal systems. The CS-1 system was adopted for three Gaz de France vessels delivered between 2006 and 2007. Since then, associated with the move toward greater fuel efficiency with dual- and tri-fuel diesel electric propulsion, variations on the Technigaz Mark III and No. 96 membrane systems have focused on reducing boil-off from as much as 0.25% per day to about 0.09% per day.

In March 2014, GTT launched its new Mark V and No. 96 Max LNG membrane containment systems, which combine elements of the existing systems. The Mark V will be available for LNG carriers constructed in 2016 and delivered in 2018. Meanwhile, the No. 96 Max, an upgrade of the existing No. 96 system, will be available in 2016. According to GTT, both systems can guarantee a boil-off rate of 0.09% of cargo volumes per day. The systems have been created in response to an increasing emphasis on more fuel-efficient propulsion systems and optimized hulls for next generation LNG carriers, which has reduced the need for and desirability of boil-off.

Shipbuilders are obligated to pay GTT a license fee—currently (2014) around $10 million per ship, or approximately 5% of the cost of a newbuilding—to use GTT membrane tanks. As South Korean yards build many LNG ships, these license payments have spurred interest in developing in-house containment systems:

- Samsung Heavy Industries (SHI) has developed the SCA (Smart Containment-system Advanced) membrane system;
- Hyundai Heavy Industries (HHI) has developed the Hyundai Membrane LNG Cargo Containment System;
- STX Offshore & Shipbuilding (STX) has completed development of its Independent Tank (ITS) containment system;
- Daewoo Shipbuilding & Marine Engineering (DSME) has developed ACT-IB: Aluminium Cargo Tank Independent Type B;
- The KC-1 system is a new membrane containment system, employing Korea Gas technology, developed with three South Korean yards—DSME, HHI, and SHI; and
- General Dynamics NASSCO has also unveiled a new IMO Type B LNG containment system.

It is difficult for new containment systems to win market share from established designs due to the conservative nature of the LNG industry and the reluctance to adopt new technology, which in this case might only result in marginal savings with unknown risks of performance and longevity. The South Korean designs will likely be first employed in newbuildings ordered by South Korean owners and built in South Korean yards. For example, Korea Gas plans to employ the KC-1 system in the two ships that will transport its FOB LNG from the Sabine Pass liquefaction project under construction on the US Gulf Coast. These new ships may cost less but they do not achieve the new lower boil-off rates now guaranteed by GTT.

Moss containment technology

Moss Rosenberg Verft patented the Moss containment technology. When the shipyard closed down, the Moss Maritime engineering company[3] retained independent ownership of the design. The Kvaerner Masa[4] yard of Finland obtained a license to build the design and gained the rights to market the product on behalf of Moss Maritime, although it was not until 1993 that the Finnish yard received newbuilding orders, with the first Moss ship delivered in 1996. In July 2001, Saipem, the offshore engineering subsidiary of Italy's ENI group, purchased Moss Maritime and with it the Moss containment patent. Sensing that the market was shifting to the more competitive Asian shipyards, Moss Maritime and Kvaerner Masa promoted the technology and sold licenses to a large number of yards.

Japan dominated the shipbuilding market from the early 1980s to the late 1990s. Since three out of the five Japanese yards capable of building LNG tankers have Moss experience, by the end of the 1990s the Moss spherical LNG tankers were the most popular newbuildings, and their total capacity outsized all other containments combined. There are other reasons why Japanese yards opted for Moss licenses, including the wider use of automation in building Moss vessels, and the perception that tankers with this containment system could be safer and stronger than membrane ships. Safety is a big concern for LNG ships calling and operating in the Tokyo Bay area. Moss tankers also had lower boil-off as they are easier to insulate.

There has been little in the way of innovation with the Moss tank concept, although the insertion of a cylindrical ring section at the tank's equator has slightly increased cargo capacity for the same diameter of tank. The Moss tank design has drawbacks, including higher Suez Canal transit costs (based on Suez Canal net tonnage), greater air draft, which limits the ability to trade into certain ports, and less maneuverability in high winds. Recent developments offer streamlined tank covers and improved maintainability of the flying bridge structure across the tank domes. The latest Japanese design, Sayaendo, or pea-pod configuration, is being employed in eight ships being built in Japan, with deliveries scheduled between late 2014 and mid-2017.

Membrane systems have come to dominate LNG transportation, with 92% of vessels currently on order incorporating this system. Membrane proponents argue that vessel owners are attracted by the greater carrying capacity relative to hull size, leading to cost savings. Membrane tanks are built within the inner hull, while Moss tanks represent a "ball in a box" design, which leads to large void spaces in the hull. Moss vessels now account for approximately 30% of the existing LNG carrier fleet.

Cargo handling systems

In addition to having safe containment systems, LNG tankers must be able to load, discharge, measure, and monitor their cargoes. This requires a whole range of cryogenic pumps, compressors, temperature sensors, gas detectors, and liquid gauges. Much of the cryogenic equipment was developed as an offshoot of the American Apollo space project.

Typical LNG Shipbuilding Project

Before contract signing

Shipowners must take into account a number of issues before ordering an LNG tanker. Aside from the project's financing structure and shipyards' quoted prices, owners must evaluate which cargo containment system, propulsion system, and specifications are best suited to their commercial and quality needs, as well as the requirements of the specific trade that the tanker will serve.

After reviewing the design with classification societies, shipyards, and equipment/machinery makers, the owner selects a yard. The selection process can take several forms, but the most common means of selection is to award a contract after a competitive tender and short-listing procedure or choose a yard on the basis of personal preference (whether it be a close historical link with the yard, a known first-class builder, or a predilection toward the builder's specifications, terms, and conditions). Only then will the final proposal be submitted to all concerned parties for approval prior to the contract signing, construction, and delivery.

Project scheduling

As an increasing number of ships are ordered, yards strive to become more efficient. Today, however complex a ship's specification may be, a yard familiar with the construction of LNG carriers can provide an exact delivery date more than two years in advance.

Construction of an LNG ship, like other specialized ships, requires the necessary yard infrastructure, highly sophisticated machinery, a professional workforce, a well-structured project schedule, and strict production management. Figure 10–3 illustrates the activities undertaken during a typical LNG shipbuilding project. In general, for a 145,000 m³ membrane tanker, the project takes about 30 months from contract signing to completion and delivery. Actual construction time is roughly 18 months from steel cutting to delivery.

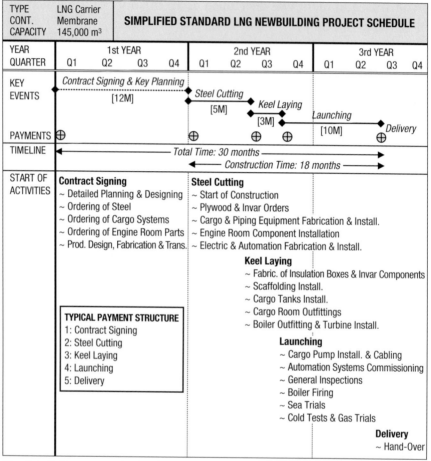

Fig. 10–3. Representative LNG membrane vessel newbuilding project schedule. (*Source:* Poten & Partners.)

During the construction process, the shipowner generally sends a site supervision team (naval architects, structural and cryogenic engineers, paint inspectors, etc., either employees or independent contractors) to represent its interests at the shipyard. Many site teams also include an owner's health, safety, security, and environment

(HSSE) advisor, especially if a series of sister vessels is being constructed and work is being carried out in several locations. The team resides on-site throughout the construction period, although the composition of the team may change as work progresses. Among other responsibilities, their role is to oversee the day-to-day activity, perform frequent inspections, and closely monitor building methods and progress to ensure that the vessel is being built according to the agreed specifications.

(*Note:* The above time frame is only indicative, as each yard has a slightly different building practice; fabrication and installation of cargo containment systems are different and alter the critical path procedures. The building period also varies in accordance to the ship's specifications required by the owner or project.)

The time frame for acquisition of a membrane containment vessel in the previous example also applies to a Moss vessel. These schedules, however, very much depend on the individual yard's capacity, LNG building experience, and order book for other merchant ships. Membrane and Moss ships have different critical paths during the building process, as they differ in technical and physical details. For example, the spheres for a Moss ship can be built separately from and at the same time as the hull, and are then inserted into the hold spaces, usually in two halves which are welded together in situ. In a membrane ship, by contrast, construction of the tanks is carried out inside the hull and cannot commence until the hull is substantially complete. However, this does not necessarily extend the schedule for membrane construction. On a Moss vessel, the piping systems and structure on deck can be mounted only after the installation of the spheres. For a membrane ship, the construction of the tanks is done through openings in the side of the ship's hull, so that the deck and all its piping structure can be installed at the same time. Another example is welding the containment system. For Moss spheres, 90% of the welding is automated. The same is more or less true for the GT-96 membranes, but much less for the TGZ Mark III, which requires more manual welding of the corrugated stainless steel membranes.

Standards of construction

In 1983, the International Maritime Organization (IMO), a United Nations agency, adopted the International Code for the Construction and Equipment of Ships Carrying Liquefied Gases in Bulk (IGC Code) to provide an international standard of construction for vessels built on or after July 1, 1986. The code's requirements include design criteria of the cargo tanks, accepted construction material, location of tanks, piping arrangements, boil-off and cargo handling requirements, and safety arrangements. Today, government bodies and classification societies of every major maritime country administer and enforce the IGC Code.

The superior safety record of LNG carriers confirms the value of the IGC Code and its enforcement, especially at a time when the LNG fleet is expanding rapidly. The IGC Code is kept under review, taking into account experience and technological development. In keeping with this philosophy, a substantial revision of the IGC Code has been undertaken and is due to enter into force in July 2016.

Typical Parameters of LNG Ships

LNG ship sizes have increased in capacity and capabilities over the last 40 years. Today's standard LNG carriers are more than five times the size of the first two *Methane* sister ships, and the Q-Max ships are nine times as large. There is a slight difference between Moss and membrane vessel sizes. For the same cargo-carrying capacity, a Moss vessel is physically larger because of the inefficient use of the hull cargo space, which is occupied by spherical tanks. Table 10–2 provides a general example of the LNG tanker specifications based on size and tank technology.

In the crude oil tanker market of the 1970s, following the closure of the Suez Canal, longer trade routes and larger ships meant economies of scale and lower dollar-per-ton-per-mile costs. The same principle has applied to LNG tankers, although the trend toward more widespread trading of LNG has resulted in the market settling for a more flexible 170,000 m^3 capacity as the standard vessel rather than the 216,000 m^3 Q-Flex and 266,000 m^3 Q-Max vessels.

Table 10-2. Main LNG tanker specifications (categorized by size and vessel type).

	ST	DFDE	TFDE	TFDE	MEGI	SSD
	Representative Specifications of Membrane Vessels					
Status	Modern	Modern	Newbuild	Newbuild	Newbuild	Modern
Cargo Size (1,000 m³)	145	155	160	174	174	216
Service Speed (knots)	19.5	19.5	19.5	19.5	19.5	19.5
L.O.A. (meters)	285.4	288.4	280.5	294.9	291.4	315.16
Draft (meters)	11.52	11.5	11.5	11.5	11.45	12
Beam (meters)	43.4	44.2	43.4	46.4	46.4	50
Boil-Off (%/day)	0.15	0.15	0.1	0.1	0.1	Reliq
Fuel Cons. (tons/day)	180	143	140	123	110*	173**

	ST	ST	ST	UST	RHST
	Representative Specifications of Moss Vessels				
Status	1st Gen	Modern	Modern	Newbuild	Modern
Cargo Size (1,000 m³)	126	137	147	155	177
Service Speed (knots)	18.5	18.5	19.5	19.5	19.5
L.O.A. (meters)	285.3	290	288	288	300
Draft (meters)	11.53	11.4	11.2	11.6	11.5
Beam (meters)	43.7	46	489	48.9	52
Boil-Off (%/day)	0.2	0.15	0.15	0.1	0.1
Fuel Cons. (tons/day)	198	163	176	134	169

(*Source:* Poten & Partners.)
Notes: Carrying capacities/efficiencies are estimates and vary with different designs. ST = steam turbine; DFDE = dual fuel diesel electric; TFDE = tri-fuel diesel electric; MEGI = M-type electrically controlled gas injection; SSD = slow speed diesel.
*Assumed and includes estimated hotel load consumption.
**Includes estimated reliquefaction consumption.

Ship Operations

Cooldown

Regardless of the containment system, after a ship leaves the yard its systems and tanks are gas free. The ship's tanks and piping systems must be gassed up and cooled down prior to loading its cargo unless it retains some LNG (known as "heel") after the last discharge. This will minimize vaporization of the cargo as it loads and avoid thermal shock to the tanks. Traditionally, a small volume of LNG is purchased at the load port and is used to gas up and cool down the containment system until there is liquid in the bottom of all tanks. Once the tanks are cooled down, it is normal to keep them cold throughout a vessel's trading period, although this may not always be possible for vessels in spot-trading service. They are warmed up again only for the next scheduled refit or if earlier access is required for essential maintenance. At each cargo discharge, the pumps are stopped before the tanks are completely empty, to retain the heel. The quantity of heel is designed to maintain the tanks in the cold condition until arrival at the next load port; boil-off from the heel also provides a portion of the fuel required for propulsion on the ballast voyage. This procedure avoids the time and cost associated with cooling the tanks down before each loading and avoids thermal cycles (warm–cold–warm) which stress the containment system. With the increase in traded cargoes, there have been many more instances of ships "heeling out," particularly when coming off a short duration time charter. This can result in additional time being required for cooldown prior to the next loading. Berth availability and cargo loading windows have to be carefully scheduled to accommodate cooldowns when required as the process can take many hours and generate large quantities of excess boil-off.

Boil-off

Any influx of heat causes the cargo to boil and generate vapor. This boil-off gas is used as fuel in the boilers of steam turbine-powered vessels, and also as fuel on dual- and tri-fuel diesel electric engine (DFDE/TFDE) LNG carriers being built today. (For a discussion of propulsion systems, see chapter 6.) By removing this boil-off gas, latent heat is removed from the cargo, and the cargo temperature maintained.

Even with today's advanced technology, insulation of the tanks is not perfect, although it is very close to it. The temperature inside the tanks increases very slightly during the sea passage and creates small quantities of boil-off. The rate of boil-off can increase during a storm, while the cargo is sloshing around inside the ship. Typically, the daily boil-off gas represents 0.1%–0.25% of the ship's cargo-carrying capacity, though the latest designs are claiming daily boil-off rates of 0.09% (see above). Larger new ships have lower boil-off rates than smaller and older ships owing to improved insulation systems. Boil-off exerts increased pressure in the tanks. If nothing is done to relieve this pressure, it eventually causes the tank relief valves to open, venting cargo into the atmosphere.

Until recently, almost every LNG vessel had steam turbine propulsion. In this case, the use of boil-off as fuel is standard practice. It greatly reduces the ship's bunker fuel requirements[5] and provides an environmentally clean fuel for the boiler, while reducing boiler maintenance costs. If the propulsion system cannot handle the boil-off gas, the alternative (other than venting, which is prohibited in many ports) is to return the boil-off to a liquid by reliquefaction and inject it back into the cargo tanks. Since onboard LNG reliquefaction facilities are expensive to construct and operate, it is a hard decision to reliquefy the gas on board.[6] The Q-sized ships built for the Qatari projects were built with reliquefaction units because their slow speed diesel engines were unable to burn the boil-off as fuel. Steam turbine vessels were able to manage tank pressures very effectively by "steam-dumping"—a process where excess boil-off gas could be burned in the boilers firing at full rate, and the produced steam condensed back to boiler feed water through the main condenser. DFDE/TFDE LNG carriers, and those using slow-speed diesel engines that were incapable of burning the excess boil-off in such quantities, are fitted with a gas combustion unit (GCU) that allows excess boil-off to be burned instead of being vented into the atmosphere. No one likes steam dumping because it reduces the volume of tradable LNG.

Cargo loading and discharge

Cargo operations are well described by Malcolm W. H. Peebles, in his book *Natural Gas Fundamentals*:

> *During loading any excess of boil-off generated on the ship is returned to the shore by compressors installed either on the ship or shore, where it may be used as fuel in the shore power plant or re-liquefied and returned to the shore storage. The vapor return flow rate will normally be the highest during the initial stages of loading when the pipelines and receiving tanks are relatively warm. During this phase the loading rate will be restricted so that the vapor return flow rate is limited to the available capacity of the compressors. As the loading system temperature stabilizes,*

the vapor flow rate will reduce and the LNG loading rate increase. Loading systems using dedicated ships are usually designed to allow a loading operation of a large LNG ship to be completed within fourteen hours. Discharging operations are essentially the reverse of loading and take approximately the same time.

In general, the description above, published in 1992, still stands. Most LNG carriers are able to load and discharge within about 14 hours, although the loading rate achieved is often dependent upon onshore facilities—number of loading pumps, hard arms, ability to handle return gas, etc.—rather than constraints on board the ship. Similarly, discharge restrictions may be based on the number of hard arms, distance to (and height of) shore tanks, and the ability to handle return gas. Ship-to-ship transfer to floating regasification vessels results in slower cargo transfer rates due to the use of smaller diameter flexible hoses and greater constraints on return gas transfer. In such cases, the transfer rate may be about 6,000 m³ per hour instead of the more normal 12,000 m³ per hour associated with land-based terminals.

LNG Ship Safety and Accident Records

To date, LNG shipping boasts an exemplary safety record. This reflects the industry's professionalism, the small number of shipowners, the value of the International Code for the Construction and Equipment of Ships Carrying Liquefied Gases in Bulk (IGC Code), and the success of bodies such as the Society of International Gas Tanker and Terminal Operators (SIGTTO), Oil Companies International Marine Forum (OCIMF), and others advocating the highest standards in all safety aspects, including design, construction, operation, crewing, and maintenance, both on ships and in the terminals that they serve. Although accidents have occurred, very few serious incidents have been recorded, and none have resulted in serious breakdowns of the cargo containment system or any significant loss of cargo.[7]

A case in point is the discovery of latent defect problems associated with the secondary membrane in a number of Technigaz Mark III vessels built in the early 2000s. There was increased nitrogen migration across the secondary barrier in one of the newly built

LNG ship's tanks that exceeded manufacturer specifications. This meant that the secondary barrier was leaking slightly. However, no LNG ever escaped the tank. It was later established that there was a problem with an epoxy adhesive administered to the triplex layer between the primary and secondary insulation.

Other LNG tanker operating incidents include collision, groundings, and mechanical breakdowns, most of which are common to all sectors of the shipping industry. On the mechanical side, however, there was a series of defects that occurred on a number of LNG carriers built in the mid-1990s. These vessels were steam-turbine driven, with the high-pressure and low-pressure turbines driving the propeller through a reduction gearbox. Routine inspection of gearing on one manufacturer's gearboxes identified heavy pitting and fractures on the gear teeth and some metal fragments in the gearbox oil. This same pitting and fracturing was then found on a number of vessels with the same manufacturer's gearbox, affecting several different projects. The problem was addressed initially by slow steaming the ships to minimize further damage and subsequently taking ships out of service for gear replacement, once these became available.

TFDE vessels—those capable of burning heavy fuel oil, diesel oil, and gas—offer much better operating flexibility and ability to optimize efficiency at various speeds but have had some initial teething problems, including generator blackouts.[8]

Maintenance, life extensions (ship longevity)

The industry has demonstrated that the useful service lives of LNG tankers are well in excess of the 25-year life normally associated with conventional oil tankers and originally assumed for many LNG projects. For example, Japanese buyers, who are extremely concerned about safety, commissioned an extensive study of one Pacific Basin fleet's condition and life expectancy and agreed to accept the tankers for continued service for another 20 years following an initial 20-year LNG supply contract. Industry experts now expect an LNG vessel's service life to be around 40 years, due to tankers being designed for a minimum 40-year North Atlantic fatigue life and assuming high standards of maintenance and a comprehensive life extension program. Older tankers have historically proved competitive, especially in shorter-term trades. However, advances in engine and propulsion technology, increased size, and improved insulation are such a step forward that older ships are finding it harder to compete in a market where longer-haul shipping prevails.

Notes

1. Malcolm W. H. Peebles, Natural Gas Fundamentals (London: Shell International Gas Ltd., 1992).
2. Ibid.
3. Moss Maritime is a leader in marine technology. Their expertise lies within the fields of special-purpose vessels, platforms, and floaters for the offshore industry.
4. Today, the Kvaerner Masa yard is owned by AKER Kvaerner ASA.
5. The cost of the calorific value (MMBtu) of boil-off has historically been less than that of fuel oil. The method of valuing boil-off has always been disputed. Some will argue that since it would otherwise be lost, there is no value, merely a savings on fuel costs. Others impose a fixed or variable cost based on LNG's FOB value or based on the resale value of the cargo at its destination.
6. Ships with an onboard reliquefaction facility currently trading include the 135,000 m^3 LNG Jamal, Nakilat's Q-class ships, and BG's TFDE ships.
7. The closest the industry has come to the breach of a cargo containment system was in 1978, when the fully laden El Paso Paul Kayser grounded near Gibraltar, causing serious damage to the bottom plating. However, the membrane containment system remained intact and no cargo was lost.
8. Anita Odedra, "The Importance of Shipping to a Global LNG Business" (paper presented at the 17th International Conference & Exhibition on Liquefied Natural Gas, Houston, Texas, April 16–19, 2013).

11

LNG Import Terminals

Introduction

The LNG receiving, or import, terminal is the final link in the LNG chain, connecting it to the transmission and distribution systems that serve consumers. While liquefaction plants act as enormous refrigerators to convert natural gas into a liquid, LNG receiving terminals turn the liquid back into gas by warming it, then send it into the pipeline system. An illustration of an onshore LNG import terminal is depicted in figure 11–1.

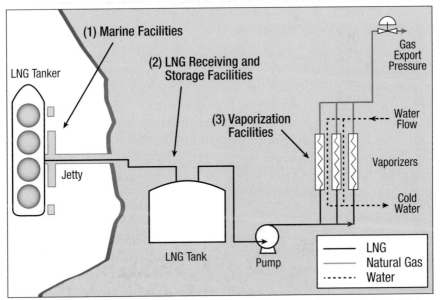

Fig. 11–1. Illustration of an onshore baseload LNG terminal. (*Source:* BP LNG.)

Since the LNG arrives in a high state of purity, the regasification and sendout processes are relatively straightforward. The marine infrastructure and storage tanks of an LNG import terminal and a liquefaction plant are essentially identical. However, while all LNG import terminals have generally similar components, their specific designs and layouts vary depending on such factors as: the nearby pipeline takeaway capacity; the size of LNG tankers expected to use the terminal; proximity to deep water; the topography, geotechnical characteristics, and surroundings of the site; the applicable regulatory regime; and the choice of vaporization technology.

Storage and Sendout Capacity

A terminal's storage and sendout capacity is a function of many variables. Storage capacity will be dictated by available land area, LNG tanker sizes, market profile (including seasonality, interruptibility, and access to alternative gas supplies), regulatory considerations, and commercial drivers for the terminal owner. Vaporization capacity has similar considerations. For example, if the terminal is located near existing gas pipelines, its ultimate capacity will be a function of the takeaway capacity of nearby pipelines and the ability to expand them. If the market is highly localized, the capacity is a function of the local market.

For example, the Sabine Pass LNG receiving terminal on the US Gulf Coast, now being converted to LNG export, features storage of 0.8 million m^3 supporting sendout capacities of up to 4.3 Bcf/d (44.5 Bcm/y), given its location in a primary gas supply area and its proximity to large-capacity gas transmission lines. LNG receiving terminals in Japan and South Korea have similar throughput capacities but much larger LNG storage of up to 2.9 million m^3 over as many as 20 storage tanks, given the complete reliance on LNG to supply the market and on the terminals to store the liquid.

There are very large regasification terminals in Europe as well. For example, South Hook in Wales, one of Europe's largest LNG import terminals, has a throughput capacity of 2.0 Bcf/d (21 Bcm/y) and five 155,000 m^3 storage tanks. Of comparable size, Grain LNG, also in the United Kingdom, has similar throughput capacity and five 55,000 m^3 and four 190,000 m^3 LNG storage tanks.

An example of a midsize terminal is GDF Suez's facility at Everett in the northeast US state of Massachusetts, which has 165,000 m^3 of storage available for tanker discharge and an average regasified LNG sendout capacity of 0.5 Bcf/d (5 Bcm/y), with peak capacity in excess of 0.7 Bcf/d (7.2 Bcm/y). The Everett terminal is located in a gas market, where the transmission lines are smaller than they would be in the Gulf of Mexico, but Everett is also connected to a 1,600 MW power plant, which expands its throughput capacity significantly. There are numerous midsized terminals dotting continental Europe's coastline that are sized to serve regional markets.

By comparison, the Caribbean terminals in the Dominican Republic and Puerto Rico are quite small at 0.14 Bcf/d (1.4 Bcm/y) and 0.18 Bcf/d (1.9 Bcm/y) respectively, as they are limited to serving modestly sized power-generating plants.

FSRUs, a relatively recent development, are deployed at locations where it is difficult to site onshore facilities, there is a need for fast-track timelines, or to supply small nascent markets. Capacities of shipboard regasification systems typically range from 0.4 Bcf/d to 0.7 Bcf/d (4.1 Bcm/y to 7.2 Bcm/y). These have proved to be very popular in South America, Southeast Asia (particularly in archipelagic nations without a developed pipeline grid), and the Arabian Gulf (initially to supply seasonal markets), as well as Europe (Italy, Lithuania, and the United Kingdom) and Israel.

Onshore LNG import terminals can be expanded with relative ease. The simplest and most cost-effective way is by adding vaporizers. However, if a significant quantity of new vaporization capacity is added, the LNG terminal developer may be faced with the expense of building one or more new storage tanks. This will not be a major problem as long as there is sufficient space on the terminal site for the new tank. However, LNG storage tanks take a comparatively long time to build and are the most expensive component of a regasification facility. An LNG terminal owner may also need to build more marine unloading facilities to handle increases in LNG tanker traffic. Before doing so, the owner must seek the consent of the relevant port and regulatory authorities, as well as ensure that the increase in LNG tanker traffic does not interfere with existing port operations.

The Everett, Massachusetts, terminal illustrates the incremental expansion approach. When originally designed and permitted, the terminal's vaporization capacity was limited; its throughput was around 0.125 Bcf/d (1.3 Bcm/y). By adding more vaporization, pipeline connections, and connecting to the adjacent power plant, the terminal's throughput capacity was expanded almost five times, and can handle more than 0.7 Bcf/d (7.2 Bcm/y) on a peak day—all with no increase in storage.

Grain LNG is another example of incremental expansion. The terminal, which is on the Isle of Grain on the River Medway, Kent, 30 kilometers east of London, began life in 1982 as an LNG peak shaving facility. In July 2005, the owner, National Grid, converted the facility into an LNG import terminal. After an open season, the initial capacity of 0.42 Bcf/d (4.3 Bcm/y) was leased equally by BP and Sonatrach. After another open season that resulted in awarding capacity to Centrica, GdF Suez, and Sonatrach, construction of the 0.87 Bcf/d (9.0 Bcm/y) Phase 2 expansion was completed in 2008. Following further open access bidding, Phase 3 capacity of 0.69 Bcf/d (7.1 Bcm/y) was awarded to Centrica, E.On Ruhrgas, and Iberdrola, with commissioning in 2010.

Design and Construction

A company that intends to build an LNG import terminal has to clear several hurdles before breaking ground. It must ensure that the proposed location is consistent with local, state, and national environmental and safety guidelines, and must go through an often rigorous licensing and permitting process. Since receiving terminals are more likely than liquefaction plants to be located near population centers, they are generally subject to a higher degree of regulatory oversight. The sponsor must decide on the proposed facility's design and operational parameters. In the case of utilities developing LNG terminals, the decision to add a facility or expand an existing one may also be subject to an economic review by the regulatory authorities. This will determine whether the new capacity is necessary for the utility to meet its customers' requirements, and will then permit the utility to include the terminal in its rate base and allow the recovery of costs and investment returns. The same approach also holds for third-party open-access terminals, although the standard of customer need is often achieved simply by entering into contracts with parties wishing to make use of the facility to unload and process their LNG.

Generally, after an initial feasibility study, the sponsor hires a recognized engineering firm to perform detailed FEED work, to create a design for an LNG terminal that fits their requirements and provides sufficient detail for regulatory review and approval. Once the design has been agreed on and the requisite licenses obtained from the regulatory agencies, the sponsor commissions an EPC contractor to build the marine facilities, vaporization system, and LNG storage tanks. The permitting and licensing schedule can vary among different countries and within the same country, depending on the degree of central versus regional involvement, political support or opposition, and the complexity of the site. Licensing can take between one and five years following formal submission of the application, and the feasibility and application preparation can easily add another one to two years ahead of that.

The time needed to build an LNG import terminal does not generally vary with the size of the facility. Rather, it is determined by the construction schedule for the storage tanks, which are a terminal's most time-consuming and expensive component. Aboveground LNG storage tanks usually take between two and three years to build, while inground tanks may take up to five years to construct. Building a floating import and regasification terminal can cut this timeline in half, even more if an existing regasification ship is available. The regulatory review process is often less time-consuming as well.

Large gas and electric utility companies, especially in Japan, were instrumental in developing many of the concepts of LNG import terminal design. Their resulting design philosophy reflected several factors: the generally conservative nature of these utility companies, the critical role of LNG in the company's and country's energy supply, and the magnitude of the downstream investments in gas distribution systems and power generation facilities. This design philosophy emphasized the importance of large storage capacities relative to throughput rates and duplication of critical components, aimed at ensuring the availability of 100% of the terminal's design throughput except under very unusual circumstances. However, in places where LNG plays a less critical role in the overall energy supply picture, such as in Europe and the United States, the need for maximum redundancy and excess storage capacity is less valued and has led to the adoption of a less conservative operating design philosophy. In all circumstances, the safety and security of the facilities have been paramount and usually result in the installation of multiple (redundant) safety systems that have proved to be highly effective.

The design parameters for an onshore baseload LNG import terminal must accommodate two primary modes of operation: vaporization, in which LNG is pumped from storage, vaporized, treated, and distributed to customers based on market demand; and LNG tanker unloading, while maintaining normal vaporization and sendout rates. LNG tanker unloading typically takes about 24 hours; maximum LNG boil-off is generated during this operation. This, in turn, establishes the design of the terminal vapor recovery system. The sizing of individual equipment components and the degree of redundancy and operating flexibility depend primarily on how the regasified LNG is utilized, as well as on variations in the daily and seasonal load demand pattern, the size and frequency of the LNG tankers calling at the terminal, downstream gas quality considerations, regulatory requirements, and on the physical location of the terminal.

A growing number of LNG import terminals supply regasified LNG for power generation. In these projects, natural gas–fired CCGT power plants are located adjacent to the receiving terminal, thereby providing an anchor market for the LNG terminal itself. Depending on the specific economics and operating rules of the power market, these plants often operate in the midmerit dispatch order, running at 50%–70% load factors.

Even if the power plant is used for baseload supply, regasified LNG demand is subject to both seasonal and daily demand variations. Adequate LNG storage must be provided to balance average annual import supply with short-term changes in demand, especially if no pipeline gas supply is available, as is the case in Japan and Korea. Vaporization equipment must also have sufficient capacity to meet peak demand, which can be highly valued in deregulated markets.

Cost

More than 60% of the cost of an LNG receiving terminal is associated with the construction of LNG storage tanks, marine and off-loading facilities, and safety systems. It is difficult to give a meaningful figure for the average cost of an LNG terminal. Factors unique to the location of the facility play a disproportionate role in determining the final construction costs, including the following:

- **Local geological considerations (such as soil stability and seismic activity) and the need to tailor infrastructure to them.** For example, LNG import terminal sponsors that plan to build a facility in an earthquake-prone area may want the LNG storage tanks to be buried. This would lead to significant extra costs, as additional civil work is required to isolate the tanks from the surrounding berm and from the incursion of water.

- **The cost of real estate.** In Japan, where waterfront space is tight, terminals are often built on filled land, which is not only expensive, but also entails additional engineering considerations to stabilize the soil for tank and equipment foundations.

- **Site layout, regulatory, and safety considerations.** These will dictate the choice of single containment, full-containment, or below-grade storage tanks, which can result in costs varying by a factor of four to five times for the same storage capacity.

- **Local and regional labor and construction costs.** For example, manpower costs are generally higher in the United States or the United Kingdom than in the Dominican Republic.

- **The choice of vaporization technology.** Open-rack vaporizers are more expensive than gas-fired vaporizers but cost less to operate.

- **The use of local power supplies or the development of dedicated power generation within the terminal facility.**

- **The need for downstream facilities to tie into the pipeline grid, including pipelines and gas treatment and odorization equipment.**
- **The marine environment.** For instance, the need to dredge the channel and berth to accommodate LNG tankers and disposal of the dredge material can impose significant costs; also, the need to place the berth away from the storage tanks can result in much higher marine costs.

Additional project costs could be incurred by licensing and permitting activities needed to accommodate local residential/ environmental concerns about the project. Upgrading or building infrastructure—such as roads, pipelines, or electric transmission—at the designated site could also add to the project cost.

Main Components of LNG Import Terminals

All baseload onshore LNG import facilities, regardless of terminal sponsor or design contractor, basically feature the same components:

- Tanker berthing and unloading facilities
- Storage tanks
- Regasification system
- Facilities to handle vapor and boil-off gas
- High-pressure LNG pumps
- Metering and pressure regulation station
- Gas delivery infrastructure
- Gas odorization, calorific value control (some terminals, mainly in Japan, are equipped with LPG unloading, storage, and vaporization facilities enabling LPG blending to raise the heating value of the vaporized LNG; in other regions, nitrogen or air injection equipment is used to achieve the opposite effect), and LNG tank truck loading facilities and/ or LNG bunkering (at some terminals).

A schematic of an LNG import terminal is presented in figure 11–2.

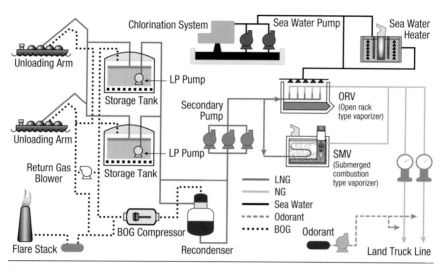

Fig. 11–2. LNG import terminal gas flow. (*Source:* M.W. Kellogg, "Pre-feasibility Study.")

Marine facilities

An LNG import terminal must have good port access to facilitate the unloading of LNG carriers. Typically, one or more berths will be built alongside the LNG import terminal, provided that sufficient water depth is available (or can be dredged) to receive and unload the LNG tankers. Alternatively, a jetty is built to connect the berth to shore. The minimum water depth required at the berth is 12 meters for a 130,000 m³ LNG tanker. Deeper water is needed to accommodate larger LNG tankers, which have a bigger loaded draft; for example, a 266,000 m³ Q-Max LNG tanker needs a water depth of 13 meters.

Each unloading berth is equipped with several arms that connect the LNG tanker's manifolds with the unloading line and the vapor return line. They incorporate articulated joints to allow for different sizes of LNG tankers and for ship movements as the tide rises and falls. The arms are equipped with automatic coupling releases, which ensure rapid disconnection in the event of an emergency. Unloading lines run from the arms to the storage tanks, where the LNG is stored until it is vaporized. These dedicated lines are usually made of stainless steel or some other material that can safely withstand the extremely low temperatures of LNG. They are insulated to reduce heat transfer from the atmosphere to the LNG as much as possible. Unloading lines also include expansion loops which permit the lines to expand and contract without overstressing as they warm up and cool down.

Most terminals have more than one unloading line, so that the lines can be kept cold between cargo deliveries by continuously circulating LNG from storage to the berth and back again. If only one unloading line is installed, then a separate circulation line is required to feed LNG into the unloading line to keep it cold. Separate vapor return lines carry LNG vapor back to the LNG tanker's cargo system in order to ensure that the vapor pressure in both the shore and LNG tanker's storage tanks is maintained within the operating design parameters for these tanks. The marine facilities may include a vapor return blower to maintain the pressure in the LNG ship's cargo tanks. LNG is unloaded from the tankers into the storage tanks using the LNG pumps installed in the tanker's cargo tanks. Where the unloading line is very long, additional booster pumps may be installed onshore between the tanker and the shore tanks to assist in the cargo pumping operation.

LNG offloading typically requires 24 hours. While the actual cargo pumping operations may only require 12 to 16 hours, the mooring of the LNG tankers, connection of the unloading arms, and the disconnection and unmooring prior to departure all add to the time involved. To this can be added the time needed to conform to harbor and port restrictions at LNG import terminals, which compete with other established, often high-density, port traffic. For example, countries such as Japan impose a nighttime transit ban on tankers carrying hazardous cargoes. All such tankers arriving at the discharge port in the evening must wait outside the bay or harbor channel before traveling to their discharge ports in a convoy—with the requisite safety and security distance between each LNG tanker in the convoy. Consequently, port transit restrictions can add several hours to the LNG tanker's overall offloading time.

Storage tanks

Design. LNG storage tanks maintain the gas in liquid state at $-163\,°C$ before it is vaporized and delivered into the local gas market. The tanks at an import terminal are exactly the same as those at a liquefaction plant (see chapter 9). An LNG import terminal developer is faced with two important decisions related to storage:

how much to build and the type of storage tank required. Because an LNG import terminal is often the sole or primary source of gas supply for a region, it is essential for its owner(s) to design the facility's storage requirements to meet unexpected surges in demand as well as unforeseen events such as a delay in cargo delivery. The selection of a tank design and the associated foundation design will be influenced by several factors, especially the geology, topography, and soil conditions of the site; seismic considerations; the quantity of land available; regional safety regulations, especially with regards to vapor dispersion and exclusion zone requirements; and aesthetic considerations.

LNG storage tanks may be above ground or partially or fully buried. LNG terminal owners must decide whether to build single- or full-containment storage tanks. Where land is at a premium, the developer may opt for a full-containment storage tank, thereby eliminating the need to build an impoundment and berm around the tank. As a rule, this impoundment will be capable of containing the capacity of one full tank plus a reserve margin in case of a leak or spill involving the entire contents of the tank. Full-containment tanks eliminate the need for impoundment and reduce the tank's footprint and, consequently, the amount of land needed to build the terminal. However, US regulatory policy now requires the use of an impoundment around full-containment tanks. While full-containment tanks are considerably more expensive than single-containment tanks, they are generally considered safer and have much smaller thermal and vapor exclusion zones than other tanks, allowing them to be placed in smaller sites.

Buried (inground) tanks are usually membrane tanks and consist of a prestressed concrete outer wall and an inner layer constructed of load-bearing foam, usually made of polyurethane and polyvinyl chloride (PVC), over which is laid a thin, cryogenic steel membrane that is in contact with the LNG. The insulation system allows transfer of the membrane loads to the outer wall. To prevent an influx of underground water, a slurry wall is first installed with the main tank built inside it. Most examples of this design are found in Japan.

In-pit tanks are similar to full-containment, aboveground tanks but are constructed below ground in a concrete-lined, excavated pit. The inner tank is constructed of 9% nickel/steel, and the outer tank is constructed of prestressed concrete. This design is particularly applicable to locations subject to high seismic activity, since it can essentially be a freestanding structure, while retaining the safety benefits of an inground installation. Tanks of this design have been constructed in Belgium, Greece, and Japan.

Inground tanks are considerably more expensive than aboveground tanks, with soil conditions, seismic considerations, and other environmental factors having a great effect on the cost. If the LNG terminal is close to populated areas, which is true of many in Asia, addressing the nearby community's aesthetic sensibilities may also factor into a decision to bury the tanks. Ironically, inground tanks may offer a lower measure of protection to their surroundings since a full tank fire, which might result in a total roof failure (often the worst case assumed by the regulatory authorities), has a greater thermal impact for an inground tank than for an aboveground tank.

Safety. For aboveground tanks, secondary containment is built around each tank in order to hold the whole tank's contents if it fails for some reason (the inner tank being the primary containment). These walls are capable of withstanding extremely low temperatures. In the case of a full-containment tank, the outer, prestressed concrete wall fulfills this function; for a single-containment tank, the outer wall made of steel cannot withstand the cryogenic temperatures, and the impoundment area and berm act as the containment. For obvious reasons, retaining walls are not necessary for tanks that are totally inground. All types of tanks must be surrounded by gas and fire detectors.

Foundations are designed based on local earthquake criteria and soil conditions to avoid foundation or tank failures in extreme earthquake emergencies. All tanks are equipped with temperature monitors to ensure that any failure of the insulation can be measured and rectified. Tank foundations are equipped with heaters to ensure the ground remains at a constant temperature and to avoid frost heaves, which could damage the bottom insulation or foundations. Tanks are also equipped with level devices that detect any shifts or settlement of soils that could also threaten the tank's integrity.

Capacity. More conservative utility companies favor large LNG storage volume relative to annual throughput. As a basic rule, LNG receiving terminals must have sufficient storage capacity to accept a full cargo of LNG from the largest LNG tanker expected to call at the terminal, plus a margin to allow for delays in shipping schedules. If there is a large variation in sendout rates between summer and winter and the LNG supply portfolio does not accommodate this fluctuation, then the terminal may require additional storage capacity in order to cope with winter demands, including extreme cold weather. In addition, the terminal's storage capacity must deal with fluctuations in cargo deliveries and sendout rates.

Also, storage capacity may be added to allow for an early cargo delivery or to accommodate a lower-than-expected sendout rate. The tanks must also be sized to hold a heel—that is, the level of LNG that is needed to maintain the operation of the tank's primary pumps, which are set in wells in the floor of the tank. Beyond the storage capacity designed to accommodate LNG tanker sizes, seasonal storage rates, and the heels, adding storage capacity becomes less of a science and more of an exercise of judgment and commercial objectives. In a world where LNG may become a more readily traded commodity, LNG storage could take on more opportunistic and optional value, which may justify the addition of storage for reload and trading opportunities. Conversely, the increasingly LNG-traded world may allow Asian utilities to meet seasonal LNG demands with less storage than has been the case, by using short-term purchases as a substitute.

Low-pressure LNG pumps are installed inside or close to the LNG storage tanks. LNG delivered by these pumps is circulated through the unloading lines to maintain cryogenic temperatures when no LNG tankers are unloading. However, the primary function of these pumps is to feed LNG to the high-pressure pumps, which raise the pressure of the liquid to a little above the desired gas sendout pressure before sending the liquid to the vaporizers. Because the capital and operating costs of LNG pumps are much lower than those of gas compressors, LNG is pumped at high pressure, typically 50–80 bar, into the vaporizers. The gas then needs no further compression before it is sent into the pipeline grid or to a power plant.

Quality concerns. Variations in LNG quality in the storage tank may occur as a result of weathering (changes in LNG composition when the LNG sits for extended periods of time in the tank due to the preferential boil-off of light components such as nitrogen and methane) or offloading LNG of a different quality (density) than the LNG in the storage tank. This is usually the result of receiving LNG from different supply sources, which is becoming a growing issue as LNG trading grows.

If LNG cargoes of different densities are stored in the same tank and are not adequately mixed, stratification (separation into distinct layers of different densities) may occur. If it does, it creates an opportunity for tank rollover—the sudden mixing of the two layers as their densities equalize because of heat leaking from outside and the tank dynamics. Rollover is a safety concern because it generates a large amount of boil-off gas rapidly and may cause structural damage to the tank, especially if the boil-off handling systems are overwhelmed by the sudden increase in volumes.

Several LNG buyers, mainly in Japan, avoid this problem by segregating LNG storage according to supply source. In other words, they dedicate storage tanks to specific suppliers: Brunei, Indonesia, Malaysia, and so on—thereby obviating the possibility of mixing LNG supplies with different physical characteristics.

However, not every LNG terminal owner has the economic means or sufficient space at the facility to do this. LNG import terminal owners with only limited storage at their disposal can fill their tank(s) from either the top or the bottom and can use the in-tank pumps to promote mixing by recirculating LNG through the pumps within the tank, ensuring that stratification does not occur.

Vaporizers

The vaporizers at an LNG terminal transform LNG back to its gaseous state by warming the liquid so that the gas is at or above 5°C (41°F). There are several types of vaporizers in common use, but they all employ the same general principle—extracting heat from water or air to warm the LNG.

Open rack vaporizers (ORVs). ORVs pump seawater from the adjacent harbor, and let it flow down the outside of hollow panels that heat LNG flowing in the opposite direction through the interior of the panels (fig. 11–3).

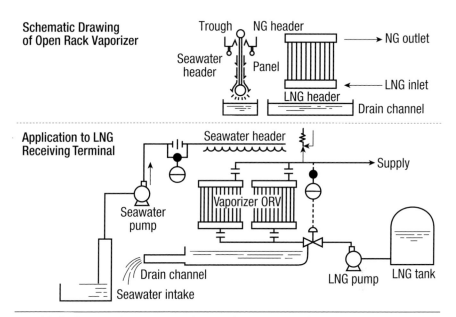

Fig. 11–3. Open rack vaporizers (ORVs). (*Source:* Kobe Steel.)

The seawater and the vaporization systems must be treated with chemicals to avoid the fouling of the vaporizers with marine growth, and the resulting water discharges carry these chemicals back to the ocean. The discharge of cold seawater into the sea may also cause environmental concerns since water temperature fluctuations may adversely impact marine flora and fauna. To mitigate this, the seawater is sometimes warmed before being discharged. ORVs are commonly found in Asian terminals but are not used in US terminals, owing to the difficulty of permitting the discharges under US environmental regulations. ORVs are usually much larger and more expensive than gas-fired vaporizers, since the ambient temperature of the seawater requires a large surface area to warm the LNG. However, their operating costs are much lower since they use no fuel other than the energy required to pump the seawater.

Submerged combustion vaporizers (SCVs). SCVs typically use natural gas to heat the LNG (fig. 11–4). They operate by flowing the products of combustion into a bath of water containing a bundle of tubes through which the LNG flows and is converted into gas. In this design, the LNG and heat flow in the same direction. SCVs use more energy than ORVs and thus are more expensive to operate, although they create no water discharges and are generally less costly to install. However, gas-fired vaporizers do create air emissions, which can raise other environmental and permitting issues. Typically, gas-fired vaporizers will consume 1.5%–2.5% of their throughput as fuel.

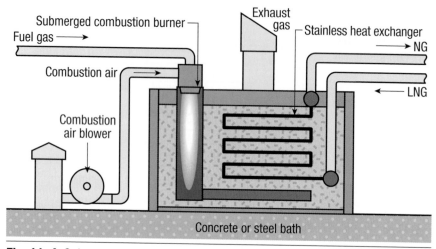

Fig. 11–4. Submerged combustion vaporizer (SCV). (*Source:* TradeKorea.com.)

Both ORVs and SCVs can be regulated to vary sendout rates over a wide range, and are very useful in combination with power plants which are subject to fluctuating dispatch rates.

Shell and tube vaporizers (STVs). STVs, generally smaller in size, are less expensive than ORV or SCV systems, but they require an external fluid heat source (fig. 11-5). Heat is usually supplied to the LNG vaporizer by a suitable heat transfer medium, typically a water/glycol mix, heated in a conventional boiler and flowing in a closed circuit. These vaporizer systems usually require a stable LNG flow at design and turndown conditions, with provisions to prevent freeze-up within the vaporizer at low flow rates. In LNG terminals, variable flow rates can be achieved by using multiple STVs or by combining STVs with SCVs. STVs are now being used for specialized applications, particularly where an alternate source of heat is available, such as from a power plant or a cold energy utilization process. In these applications, an intermediate fluid between the heat source and the vaporizer will be used. STVs have similar fuel consumption requirements as SCVs, but their air emissions can be much lower.

Because of their compact footprint, STVs are also the vaporizers of choice for FSRUs.

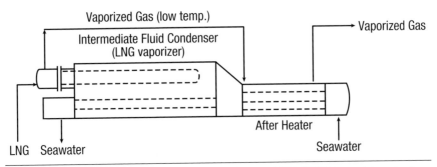

Fig. 11–5. Shell and tube vaporizer (STV). (*Source:* KBR, "LNG Import terminals–Recent Developments.")

Ambient air-heated vaporizers. These utilize ambient air, in either a natural draft mode or a forced one, to vaporize LNG. Such vaporizers are best suited for warm climates. They are manufactured by conventional air-cooler manufacturers and have been used at the Petronet LNG terminal at Dahej, India, as well as the Sabine Pass terminal in Louisiana. They do require large sites as their footprint is large. The advantage is that these vaporizers have the lowest environmental impacts, with no air or water discharges and minimal energy use.

Combined heat and power unit with a submerged combustion vaporizer (CHP-SCV). In order to decrease the gas consumption of SCVs, as well as to increase the efficiency and economics of the entire regasification process, the receiving terminal can be modified to use a cogeneration (also known as combined heat and power or CHP) concept that offers energy saving and environmental advantages. This has been implemented at Belgium's Zeebrugge LNG terminal. The heart of the CHP facility is a gas turbine that generates electric power. The hot exhaust gases from the turbine pass through a heat recovery tower, which uses the heat to raise the temperature of a closed hot water circuit. This hot water is then circulated and injected into the water bath of the vaporizers to regasify the LNG.

Most onshore terminals are equipped with ORVs and SCVs. The combination and number of vaporizers are a function of the facility's expected sendout profile, the quality and temperature of the available seawater, regulatory limitations, gas prices, and other factors. For example, an LNG import terminal may use energy-efficient seawater vaporizers for the baseload component of terminal sendout and install gas-fired vaporization capacity to meet short-term demand fluctuations and act as backup to enhance system reliability.

Boil-off gas facilities

Because LNG remains in a liquid state through natural evaporation (a process known as autorefrigeration), there is always some boil-off generated in the LNG storage tanks, as well as during cargo unloading and through heat leakage in the terminal components. Generally, boil-off is about 0.05% per day or less of the entire stored volume. All LNG terminals are equipped with systems that either capture the boil-off gas and send it into a reabsorber downstream of the LNG pumps or compress the gas and export it from the plant to the local distribution system. During LNG tanker unloading, boil-off rates are elevated by the energy generated through the pumping process, and vapor must be returned to the LNG carrier's cargo tanks to maintain the appropriate operating pressures in the cargo tanks and shore tanks. If a terminal has a gas-fired electricity generator, the boil-off gas may be used to fuel the plant's own power needs. Many terminals are provided with heated vents on the storage tanks or a remote flare stack to dispose of boil-off in case of an equipment failure or in case the rate of boil-off exceeds the capacity of the terminal's boil-off recovery system.

Gas Quality and Interchangeability

The quality of the gas that comes from an LNG import terminal must be consistent with the requirements of downstream gas customers or meet the specifications of the interconnected gas transmission lines, which vary by country and by region.

- **Asia.** In the traditional markets of Japan, South Korea, and Taiwan, the marketed gas is rich, with heating value greater than 43 million joules/m³ (1,090 Btu/cf). Newer markets such as China and India are developing pipeline grids and have more flexible gas quality specifications.

- **United States and United Kingdom.** In these markets, distributed gas is lean, with a heating value usually less than 42 million joules/m^3 (1,065 Btu/cf).

- **Continental Europe.** On the continent, significant gas quality variation is permissible within the pan-European EASEE-gas common business practices governing crossborder gas transactions. Gas quality ranges are not governed by heating values, but rather by the Wobbe Index, which can range between 13.60 and 15.81 kWh/Nm^3, and gas specific gravity, which can range between 0.555 and 0.70. These limits indirectly allow heating values between 10.13 and 13.23 kWh/Nm^3 (928–1,212 Btu/scf), though this range is deceptively wide, as not all gases with these heating values will necessarily fall within the allowable Wobbe–specific gravity envelope. Individual countries still have their own gas quality specifications, though these are for the most part similar to the EASEE-gas specifications, with the most notable exception being the Netherlands' low-calorific value G-gas system, which flows a high-nitrogen gas (14% N$_2$) from the Groningen gas field.

Many international gas markets have adopted interchangeability parameters to protect end users while dealing with multiple supply sources. The current specifications in most US pipeline tariffs are based upon overall heating value, or gas calorific value (GCV). However, GCV only addresses the energy content, not the burner performance of the gas. To indicate burner performance, there must be an adjustment for the relative density of the gas. The Wobbe Index, which divides the lower heating value by the square root of the relative density, is the most widely used internationally. Generally, markets dominated by power generation and industrial customers can accommodate a wider range of heating values if the end users are given time to adjust their combustion equipment. However, once a significant number of residential and commercial consumers are served by the grid, the range of acceptable heating values narrows, since there is no reasonable way to adjust their appliances if gas quality varies.

Pacific basin

Many natural gas liquefaction plants are located in remote locations far from natural gas liquids (NGL) markets, making it commercially expedient to leave the NGL components in the LNG, especially ethane (C_2), and in some cases, propane (C_3) and butane (C_4) as well. This Btu-rich product coincided with market domination by Japanese buyers, who prefer high calorific gas and often choose to add C_3 and C_4 to boost the energy content of the gas supplied when the delivered LNG is too lean.

Japan's terminal operators managed the introduction of lean LNG, such as that from Alaska, primarily through the addition of LPG. This process is illustrated in figure 11-6. Japanese buyers with numerous LNG storage tanks at their terminals and with LPG injection capabilities are positioned to accommodate growing lean regional LNG production such as the very lean product produced at Australia's CBM-to-LNG projects in the state of Queensland and from purchases of lean LNG from the Middle East and the Atlantic basin under both spot and term arrangements.

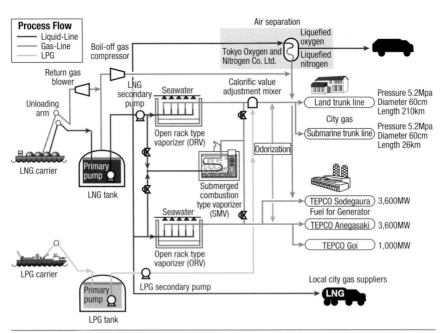

Fig. 11-6. Typical Japanese LNG terminal LPG enrichment process. (*Source:* TEPCO.)

South Korean city gas quality is Btu-rich, similar to that of Japan. To date, Kogas has controlled gas quality by blending LNG received from various sources to achieve the nominal sendout specification. While Kogas has considered introducing LPG blending several times, it has not been necessary to date. Meanwhile, the government has been lowering the city gas heating value specification to accommodate the trend towards leaner LNG project developments, together with the purchase of lower heating value LNG on the spot market.

In the absence of an integrated national gas transmission system, Chinese gas quality specifications remain regional, governed by the properties of the prevailing gas source. As a result, LNG supplies contracted to establish terminals in China show a similar wide range of properties. For example, North West Shelf and Malaysia will provide relatively rich LNG to Guangdong and Shanghai respectively, while Tangguh supplies a much leaner product at Fujian. The China gas market is dominated by industrial and power generation plants, which can be set up to accommodate a wide range of gas quality. The other main supply of gas to the east coast of China is the West–East Gas Pipeline, which will become part of a future national grid.

Similar to China, India has the beginnings of a national pipeline system in place. The Indian gas system currently handles a wide variety of gas compositions as well. The main HBJ pipeline system from the northwest coast to central northern India was deliberately designed for high and variable gross heating value (GHV) gas from the Bombay High gas field. Very rich gas is fed into the system at the coast, with the gas composition becoming progressively leaner further inland as LPGs are extracted. Gas demand is almost entirely from industrial users (mainly as ammonia plant feedstock) and power generation customers, all of whom can accommodate variable heating values. Similar to China, India will likely have considerable natural gas quality flexibility.

Atlantic basin

The issue is the reverse in the United States and the United Kingdom, where consumers prefer lean natural gas. These markets have traditionally been supplied by pipeline gas where the liquids, including ethane, have been stripped out of the supply. However, not all the NGLs are removed prior to liquefaction, making the LNG richer than local pipeline gas.

Receiving facilities located close to local producing regions can accept LNG with wide-ranging calorific values. For example, the Lake Charles, Louisiana receiving terminal can import lean LNG from places such as Trinidad, as well as rich LNG from Australia and Qatar. This is because the regasified LNG can be mingled downstream with gas production from the Gulf of Mexico and treated in NGL-processing plants before being sent on to market.

LNG import terminals located in market areas or serving buyers with specific gas quality restrictions must closely monitor the quality of LNG unloaded at the terminal as well as the regasified LNG that is delivered into the local grid. At these terminals, measures must be taken to lower the heating value of the gas. The most common approach is to inject inert gases, such as nitrogen, up to the pipeline limit for inert content, usually 2%–3%. For example, the owner of the Cove Point terminal installed Btu-stabilization equipment that uses nitrogen injection to reduce the Btu level of the regasified LNG. Air injection is also technically feasible, but is only used in areas where there is little or no water present in the pipeline gas, which might interact with air to create a corrosive environment inside the pipe. The Everett, Massachusetts, terminal treats regasified LNG using air stripped of water to meet the tariff requirements of the downstream pipelines.

LNG producers

The diversification of international LNG supply chains means that producers now plan to supply a range of countries with varying natural gas quality specifications. In the past, most liquefaction plants were designed to serve clearly identified markets under long-term supply contracts. These contracts also specified the particular LNG quality parameters that the plant had to meet. LNG producers increasingly have to decide whether to produce more than one LNG specification to satisfy all potential customers, which would require additional capital investment and operating costs. Qatar Petroleum and its foreign partners have equipped some of their liquefaction trains with front-end NGL extraction to supply lean Atlantic basin markets, while the other trains supply rich LNG to Pacific basin buyers.

Metering and pressure regulation station

Before leaving the terminal, the regasified LNG passes through a pressure-regulating and metering station to measure the gas quantities and qualities. The gas may be odorized (with a substance such as mercaptan, a sulfur-based additive) to aid detection of any leaks in the gas transportation system or customer appliances.

Safety and Security

The safety guidelines that govern the operation of LNG terminals are more stringent than for liquefaction plants, since the facilities are often located in or close to major urban areas. This became an even more sensitive issue after September 11, 2001. When potential terrorist targets were assessed, energy production, transportation, and delivery infrastructure were near the top of the list. As far as the LNG industry overall is concerned, the events of September 11 did not alter the fundamental properties and behavior of LNG and natural gas. The threat of a terrorist attack on an LNG terminal or the LNG tankers serving it may represent a new risk to the safety and security of the surrounding communities, the terminal staff, and the facility, but the hazards and associated consequences remain very much the same as from an incident occurring as a result of an accident.[1]

The design of the terminal, the equipment and materials used, and the operational procedures are geared toward risk reduction. This is achieved through the prevention of spills of LNG and leaks of regasified LNG and, in the unlikely event that they would occur, through the management of the consequences (that is, ensuring that any spills be detected, contained, and allowed to disperse safely).

Safety systems are designed to be both active and passive. Examples of active systems include gas, fire, and smoke detectors, shutdown systems, and firefighting systems. Passive systems include the installation of spill control channels leading to sumps, where LNG can be contained and its boil-off rate (and resulting vapor production) can be reduced to low levels.

The locations of and distances between the tanks, vaporizers, liquid and vapor lines, and other facilities are determined by national or local regulations and/or by recognized international codes of practice. Distances to the nearest property lines and to populations off-site from the terminal are also safely assessed and controlled. By ensuring adequate distance between the terminal and the surrounding area, in the event of a spill or fire at the terminal, the gas vapor will be diluted to below flammable limits before leaving the facility site, or any fire will be far enough away that the public would have time to move to a safe distance and avoid the risk of being burned. In the US regulatory system, the safety of the terminal is managed by the imposition of exclusion zones for hypothetical vapor and thermal spills that are calculated based on the specific design of the terminal and its site, and within which certain activities are prohibited. There are fewer prescribed standards in Europe, but regulatory authorities generally require that each facility be subject to a specific hazard analysis, also known as a quantitative risk assessment (QRA) developed just for that facility.

All LNG terminals have security systems to ensure that unauthorized people do not have access to the facility or LNG tankers berthed there. The arrival, departure, and berthing of LNG tankers will normally be controlled by the local harbor authority or by the operating company, as appropriate, and each port will impose differing levels of safety and security for the LNG tankers. The result of this focus on safety has been an impressive operating record, with no incidents at LNG terminals ever resulting in any injury to the public or to adjacent property since the Cleveland accident in 1944. Fatalities and injuries to terminal staff have been few and far between and rarely have had anything to do with the LNG itself. The same has also been the case for the LNG tankers in the arrival port and waterways.

Peak Shaving LNG

All import terminals are designed to send out gas throughout the year in baseload service, although daily sendout rates can vary significantly. A peak shaving facility sends out gas only for a few months, weeks, or even days of the year. Some LNG import terminals have facilities for loading trucks to deliver LNG to smaller, satellite storage and regasification plants in the marketing area.

Whereas LNG terminals must be sited on or close to the coast, peak shaving plants are located where they make most sense from a supply standpoint, usually near critical areas of gas distribution infrastructure and in regions that experience extreme demand peaks, such as the northern United States. While most peak shaving plants are generally smaller than their import terminal counterparts, they employ essentially the same design and technology concepts, with the exception of the marine unloading infrastructure. LNG trucking is also used to deliver LNG to remote markets, which cannot justify the construction of new pipelines.

Floating LNG Import Terminals

Onshore baseload LNG receiving facilities represent the vast majority of the world's LNG import capacity. In recent years, however, there has been great interest in building floating LNG import infrastructure. In spite of initial skepticism, floating LNG import terminals are now an accepted technology. Poten & Partners projects the number of FSRUs operating around the world will reach 25 by 2020, an increase from the 16 units operating as of year-end 2014 (fig. 11-7). South America has 4 units operating year-round with another 10 in various stages of development. There are units offshore or moored to jetties in Italy, Israel, Lithuania, Indonesia, Kuwait, Dubai, and Egypt. Three units are being promoted in the Caribbean, where they are ideal to supply these small, isolated gas markets. Numerous floating regasification projects are in the works across Southeast Asia, from Pakistan to China. On the Arabian Gulf, two units are in operation meeting seasonal natural gas demand, and two more are in various stages of development.

While there are other floating alternatives, the floating LNG terminal segment has come to be dominated by FSRUs (also known as LNG regasification vessels or LNGRVs), which were first developed to import LNG into the United States owing to less onerous hurdles for building and permitting relative to conventional onshore terminals. Gulf Gateway was the first offshore LNG import facility in the world. Excelerate Energy commissioned Gulf Gateway in March 2005. Gulf Gateway provided the means for establishing the technical and commercial viability of FSRUs.

Fig. 11–7. Existing, under construction, or planned floating regasification projects. (*Source:* Poten & Partners.)

Since their introduction a decade ago, FSRUs' share of global LNG imports has grown steadily. The number of significant benefits associated with utilizing these units versus onshore facilities include:

- Serving new and relatively small, niche LNG markets more cost effectively than onshore terminals.

- Faster implementation, as the units usually take two to three years to build, which is approximately two-thirds the time required to build an onshore terminal; permitting is often faster as well.

- Ability to act as a bridging solution until a larger onshore terminal can be built.

- Inherent mobility also provides greater flexibility as the units can be chartered for use during peak demand seasons and then traded as conventional LNG carriers or transferred to another location during off-peak periods.

The initial FSRUs involved the construction of a buoy and riser system with a high-pressure natural gas pipeline to shore. This structure has been implemented in four locations worldwide, with one location in the US Gulf of Mexico having since been decommissioned. Following the entry of custom-built FSRUs into the market in 2005, industry participants began retrofitting decommissioned LNG carriers with onboard regasification capabilities. Additionally, the mooring arrangements moved from offshore buoy systems to include the use of jetties to which the FSRU was moored permanently or semipermanently, as well as custom-built discharge systems to which FSRUs could be moored by the bow to a structure above the water surface. As of 2014, these dockside regasification structures are being utilized in approximately 10 projects worldwide.

Custom-built floating regasification units are more expensive than converted LNG carriers, costing upwards of $250 million per unit, with construction taking between 24 and 30 months. Standard import capacities range from 2 to 3 MMt/y (2.7 to 4.1 Bcm/y) for a converted vessel with a capacity of less than 140,000 m³, while newbuildings have import capabilities of 3–4 MMt/y (4.1 to 5.4 Bcm/y) and storage capacities between 160,000 and 170,000 m³. The newbuildings are also equipped with more efficient propulsion systems, which make them more appropriate than older FSRU conversions to operate as conventional LNG carriers in trading operations (fig. 11–8).

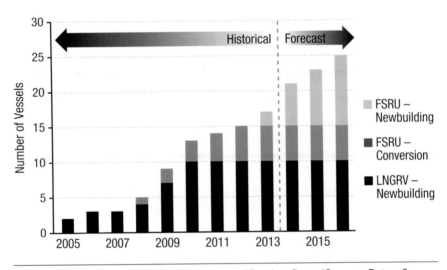

Fig. 11–8. Development of the floating regasification fleet. (*Source:* Poten & Partners.)

LNG shipboard regasification

The early FSRUs relied on Advanced Production and Loading's (APL) Submerged Turret Loading (STL) system (fig. 11–9). The STL buoy is anchored to the seabed and connected to a subsea pipeline by a flexible riser. When no FSRU is present, the STL floats submerged at a depth of approximately 30 meters. The offshore buoy system generally requires a minimum water depth of 150 feet for economic mooring (the system may have one or two STL buoys) and riser design. When ready to discharge, the FSRU connects to the terminal by pulling the STL buoy into a compartment of the FSRU (also known as a "moon pool"), where it is connected to the vessel system. The LNG is regasified on board the vessel, typically via a closed-loop fresh water circulation system using shell and tube vaporizers (STV). The recirculating loop for the STVs is heated using steam from the LNG tanker's boilers. The vaporized gas then passes through the STL system to the riser connected to a pipeline and then to the onshore gas market. The first application of the LNG shipboard regasification concept at Excelerate's Gulf Gateway terminal also required construction of a separate metering platform, as the system ties into two separate offshore gas-gathering systems. Capacity of the shipboard regasification system varies between 0.05 and 0.8 Bcf/d (0.5 to 8.2 Bcm/y).

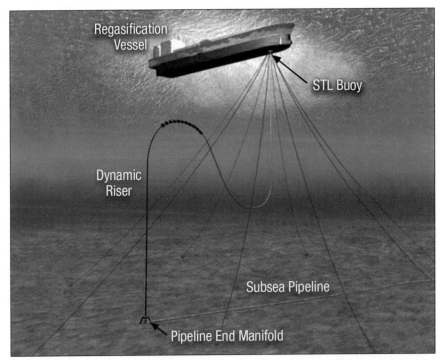

Fig. 11–9. Illustration of Excelerate's EnergyBridge™ FSRU. (*Source:* Excelerate.)

Floating storage and regasification units (FSRUs)

In concept, the FSRU can be loaded at a liquefaction plant just as a conventional tanker. The FSRU then sails to the offshore buoy system, where it connects and discharges its cargo. The process requires at least two FSRUs, one loading cargo while the other delivers regasified LNG to the market. However, this approach to offshore regasification results in several trade-offs. These include the need for a dedicated FSRU fleet with turret connections and vaporization equipment on all the vessels. The project owner must also factor in additional tonnage for a given trade, as each vessel must remain moored for several days to vaporize its LNG cargo. To maintain an FSRU system in baseload service requires more than one offshore buoy and the equivalent of more than one vessel attached to the offshore facilities at all times. Gulf Gateway also represented the first application of the STL buoy system to intermittent high-pressure gas service, and it was specifically designed for spot trading of LNG cargoes.

Concerns linger around the potential for sloshing damage to partially emptied membrane tanks due to the prolonged mooring periods required to discharge a shipboard regasification vessel. (Sloshing would not be a concern for vessels that employ a Moss-type containment system.) The potential for sloshing damage is greatest for membrane tanks between 30% and 70% full—an FSRU will spend at least three days with tanks in this range while unloading. While FSRUs can readily call at existing onshore terminals, conventional LNG tankers cannot call at the offshore terminals, raising concerns about reliability in the event there is a problem with one or more of the shipboard regasification vessels in a given trade.

In a variation of this approach, Excelerate developed a hybrid terminal first implemented in Teesport in northern England, where the FSRU connected to a shoreside manifold and discharged vapor from the vessel into the pipeline system nearby. This type of dockside regasification has become popular.

Another development was the move to offshore ship-to-ship cargo transfers, via flexible hoses in relatively calm waters. A conventional LNG tanker would be tied alongside the FSRU and discharge some or all of its cargo into the FSRU. This overcame the need to have an FSRU make the shuttle trips to the liquefaction plant, but could require the FSRU to disconnect from the offshore buoy and rendezvous with the LNG tanker in calm waters. The cargo transfer also proceeded slowly due to the limited capacity of the flexible cargo hoses.

As the drawbacks of the offshore STL system became more apparent, the industry moved to the adoption of the technology in combination with shoreside infrastructure. This approach eliminated the need for connection equipment on board the FSRU and lowered the vessels' costs. FSRUs can take two forms: a custom-built vessel or a converted LNG tanker permanently moored at the designated site (converted LNG tankers are retrofitted with regasification equipment on board, just like custom-built vessels).

Another variation is the development of FSRUs, which employ either a yoke or turret system (a tower-like revolving structure, fig. 11-10) allowing the FSRU to weathervane (rotate) depending on the prevailing water and wind currents around a structure fixed to the seabed. A conventional LNG vessel unloads to the FSRU using either a side-by-side or end-to-end connection. The LNG is stored and vaporized on board the FSRU and piped through the mooring yoke and then to shore via a subsea gas pipeline.

Fig. 11-10. PGN FSRU Lampung (Indonesia). (*Source:* Hoegh LNG.)

FSRU terminals present certain design and operational issues. One of the biggest issues is the LNG tanker berthing process, which will result in motion between the LNG tanker and the FSRU during cargo discharge operations, unlike the rigid mooring arrangements of an onshore terminal in sheltered waters. While offloading through a loading arm or some other special system for the transfer of cryogenic liquid between the LNG tanker and the terminal, the stresses on the transfer system are significant. This challenge results in a very intensive site selection process to find a calm marine environment.

Gravity-based structures (GBSs)

GBSs have been employed in offshore oil and gas production for several decades, but have been applied to only one LNG import facility. ExxonMobil, together with joint venture partners Qatar Petroleum and Italy's Edison Gas, commissioned their Isola di Porto Levante GBS LNG terminal on the Italian coast near Venice in 2009 (fig. 11–11).

GBS units are massive, concrete structures prefabricated in a graving dock located onshore. The structures are then floated and towed to the installation location, where their ballast tanks are flooded and the structure sunk to rest on the seafloor. The deck of the GBS unit is located at an elevation above the sea surface that provides sufficient freeboard for the sea conditions expected at the location. Their use is limited to areas with suitable substrates (i.e., a relatively flat, sandy seabed) and water depths ranging from 60 to 85 feet. Because GBS units sit securely on the seafloor, LNG terminal operators do not have to worry about LNG sloshing in the unit's storage tanks during rough weather.

Fig. 11–11. Illustration of gravity-based structure (GBS) LNG facility. (*Source:* ExxonMobil.)

Nonetheless, offshore facilities are highly exposed to the effects of meteorological and oceanographic forces from high winds, waves, and currents. Their fixed design also means that LNG tankers cannot weathervane to account for the prevailing wind and current conditions during docking and unloading. In addition, LNG tankers must deal with wave interaction (coupling) from close proximity to the solid, stationary platform. The LNG unloading arms must also compensate for the relative motion between the terminal and LNG tanker during the unloading operation. The relative inflexibility of this design and the limited number of areas suitable for deployment have resulted in little interest in further development since the initial deployment.

Note

1. D. Quillen, "LNG Safety Myths and Legends: Investment in a Healthy US Energy Future" (presentation given at the Natural Gas Technology conference, Houston, TX, May 14–15, 2002).

12

LNG Project Formation

Introduction

Every physical LNG chain runs from gas exploration, production, and processing, through pipeline transportation, liquefaction, shipping, importation, and regasification, to distribution. Alongside this runs a parallel chain of business ventures, governments, banks, and contracts that bind the links, apportioning the revenues to support the financing of the facilities and operations, allocating and assigning risks. Assembling this chain is often described as project formation.

The decisive event in project formation is the commitment to fund the liquefaction plant by the shareholders and the lenders, called the final investment decision (FID). Usually, the liquefaction plant's owners take FID after committing upstream natural gas reserves and securing an assured, firm, long-term revenue stream from LNG buyers sufficient to underwrite the capital investment and cover the operating expenses. In turn, traditional LNG buyers are seeking an assured LNG supply to serve their markets.

To provide these assurances, a web of intertwined and interdependent venture and fiscal agreements, resource and cost assessments, and construction contracts must be established. Shipping as well as buyer commitments must be secured. All of these must be brought together on a closely coordinated schedule. Project formation—the implementation of the set of agreements leading up to and including the sales agreements and the FID—is a complex, costly, and often lengthy process. When project financing is introduced, the complexity is further increased.

There are two keys to successful project formation. The first is to establish a supply project in which the participants bring project management capabilities and experience and align seller, buyer, host country, and buyer country interests. The second key is a creditworthy revenue commitment from the sale of LNG in order to secure the funding.

Monopoly Utilities and the Project-Chain Business Model

Through the mid-1980s, natural gas and electricity markets were largely controlled by regulated, monopoly-franchised utilities. In Japan, these were natural gas or electric utilities. In Europe, South Korea, and Taiwan, LNG buyers were national gas transmission utilities. In the United States, most LNG buyers were gas transmission companies. They bought LNG CIF or DAP, or if they controlled shipping, FOB. Each utility managed all the downstream logistics, operations, and commercial transactions in its own market, including importation and regasification, inland transportation, and demand aggregation. They sold the natural gas wholesale to electric generation, industrial, and gas distribution companies or distributed gas or electricity directly to end users.

The natural commercial structure for an LNG project in this environment was the project-chain model, in which the LNG supply project was the seller, and the utility was the buyer. To secure funding, the supply project required utilities to make a long-term purchase commitment and receive the LNG through import terminals they controlled. The supply project committed to funding and construction of natural gas production facilities and the liquefaction plant only when its design capacity was completely sold. Utilities could afford such commitments, because they could recover their gas supply costs through regulated tariffs. Shipping was dedicated and controlled either by the seller (CIF or DAP sale) or the buyer (FOB sale), but, since there was little opportunity for trading, the control of shipping was not a strategic issue at the time. In the absence of competitive wholesale natural gas markets, LNG sales prices were indexed to the fuel it replaced—typically crude oil prices in Asian markets and fuel oil prices in Europe.

LNG sales evolved as a layered sequence of bilaterally committed trades. Each new LNG project had to find markets in advance.
This model supported the financing of LNG projects, but the
rigid commercial structure carried costs. Inflexible contract terms
prevented buyers from managing the risk of mismatched supply and
demand. Sellers were precluded from selling to markets that offered
higher prices, except on rare occasions when there was alignment
of uncommitted excess production, shipping capacity, and access to
the import market.

The buyers' emphasis on reliability (an outcome of the regulatory
bargain described in chapter 2) resulted in excess capacity being
embedded along the entire chain. The inefficiency was costly, but
the buyers had no economic motive to support more efficient and
profitable commercial arrangements; gas and LNG supply was not a
profit center. Finally, the financing arrangements were often underwritten by the importing country's export credit agencies and typically restricted the project to use that country's EPC contractors and
shipbuilders. Limited competition resulted in higher costs, but again
there was no incentive for purchasing utilities in the consuming
nation to break the pattern under supply cost recovery regulation.

The Emergence of LNG Merchants

In the 1980s and 1990s, three converging forces started changing
markets, costs, and financing requirements for LNG supply projects:

- The commoditization and liberalization of North
  American wholesale natural gas and gas transportation
  markets, followed by liberalization in Europe, weakened
  the monopoly-franchised utilities and their guaranteed
  tariffs, which had underwritten an integrated model of gas
  importation, supply, transmission, and wholesale gas sales.
  As a result, the changes in logistic and commercial functions
  the utilities had provided—import and inland transportation
  and wholesale merchant gas sales—emerged as both
  problems and opportunities for LNG sellers.

- During the transition to liberalized markets, wholesale gas
  prices initially rose, demand then fell, and prices dropped in
  turn. After initially experiencing steep drops in price (North

America and the United Kingdom being the most dramatic examples), natural gas demand and prices began to rebound in North America and Europe. Liberalization of electric power markets in North America and Europe contributed to this trend as gas became the fuel of choice in new power plants. The United States, the United Kingdom (with its famous "dash for gas"), and Spain were dramatic examples. Finally, the widespread availability of LNG import terminal capacity made access to these markets easier. This excess capacity had been created in the United States as a result of the failure of earlier LNG import projects and in Europe as a result of conservative design.

- Driven by the opportunities and challenges of new markets, LNG costs throughout the chain dropped by nearly half in real terms by the early 2000s, owing to improved design, increasing scale, and more competitive procurement from EPC contractors, process vendors, and shipyards. However, this trend has since reversed. Reliability became less of a concern than efficiency, as it was clear to many that the LNG chain was inherently reliable, no longer requiring (and increasingly, no longer rewarding) the earlier conservatism.

However, the impacts and outcomes differed in the Atlantic and Asia-Pacific markets. Today, for projects looking to Asian markets, monopoly utility buyers still dominate, although this is less true in South Korea and may not be the ultimate model for China or India. In contrast, liberalization has eroded monopoly franchises in Atlantic Basin markets—offering increased access to and within the markets—and in general has increased the level of competition. But to realize the full benefits of competition, changes were needed in access to LNG terminals, domestic transportation capacity, and downstream demand aggregation and marketing. Traditional LNG project ventures are ill-suited to these functions but are being forced to adapt to these changes. Security of access to markets, certainty of LNG price structures, and creditworthiness of buyers become more challenging issues in this new environment.

After 2000, a new class of LNG merchants emerged and began to step into the role classically filled by the utility buyers. The newly emergent market structure created opportunities for supply acquisition, marketing, and short-term arbitrage on both ends of the LNG chain. For companies such as Shell, BP, BG, and Repsol, these new opportunities provided a means to enhance their positions in diverse liquefaction projects by increasing their control of LNG supply, shipping, import terminal capacity, and marketing across diverse terminal positions, often by acquiring capacity in open-access terminals, or building their own proprietary terminal and acquiring ownership or control of their own, nondedicated LNG tankers.

ExxonMobil, Chevron, and ConocoPhillips formed trades from new large export projects targeted at proprietary import terminals in Europe and North America. In some cases, the energy majors and national oil companies take or retain title to the LNG through the import terminal and assume direct market risk as merchants. In many instances, these schemes have faltered on downstream market changes. On the other end of the chain, downstream players such as GdF Suez, Gas Natural, and Union Fenosa have expanded their shipping, terminal capacity, and marketing activities, and are now participating in LNG supply projects as buyers and investors.

This environment has resulted in a change in the commercial structures for new LNG sales and purchases. FOB sales from Trinidad's first Atlantic LNG train provided its customers with considerable sales and destination flexibility. Consistent with EU policy, Nigeria LNG has followed Norway and Russia in abolishing destination restrictions, although for the early trains, Nigeria LNG retained control over the shipping. Algeria had no choice but to conform its pipeline and LNG sales terms under EU pressure. However, the loosening of destination restrictions does not translate into complete freedom for buyer or seller, since the terms of sales still may require the consent of both parties to a given change in destination.

LNG supply project sponsors now buy output from the project through purchase agreements or through equity LNG for merchant sale in direct competition with the classic utility customers. New liquefaction trains in Trinidad and Egypt were structured as utility-like tolling facilities without any direct LNG sales role for the owning joint venture (though their shareholders are usually involved in the upstream and/or in LNG sales and purchases). The proposed liquefaction projects in the United States have adopted the tolling model as well, albeit with some variations (Cheniere being the most obvious example—Cheniere actually takes title to the LNG and makes FOB sales, but the commercial result is very close to a tolling model). Utilities are being required to segregate their merchant and transportation functions, and are moving upstream by taking equity ownership positions in LNG supply projects.

With short-term sales growing as a share of the overall LNG market, supply project sponsors have been increasingly willing to commit to FID when only part of the design capacity has been sold to third-party buyers. Beginning in 1999, Qatar's Ras Laffan LNG, Oman LNG, Malaysian LNG Tiga, and Sakhalin II all moved forward with less than full capacity sold. The Australia NWS Train 4 was committed with an extended buildup period in its long-term SPAs with Japanese buyers. The resulting volumes—the so-called wedge volumes because of the shape of their supply curve—were sold to Shell, an equity participant, for short-term trading.

In Japan, buyers are shortening contract durations while acquiring shipping capacity to match the very gradual liberalization of their domestic markets. This loosening of trade terms has also aided the electric utility buyers in managing shortfalls in nuclear power generation by rapidly expanding thermal generation fueled by short-term LNG purchases. Kogas has been very active in short-term trading to cover the highly seasonal demands in its market. New LNG buyers, such as Pohang Iron and Steel and GS Energy, have emerged in the South Korean market, developing their own import terminals to serve captive market loads. Petrobras has become a major LNG trader, exclusively using short-term purchases coupled with low-cost FSRU terminals to manage flexible thermal power demands during low hydroelectric production periods. Power generation companies are following suit. Flexible trading will grow in Asia-Pacific, European, and South American markets and should be further facilitated by the emergence of Henry Hub–linked LNG exports from North America.

Supply Project Structure and Formation

Increased liquidity, flexibility, and capacity throughout the chain is changing the LNG business model; however, the sale of the majority of the LNG produced by a project must still be guaranteed under long-term arrangements (often with greater destination flexibility) for a new LNG supply project to move forward.

Requirements for an LNG project

Every LNG project must have the following basic components:

- A host country environment that offers stable legal and fiscal structures for resource exploitation and business income and resource taxation

- Assurance of the quantity and expected production profile of the natural gas and condensate resource base

- Commitments by experienced, credible producers, contractors, and process vendors to build the facilities for gas production, transportation, the liquefaction plant, and, if necessary, the import terminal at a known cost for a fixed work scope, on a set schedule, and with an assured level of performance

- Ownership or contractual control of shipping, LNG import/ regasification capacity, and/or market access by the seller and/or buyer

- Guarantee of sales revenues through a mix of long-term contracts incorporating assured volume offtake, market-responsive prices, and reliable market access

- Regulatory and political conditions in the end-market country that protect the terms of purchase and market access

- Strong, credible, creditworthy, and aligned contracting parties, which may include joint ventures in many project roles, with experienced management and operating track records

- Access to adequate funds, including equity and debt funding, for project formation and construction financing and in amounts sufficient to cover unanticipated up-front investment costs and adverse developments during operations, especially sustained low-energy-price environments

Project formation

The process of assembling and orchestrating these elements is called project formation. Much like a chamber music group, there are several key players, but no single conductor directs the musicians, although often this role is assumed by the supply project venture, which has the largest investment.

Host country agreements. The first order of business for the new project venture is to conclude a set of host country agreements (if these are not already in place) that establish the fiscal, commercial, and regulatory environment for the project's construction and operation. These agreements include the commercial and natural resource exploitation laws of the host country and prior agreements relating to specific resource access development, participation, royalties, and taxation.

In addition to preexisting development terms and fiscal structures, issues to be resolved during the formation stage may include:

- Specific targeted fiscal or other government incentives associated with the LNG project development
- Access to or acquisition of lands, sometimes owned by the government
- Safety, design, and environmental standards
- The terms of importation of project equipment, personnel, and material
- Targets for the use of local content
- Labor relations, maritime regulations, and operating standards
- Financial and currency issues (including foreign currency borrowing and repayment, currency conversion, repatriation of dividends, withholding and other taxes on financial transactions)
- Regulatory stability, dispute resolution, and arbitration procedures

These can be provided through legislative or regulatory actions or through sovereign contracts when the existing legal regime is inadequate. In developing areas, the lender's environmental standards will require extensive environmental and socioeconomic baseline studies and policies to ensure the rights and well-being of both the environment and indigenous peoples throughout the development process.

Adequacy of the resource and gas supply. Typically, the project aims to monetize a gas resource that is known but not fully delineated. Compared to oil, natural gas projects require greater up-front investments and have much longer payout periods. Traditionally, neither buyers nor sellers would commit to an LNG project until sufficient resources and reservoir performance are demonstrated and certified for the economic life of the project (which can run 20 years or more). This may be an easy task, as was the case with Qatar's 900 Tcf North Field, or a more difficult and costly undertaking, as in the case of challenging offshore prospects in Australia, Russia, and West Africa. Other recent challenges include the following:

- In the Australian state of Queensland, the exploitation of coalbed methane resources will require ongoing and expensive drilling programs over the life of the projects.

- In Egypt, the Spanish Egyptian Gas Company was buying feed gas from an existing natural gas supply system, but domestic demand has curtailed feed gas availability, and the facility has shut down.

- In the United States, the liquefaction projects are also relying on feed gas purchased from the natural gas pipeline grid. At most of these projects, the companies that contracted for liquefaction capacity will be responsible for acquiring their own feed gas and are relying on the continued availability of domestic gas production at competitive prices.

Facilities construction. Construction of a project's major facilities is usually performed by an EPC contractor, acting alone or in a consortium. Separate contracts may be let for upstream, pipeline, liquefaction, and marine facilities. Marine infrastructure may be built and owned by the project or by the host country (with the government's investment recovered through port fees). The liquefaction plant EPC contract will include cryogenic storage tanks, gas turbine and compressor strings, electric power and other utilities, and the liquefaction process units and license agreements.

In the early days, the design and management of the construction project was a formidable technical and managerial challenge. EPC contracts were negotiated on a cost-plus basis and required intense cooperation between the contractor and the owners, with either independent engineering firms and/or one or more owners acting as technical advisers to the project venture.

Today, project management capability is more widespread, aided by a better understanding of the design and operating parameters of the liquefaction plants, and clear evidence of the overall robustness of the technology. The sponsor will often solicit and pay for multiple detailed FEED evaluations on different liquefaction processes. EPC contractors then submit lump-sum bids for a turnkey, performance-guaranteed construction contract based on these designs (which they may have developed). There are a growing number of EPC contractors who can bid on such contracts and a growing number of vendors of cryogenic heat exchangers and processes (see chapter 9).

However, this trend has seen some reversal in the past few years. Especially in the new (post-2008) Australian projects, the scale and risk of the EPC component has grown so large that no single contractor or even a consortium can afford the lump-sum approach and its inherent risks in cost overruns and completion delays. In these cases, the model has reverted to the old reimbursable or, more recently, open-book approach, though selected portions of the EPC contract (or at a later stage of construction the entire EPC contract) may be converted to a lump sum. Almost all these Australian projects have suffered cost overruns, and most of these additional costs appear to have been absorbed by the owners. In any LNG project, some cost increases are always the responsibility of the owner.

LNG Sale and Purchase Agreements (SPAs)

The basis of project funding (except in the United States) is the revenue generated through long-term LNG SPAs. The principal provisions of the SPA establish the LNG sales price and indexation, buyers' offtake obligations (including quantity flexibility and rights to excess production), and sellers' delivery commitments, projected start of delivery, and procedures for narrowing the start date. Like many long-term contracts in international trade, LNG SPAs contain provisions setting out the duration of the contract, the responsibilities and obligations of the parties, the technical specifications of the LNG, definitions of the sellers' and buyers' facilities including ships, scheduling the deliveries, payment terms, dispute resolution, financial guarantees, and force majeure. LNG buyers and LNG projects sign deals if both believe they have the best deal they can get and the project believes the long-term income ensures sufficient return on investment and provides a fair allocation of price risk.

LNG pricing

For markets outside the United States, the United Kingdom and northwest Europe, LNG prices are generally indexed to crude oil or oil product prices. These links are not always linear, as some contract price formulas incorporate provisions to dampen the extreme ends of the oil price range. This formulation is known as S-curve due to the resulting shape of the price curve. In the early years of the LNG industry, oil and gas were viewed as somewhat substitutable for one another (except in transportation fuels), and it made sense to link the pricing. Where there is no competitive gas market, and where the gas price is passed on to the customer, there is much less pressure to set the price on competitive terms. The buyer has less economic incentive to achieve a lower price, and oil-indexed pricing has the advantage that the regulators who oversee the utility's pricing can readily understand it. On the seller's side, pricing linked to oil makes the LNG project revenue stream look more like the revenue stream from an oil project, a concept with which the producers, their shareholders, and financiers are comfortable. Once adopted, this pricing approach has proved to be very durable, surviving for four decades in the Japanese, South Korean, and Taiwanese markets with only small variations.

Oil indexation also has the advantage of being based on a market with liquid and transparent pricing, which prevents manipulation. However, while crude oil and oil products were a significant part of electricity sector inputs in the 1970s, natural gas increasingly competes with coal, not oil, in power generation. A few sales of natural gas and LNG in Europe have included indexation to coal or electricity prices. Increasingly, term gas prices are including hub-based prices at the demand of buyers.

Term LNG sales to Japan are indexed to the Japan Customs Cleared price (also known as the JCC, or Japan Crude Cocktail). The JCC is published monthly, usually with two months' lag, and is calculated as the total import cost divided by the associated total volume of imported crude and semifinished oils. Indonesian LNG (and crude oil) sales were indexed to the Indonesian Crude Price (ICP), a basket of Asia-Pacific crudes priced at Singapore. Indonesian sales to other

East Asian buyers follow this pattern. Some Indonesian contracts have a minimum price. Indonesian LNG prices are now switching to JCC at the urging of Japanese buyers.

• A typical JCC-driven formula:

LNG Price (LNGJCC) = JCC × slope + constant (e.g., shipping cost)

Traditionally the slope has been around 0.145 to 0.148, but Poten estimates that the slope to JCC will decrease from the current 0.145 owing to supply competition from new regions such as North America and East Africa. The constant is a fixed amount, usually less than $1.00 per MMBtu.

Some contracts have an S-curve structure, in which the slope of the price is reduced outside a central range. S-curves emerged between 1988 and 1990 when crude oil prices dropped into the $10.00 per barrel range. The price uplift assists sellers during periods of low crude oil prices and is balanced by price dampening when oil prices are high. S-curves were particularly common after the Asian financial crisis in 1997–98, when weak demand strengthened the buyers' negotiating position. Prime examples included the sales by BP from the Tangguh project in Indonesia to buyers in China and South Korea, and a 2001 sale from RasGas II to India's Petronet LNG that had the price fixed at a low level for the first five years. Subsequent surges in world crude oil prices, combined with a tightening LNG supply situation, have encouraged LNG sellers to demand—and receive—significantly higher prices from customers in Asia in recent years.

LNG sales to European buyers were traditionally indexed to European oil product prices, and some are similar to the pricing of Norwegian pipeline gas. Newer LNG sales to Europe generally continue to mimic pipeline sale prices, except for the Nigeria LNG sale to Italy's Enel, which included a large coal-based component. Increasingly, European buyers have been renegotiating these terms as they come under pressure from spot gas pricing, and there have been several instances of price revisions being taken to arbitration, with the outcomes heavily favoring the buyers. Sales to the United States were made on the basis of netbacks from wholesale gas market prices, such as the Henry Hub.

LNG exports from the United States offer a paradigm shift in LNG pricing, representing the first export of LNG where the feed gas prices are set at a gas trading hub. For future LNG sales from the United States, FOB LNG prices will be linked to the Henry Hub price under formulas similar to:

$$\text{LNG FOB Price} = \text{Henry Hub} \times 115\% + \text{liquefaction cost}$$
$$\text{of } \$3.00 \text{ to } \$3.50/\text{MMBtu}$$
$$\text{(partially adjusted for inflation)}$$

LNG offtake

In the absence of accessible alternative markets, the buyer's purchase obligation provides the revenue stream that sustains project funding. In the typical project-chain business model, the buyer commits to lift (or pay for if not lifted) an annual contract quantity (ACQ). Since variation in supply flow requires increased production, storage, and shipping capacity, annual volume flexibility is limited—often to less than 10% of ACQ and cumulatively to 30% of ACQ over the contract duration. Typically, the buyer is subject to take-or-pay for a high percentage of the ACQ, ensuring cash flow for the supplier. Failure to lift the ACQ can result in the buyer paying for the volume shortfall, with limited ability to recover the take-or-pay quantities later through makeup rights. The seller and buyer are obligated to ensure the availability of their facilities and to cooperate in the development of an annual delivery program for scheduling production, shipping, and delivery.

In the United States, the liquefaction capacity holder has no offtake obligation. However, the firm must pay the liquefaction capacity charge whether or not the capacity is used. These tolling fees ensure revenues to cover operating and loan costs, plus a rate of return on the liquefaction investment. Failure to lift LNG (or use the tolling capacity) gives the customer no rights in future production and is simply forfeited.

Shipping

In the traditional project-chain business model, ships are committed to the trade and are controlled or owned by the seller in a CIF or DAP trade or by the buyer in an FOB trade. Ships are typically owned independently of the liquefaction and import facilities, and ship financing is increasingly highly leveraged with debt secured by charter payments.

Funding

Project funding comprises sponsors' equity contributions and debt finance. In the first two decades of the LNG business, project technology, costs, and performance were uncertain, and the financial community was not sufficiently knowledgeable to engage in project financing (limited-recourse lending). Projects were financed through equity and loans advanced by buyers' banks, multilateral credit agencies, and export credit agencies (ECAs) and were often guaranteed by the sponsors.

Project finance, in which the lenders' recourse for debt repayment is primarily limited to the revenues and assets of the LNG project itself, began in earnest in the early 1990s with the financing of Indonesia's Bontang train F, and it has been widely employed since then. Project financings typically consist of a mix of syndicated loans from ECAs, other multilateral agencies, public or private bond financing, and commercial bank debt. The ECAs provide credit for project procurement in their home countries and create credit enhancement for other lenders as they carry attributes similar to sovereign lending, which reduces the political risk of borrower defaults. Ras Laffan Liquefied Natural Gas Company initiated the first successful capital market financing in 1996 for an LNG project, with an oversubscribed $1.2 billion bond issue.

Project financing is costly, with high up-front fees and higher interest rates than other long-term borrowings. However, it can bring structural benefits to the project: it removes the project debt from the sponsors' corporate balance sheets and allows much higher borrowing leverage by the project. It provides third-party funding to support national oil company participation. It brings the discipline of third-party due diligence to the technical and commercial structure of the project. It may help to align participation among the sponsors and host government entities that have disparate capabilities to underwrite an all-equity participation in a project without serious dilution of their shareholding. For all but the highest credit rated sponsors (ExxonMobil, for example), project financing usually lowers the overall cost of capital associated with LNG project financing.

Evolving Project Venture Structures

This section reviews in more detail the types and evolution of project venture structures and participation. The structure and character of participation in these ventures vary depending on the sponsors, the host country, market, resource access, and technical conditions. Ownership of facilities for upstream production, feed gas pipeline transportation, liquefaction, and the port may be integrated into a single venture (or legal entity) or be distributed among several ventures. Shipping can be chartered either to the LNG seller or buyer, but is typically owned and financed separately.

Project venture participation includes companies that have production rights to the resource. One or more major international oil companies (IOCs) may join—even if they were not originally involved in the resource discovery—bringing financial, technical, project management, marketing, and operating capabilities, and usually purchasing an interest in the reserves. If the host country has a national oil company (NOC), it typically has a significant equity share. Buyers may also make equity investments in order to ensure greater security of supply, access project information, and as an investment opportunity in exchange for their purchase commitment.

Types of project ventures

LNG supply project venture structures may be classified into six types.

State-owned and financed. The existing Algerian projects are state-owned and were financed through sovereign-guaranteed debt and ECAs.

Integrated joint venture project. The simplest structure in theory is the integrated joint venture, in which the ownership of the upstream, pipeline, and liquefaction segments of the LNG train are shared equally among all project participants. Examples are the Australia NWS, Qatar's Ras Laffan LNG, and Russia's Sakhalin II project. There is no transfer price between the upstream and liquefaction, so that all project segments are taxed as a single entity. Risks—price risk, production risk, and political risk—are shared equitably among all the project participants. Setting up an integrated joint venture, therefore, requires full agreement by the partners about the project's strategy and governance—in other words, a full alignment of interests.

Norway's Snøhvit demonstrates one of the benefits of the flexibility and simplicity of the joint venture approach: two of the project participants are taking their share of production as LNG, while the other three are selling at the FOB point to the two buying partners.

Resource taxation is often assessed on an imputed netback to feed gas production. Because of Australian fiscal regimes, the NWS is organized as an unincorporated joint venture, with each investor responsible for making parallel decisions; a sales contract with the NWS is, in fact, six separate but identical sales contracts with each of the six participants.

Split-revenue ventures. The upstream and downstream facilities are owned and operated by different ventures, and the LNG sales revenue is split according to an agreed formula between the liquefaction and production interests. This structure allows different shares in the upstream and downstream ventures. In some projects, buyer representatives and the NOC may participate only in one segment. Examples of split-revenue ventures are Brunei LNG, Malaysia LNG, Atlantic LNG Train 1, and Nigeria LNG. In this structure, different fiscal regimes typically apply to the distinct segments of the project. Other variations include further subdivision of interests, as is the case in Trinidad, where the National Gas Company also provides gas transmission services between the gas-producing fields and the liquefaction plant.

Fixed-price feed gas ventures. In this structure, the feed gas is owned and produced by the government or a government-controlled enterprise. Examples are ADGAS, Qatargas, Oman LNG, SEGAS, and Angola LNG. Feed gas payments may involve some LNG sales revenue sharing. For some projects, large associated condensate production provides significant revenue.[1]

Cost-based liquefaction. In Indonesian projects, the liquefaction facility, owned by the NOC, Pertamina, is 100% debt financed, and the borrowings are repaid on a cost-recovery basis. LNG sales revenues are deposited in a trustee bank, which then allocates the revenue to debt-service, liquefaction plant operating costs, and taxes, with the remainder going to the production-sharing contractors. The LNG plant makes no profit. This structure has the effect of pushing profits into the upstream, where they are subject to higher taxation.

Toll-based liquefaction project. The emergence of merchant trading by major oil and gas companies has led to a new structure, in which the producers retain ownership of the gas through the liquefaction facility, which is owned by the producers, buyers, and NOCs on some agreed split (which may not reflect the share of upstream production or purchase rights). The gas producers pay a fee to the tolling venture and sell the LNG on their own terms without reference to each other. A commercial structure that treats the liquefaction segment as a tolling facility provides a steady income for liquefaction and pushes economic rent into the upstream.

In this structure, the allowable rate of return (RoR) for the LNG trains is fixed (usually somewhere between 8% and 12%), so that the liquefaction part of the project receives a set portion of the project revenue to cover operating costs and investment returns. Feed gas price is based on a netback from the LNG sales price. The liquefaction segment, therefore, gets back its costs plus a modest return. Within this structure, the owners and operators of the liquefaction venture are less exposed to price and production risk. Their share of revenues is essentially guaranteed as long as the project has an after-tax cash flow that is equal to or greater than the allowable RoR. However, there is limited upside for the liquefaction segment, usually through earning additional fees through higher-than-guaranteed production and through reimbursements for common facilities in expansions. This limited sharing does ensure the commercial interests of the gas producers, LNG offtakers, and the LNG plant owners are aligned. Any economic rent over and above the RoR set for the project goes to the upstream. Nearly all of the risk on the project is therefore borne by the production segment.

Pushing rents upstream is a good way to remunerate the upstream participants for the risks they have taken in exploration and development of the gas fields. However, this system exposes more revenue to upstream taxation, which is typically higher than for liquefaction. The tolling-facility model is best applied in situations where either the liquefaction segment is controlled by a government entity or NOC that is not involved in the producing fields or where gas is sourced from a number of different fields, with different owners, some of whom may not participate in liquefaction.

Trinidad's ALNG is a good example of a downstream tolling structure. Ownership of ALNG's first train is split among five participants, including the national gas company, while only three IOC partners participated in Trains 2 and 3, and the state-owned gas company rejoined Train 4. In each case, the gas is sourced from different fields operating under different license agreements or PSCs. The liquefaction segment operates on a cost-plus-return system, while the feed gas price is based on a netback from the FOB sales. A similar structure is used for the LNG facility in Idku, Egypt, where the partners have built two trains, each liquefying gas from different Egyptian sources on a tolling basis.

LNG Industry Trends: The Emergence of Commercial LNG

Changes in natural gas markets have impacted the commercial structure of the LNG supply business. When monopoly-franchised utilities dominated natural gas and electricity markets, they controlled market access and the downstream sales of a project's LNG. The project-chain business model supported the financing of the supply project—but at a potential cost in lost demand due to uncompetitive LNG prices.

In the Atlantic Basin, utilities' monopoly franchises have been heavily eroded at the wholesale level. In North America, pipelines have split their transportation services and merchant functions, resulting in competitive markets for both the natural gas commodity and its transportation. This split has been furthered by the use of master limited partnerships (MLPs) as the ownership vehicle of choice in the energy infrastructure market, given their tax advantages as compared to conventional corporate structures. Virtually all interstate gas transmission pipelines in the United States are now owned by MLPs. Since MLP shareholders seek stable returns, virtually no MLPs have commodity trading operations, though some of the general partners of the MLPs do. The United Kingdom has a single gas transmission provider, but the wholesale and retail markets are fully competitive. In continental Europe, a parallel process is underway driven by EU market liberalization policies, although national champions continue to play a much larger role, often encouraged by their host governments (who are their former shareholders).[2]

Where there is effective open access through import terminals and domestic transportation systems, the LNG seller may have to acquire access and transportation all the way to the electricity generators and gas distribution companies or turn to independent energy merchants to fill this role. Ironically, the move to liberalization has resulted in a decrease in the transparency of and access to LNG terminals. In North America, LNG terminal access is negotiated between the terminal owner and parties seeking capacity, and there is nothing to prevent an LNG terminal owner from retaining the entire capacity for its own purposes. The European Commission has been granting exemptions to open access for many, if not all, new LNG terminals. For example, South Hook, Dragon, GATE, and Dunkirk are not subject to regulated third-party access.

LNG merchants are emerging to take advantage of the opportunities and challenges that are opening up. These are large companies capable of owning and managing a portfolio of diversified positions throughout the chain: equity in supply projects, control over shipping, and import terminal capacity. They engage in commercial LNG—that is, a mix of long-term sales and purchases under increasingly flexible conditions, and short-term trading.

In this emerging structure, the role of the LNG project as a merchant seller recedes. In some projects, the LNG is sold to one or more of the sponsors, who assume the downstream merchant function. In others, the liquefaction plant is a tolling facility, and upstream gas producers retain title through the chain. In the United States, the tolling customers arrange feed gas supply (except in Cheniere's projects) and market the LNG. There is no common pattern, and each project is developing individual commercial characteristics that reflect its shareholder makeup, financing requirements, and target markets.

In Taiwan and China, utility buyers still prevail, encouraged by government policies. In Japan, the utilities still dominate as well, but they are seeking to broaden and increase the flexibility of their supply portfolios to enhance security of supply and trading options. Opening access in South Korea and domestic merchant buyers in India are enhancing LNG competition and the value of flexible trading. Shrinking LNG demand in the United States and Europe has also encouraged the growth of commercial LNG as it is rerouted to more lucrative markets.

The emergence of merchant trading has encouraged industry entrants who do not fit the classic roles of the club. Statoil took an import position at Cove Point and began trading through the terminal before it began construction of the Snøhvit project. Financial institutions, such as Citigroup, Merrill Lynch (now owned by Bank of America), and Morgan Stanley, added LNG to their commodity-trading operations.[3] Gazprom, expecting to become an LNG exporter, entered into LNG trading through gas and LNG swaps in both the Atlantic and Pacific basins. More recently, traditional commodity trading houses such as Gunvor, Vitol, Koch, and Cargill have entered LNG trading. Many of these new participants own no liquefaction, ships, or import terminals. Meanwhile, some firms like Cheniere Energy invested in LNG import terminals and leased capacity to traditional LNG producers. As US shale gas production rose, LNG imports declined, terminal owners moved to add liquefaction capability, utilizing existing marine infrastructure and LNG storage, to benefit from the expansion in US unconventional gas production and the wide price gaps between regional markets. The results for the pure traders and banks have been mixed, and the banks especially are under pressure to exit commodity trading to meet regulatory capital requirements.

Project Finance

As described earlier, there are many models for the commercial and legal structure of the LNG chain components. However, the most common structure has been the joint venture between the participants in the liquefaction plant, representing arguably the most critical and certainly most capital-intensive component of the chain. The joint-venture structure accommodates equity participation by a variety of shareholders—typically IOCs, NOCs, buyers, and other entities, such as the Japanese trading houses and downstream utilities that may have some original stake in the resource or bring marketing and financial expertise.

A joint venture has an independent financial structure that often lends itself most readily to limited-recourse debt financing secured by the revenue and assets of the venture. Initial funding during the project feasibility and development stage comes from equity provided by the sponsors. Ultimately, the project venture may contract for debt financing from some combination of commercial banks, export credit agencies (ECAs), multilateral agencies (such as the World Bank or Inter-American Development Bank), and bonds floated in capital markets.

The term "project finance" usually refers to debt ultimately secured solely by the revenues and assets of the joint venture and with only limited recourse to the equity sponsors. This financing is generally put in place before construction of the project begins, and the loans are drawn down and equity contributed as the construction proceeds. In such a structure, the lenders will not take project completion risk, and the equity sponsors will guarantee debt service until completion and performance testing are demonstrated to previously agreed standards—hence the term limited recourse.

In most projects, the sponsors require EPC contractors to assume a measure of completion risk and performance guarantees, along with bonuses for early completion and production above design standards, through lump-sum, turnkey EPC contracts. This form of risk allocation reduces the sponsors' completion risk but clearly requires EPC contractors with strong balance sheets of their own. Where other EPC contract structures are used, the sponsors may need to assume a much greater degree of completion risk.

History and motivation

Project finance has a long history but is a relatively recent development in LNG. It requires sophisticated lenders and borrowers and entails a complex series of agreements and commitments between a wide group of participants to secure and allocate the cash flow, risks, and contingency obligations. Project finance lenders also require specific analysis and oversight of the technical and commercial viability of the project through the loan syndication process (where the loans are shared among many banks).

In the early history of the LNG business, liquefaction plants were financed through sponsors' equity or sponsor-corporate loans. NOCs were rarely involved in the upstream projects (Algeria and Nigeria are exceptions), which were equity financed by the gas producers. This was the case with the projects in Alaska, Brunei, Abu Dhabi, and Malaysia LNG Satu. In Indonesia, the liquefaction plants are owned by Pertamina. The first five trains of both the Arun and Bontang projects were financed through Japanese loans. The sixth Arun train was financed by a loan from Mobil, the upstream producer.

One of the earliest limited-recourse loans in the LNG business was made for Woodside Petroleum's participation in the NWS project. Since the project was unincorporated to permit the pass-through of early tax losses, the loan was secured on a limited-recourse basis by Woodside's one-sixth interest in the project. Subsequently, project finance was used for Indonesia's Bontang Train F; Malaysia LNG Dua; both the upstream and downstream ventures of Qatargas; Trinidad's ALNG Trains 1, 2, and 3; Oman LNG; and the integrated Ras Laffan project, which floated LNG's first capital market bond issue. Limited-recourse project finance has become commonplace in subsequent projects and those already under construction.

Typically, projects considering or seeking project financing will retain a financial adviser in the early stages of the project to support the financial analysis and provide the lenders with a degree of assurance that the project has been well-structured and is expected to provide financially attractive returns to the shareholders during its life. Advisers are usually units of commercial banks, such as Société Générale (SocGen), Citigroup, and the Royal Bank of Scotland (RBS), or investment banks, such as Credit Suisse. The advisers also play a key role in validating the project's financial projections, identifying the optimal borrowing structure and sources of funds, and helping conduct negotiations with the lenders.

Advantages of project finance

Project finance is expensive, typically requiring millions of dollars in up-front fees and the set aside of financial reserves that cover several months of debt service while carrying a higher interest rate than other forms of debt finance. Nevertheless, it brings several advantages.

- Project finance does not appear on the participating sponsor's corporate balance sheet. Projects are usually financed with much higher leverage than corporations, with strong projects able to borrow in the range of 60%–80% of their capital requirements. Only the sponsor's share of the equity investment in the project is shown on their balance sheets. Project financing allows the sponsor to maintain a stronger balance sheet (lower debt-to-equity ratio) and gives it improved corporate financial flexibility, especially where the share of the project would represent a very significant investment exposure for the sponsor. During construction and acceptance testing, when lenders still have recourse, total investment and funding may require disclosure in the sponsors' financial statements; although, to the extent the completion risk has been shifted to the EPC contractors, this disclosure may be limited.

- Project finance may support larger, politically required participation by the host NOCs—who may be hard-pressed to provide equity funding—and may therefore avoid delays or other complexities arising from the need for NOC funding. It also facilitates participation of smaller IOCs, who may be less financially robust than the majors. Project finance can help align the interests of sponsors with very diverse financial capabilities.

- The need for third-party due diligence, which lenders require for the entire project chain, provides checks and balances in the project structure and can help uncover and correct issues that could later prove problematic. It is often more politically acceptable for NOCs to accept funding with commercial and legal conditions laid down by independent lenders than to accept such terms from IOCs who are the partners of NOCs in the upstream investments.

- The participation of multilaterals, commercial banks, and ECAs may deter host countries from expropriation or other significant legal or regulatory changes, such as taxation or currency convertibility after start-up.

Risk mitigation

Key risks lenders will seek to minimize include:

- Management of design, permitting, construction, and operations—especially the risk of cost overruns or misestimating. This includes the responsibility for future capital investments to maintain the plant or comply with legal or regulatory mandates, as well as local labor agreements.

- Experience and reputation of the borrowers/ sponsors and the other counterparties to the major commercial agreements.

- Legal rights to the gas resources and to the land where the plant is to be built.

- Technical performance of the gas fields and pipelines.

- Fiscal, regulatory, and commercial frameworks of the host country.

- Market revenue risk—both price and volume.

- Contract rights that can lead to termination or suspension of key agreements and halt LNG sales or operations. This includes mismatches between the terms of key agreements in the supply chain that could give rise to claims by parties in the future.

- Appropriateness of insurance coverage at each stage of the project development and commercial operation, with the lenders' interests secured for loss of assets or business interruption.

- Diminution of lender's rights after financial closing, such as during subsequent plant expansions, which could access the secured assets or impinge on them.

- Project-on-project risk, such as when a third party is required to provide shipping or import terminal capacity, which in turn may be contingent on their own commercial, legal, financing, and funding agreements.

The lenders make their financial commitment after all the elements of project formation are put in place. Among the agreements and undertakings the lenders will require are the following:

- Joint venture agreements defining participation, funding obligations, shareholder rights, decision making, corporate governance, future expansion rights, and rights to acquire and dispose of assets.

- Host country fiscal, regulatory, and operating agreements, including any necessary licenses, rights-of-way, and title to property required for the plant.

- Reserves delineation and certification, and confirmation of expected upstream development plans and production forecasts by independent petroleum engineers.

- Binding agreements for the supply of gas, provision of shipping, and the sale and purchase of LNG.

- An EPC contract that provides appropriate assurance that the construction will be completed on time, within budget, and to the specified level of performance.

- Independent verification of the estimates of future gas prices and buyers' market access and development.

- Assurance of the availability of sufficient shipping and import terminal capacity to secure the project's LNG offtake.

- An appropriate package of insurance coverage, including commitments to maintain agreed coverage during the project's construction and operation.

The lenders will seek to understand whether the project represents a financially viable undertaking for all parties, not just sponsors. An LNG SPA that proves uneconomic for the buyer will not likely survive the full term, as history has demonstrated. Venture participation must include one or more companies with credible technical, financial, and management capability to form, construct, and operate the project. Significant participation by the NOC, where present, should align sovereign interests with the project.

Revenue security is enhanced by the SPA terms, which allocate market pricing and volume risks between the seller and buyer. Recent developments in destination flexibility should reduce buyers' and sellers' volume exposure, though these may not receive full credit from the lenders, as they are not yet time-tested. In many projects, buyers' payments for LNG are made to a third-party bank, which disburses funds for operating expenses, debt service, upstream gas suppliers, taxes, and, finally, to the shareholders according to an agreed prioritization process.

Structure and participants

Typically, the package of bank loans to a project may have a tenor (term) of 8–12 years after completion, with an initial debt service grace period of several months. A reserve account covering up to 6 months of debt service is set aside to cover debt repayments in the event of an unexpected interruption or diminution of the project's cash flow. Considerable attention is paid to structuring the priority of claims on project cash flow, known as the "cash flow waterfall." Innovations in recent project financings include enhanced credit support from the sponsors to protect against low prices at RasGas and subordination of feed gas payments by Oman LNG. Features such as this can permit a higher degree of leverage than might be the case in their absence. The loan packages themselves will typically be the most lengthy and complex agreements negotiated and executed by the project.

The principal sources of project finance have been ECAs and major international commercial banks, including ABN Amro, BNP, Barclay's, Citigroup, Credit Lyonnaise, JPMorgan Chase, Schroeder's, and Société Générale. Large Japanese banks, such as the Industrial Bank of Japan and the Bank of Tokyo, have traditionally participated in LNG project finance. Multilateral development banks—such as the Inter-American Development Bank, the Asian Development Bank, and the African Development Bank—may participate. These banks are often focused on financing social infrastructure development tied to the export project.

Export credit will usually be provided by export-import (ExIm) banks (also known as ECAs) for equipment and services from the lending country. Some ECAs provide direct loans, while others provide the insurance, principally involving trade with other OECD countries that allow commercial banks to lend. Prominent ECAs are the US Export-Import Bank, the Japan Export-Import Bank, the Korean Export-Import Bank, the UK Export Credit Guarantee

Department, France's Compagnie Française d'Assurance pour le Commerce Extérieur (COFACE), and Italy's Servizi Assicurativi del Commercio Estero (SACE). ECAs are constrained to lending to OECD countries and often cooperate with one another. Some ECAs provide untied loans on the grounds that the LNG will flow to their country. Project shareholders have also provided substantial loans to credit-constrained partners; for example, ExxonMobil made loans to its partners in Papua New Guinea.

Credit enhancement in the form of sovereign risk insurance is provided by official credit enhancement agencies, such as the US Overseas Private Investment Corporation (OPIC), which may also act as lenders in their own right. Commercial banks are increasingly using political risk insurance (PRI), including coverage against currency inconvertibility or nontransferability, expropriation of assets, stock or bank accounts, and political violence, including war, terrorism, and civil strife.

Performance

For the most part, LNG projects have performed extremely well from a financial point of view. Sustained high double-digit returns on equity are not uncommon, and the scale of the projects can have a meaningful impact on even a major company's financial performance. During the mid-1980s, the Arun project accounted for as much as 40% of Mobil Corporation's worldwide after-tax income, and Shell is believed to have sustained after-tax returns in excess of 20% on its LNG investments over decades.

All trades to East Asian buyers have proceeded without financial interruption or distress, although contractual shortfalls from the Bontang project since 2005 somewhat undermined this record. The fixed-pricing terms for the early Alaska and Brunei projects were renegotiated in the mid-1970s oil price shock era. Difficult restructuring of pricing and offtake terms between buyers and sellers was completed without service interruption through the turbulent price and market conditions of the 1980s.

There have been two significant commercial and financial interruptions. The Algerian–US trades were interrupted by regulatory changes that exacerbated the commercial failure of the underlying

SPAs in the face of high international oil prices, collapsing US gas prices, and the deregulation of the US gas market. The Dabhol project in India was never economically viable given the high price of the project's electricity, which made the power unmarketable, and it stalled completely when sovereign support underpinning the power purchase agreements was withdrawn.

More recently, large LNG import facilities were built in the United States and the United Kingdom as part of Qatari megatrain schemes. But when demand diminished or disappeared altogether, under the flexible project model adopted by the Qatar projects, LNG volumes were diverted to premium outlets in Asia without any loss of export volumes, though the venture had to scramble to arrange different shipping arrangements. Similarly, the collapse of the US import market in the face of the "shale gale" did not result in the financial failure of any of the terminal owners, as the terminal costs were largely underwritten by large capacity holders who had the ability to divert their LNG to higher value markets elsewhere. The same is also true of the European terminals developed post-2000.

It appears that new commercial structures for projects, with their enhanced flexibility, which opens up new opportunities, can help diminish project risks through changes in the traditional risk allocation in the LNG chain.

Notes

1. Condensate, or liquids, credits represent income accrued to the project from the production of associated condensate, LPGs, or (in some cases) crude oil. These credits essentially lower the per-MMBtu cost of producing LNG.
2. This is despite EU gas and power sector liberalization measures. For example, in the spring of 2006, the European Commission raided the offices of 20 gas firms across five countries as part of an inquiry into anticompetition practices. Germany's RWE, France's Gaz de France, Austria's OMV, and Belgium's Fluxys said they were raided. Reports said firms in Italy and Belgium were also involved. The Commission said it was an initial step into "investigations into suspected anticompetitive practices." Separately, the Commission said it also raided power companies in Hungary. The EU has recently expressed concern about the need for more competition in energy markets across Europe, after a cold winter in which political problems led to disruption of supplies. The Commission was particularly worried about the way firms granted or restricted access to pipelines. See "EU Probes 20 European Gas Firms," *BBC News* (May 17, 2006), http://news.bbc.co.uk/2/hi/business/4990292.stm.

3. In March 2006, Sempra LNG executed a terminal use agreement with Merrill Lynch Commodities, Inc., a subsidiary of Merrill Lynch & Co., Inc., for capacity at the Cameron LNG import terminal in Louisiana. The 15-year, full-service capacity agreement provides Merrill Lynch Commodities with the capability to import 0.5 Bcf/d (5.2 Bcm/y).

13

Upstream Gas Supply Agreements

Introduction

The LNG chain begins with the natural gas reserves that will be produced and delivered to the liquefaction plant as feed gas, including fuel for the production of LNG. The terms and conditions for the delivery of gas to the LNG plant are generally set out in a gas supply agreement (GSA). Its specific terms can be heavily influenced by the energy policies of the host government, the agreements that give the upstream venture the right to produce and sell gas to the LNG plant, the ambitions of the NOC, and the LNG business structure established by the project sponsors in consultation with the host government.

The purpose of the GSA is to formally establish the conditions under which gas will be supplied to the liquefaction plant. Key issues that must be agreed upon include:

- **The party(s) responsible for supplying gas to the LNG plant (the gas supplier[s]).** That party may be the government (or national oil company) of the host country or a producer that has been awarded contractual rights, often under production sharing contracts (PSCs), to produce, deliver, and sell natural gas owned by the host country.

- **The quantities of natural gas to be delivered to the plant, the reserves committed to the LNG project, and the daily and annual production rates that the upstream venture**

will maintain for the feed gas supply and the permitted variability of deliveries. Natural gas reserves and production commitments are essential to assure the LNG project sponsors, their buyers, and their lenders that the project has sufficient natural gas resources to fulfill all the LNG delivery commitments and allow for some reserve margin.

- **The duration or term of the GSA.** The starting date and the duration of the gas delivery commitment must correspond to the obligations to deliver LNG, as well as provide initial gas supply during commissioning of the liquefaction plant.

- **The price of natural gas supplied to the plant.** This is a key element that establishes the allocation of project revenue between the liquefaction and upstream ventures. Project participants should expect this transfer price to be heavily scrutinized, because the LNG revenue allocation will have a major impact on the government's tax receipts. In most countries, taxation rules for manufacturing and processing industries, such as liquefaction, are quite different (and usually less onerous) from the fiscal regime that applies to upstream natural gas production.

- **The quality of the natural gas to be supplied.** This includes upper limits on constituents such as moisture, CO_2, and H_2S, which must be removed during the pretreatment process. The quality specification may also permit the delivery of higher Btu hydrocarbons, and the disposition of these liquids, if separated from the LNG, must also be addressed.

- **The natural gas delivery point.** This determines whether the upstream or downstream venture has the responsibility to construct, operate, and maintain a pipeline from the gas field(s) to the liquefaction plant.

- **The responsibility for metering, measurement, and billing, as well as any specific payment support (such as a letter of credit) in favor of the producer.**

- **Any approvals and authorizations, which must be obtained by the parties prior to the GSA coming into force and effect.** Examples might include government authorizations for the upstream development and the liquefaction plant, shareholder votes to proceed with the construction of the necessary facilities, and requirements associated with financing.

- **The amount and allocation of damages in the event of a failure either upstream or downstream of the liquefaction plant.**
- **The choice of law and dispute resolution provisions.** This often includes expert determinations and arbitration.
- **The applicability of force majeure.** This excuses one of the parties from performance under the GSA owing to events beyond the affected party's control.

LNG Business Structures

While all GSAs are expected to address the issues listed above, the business structures that can be chosen for the LNG chain present different issues to be addressed in the GSA(s). The business structures currently adopted in LNG projects can be grouped into three categories (as described in chapter 12):

- Separate upstream and downstream ventures
- Fully integrated ventures
- Natural gas tolling arrangements

Separate upstream and downstream ventures

Most LNG projects are organized into separate upstream and downstream ventures, a structure that requires the most comprehensive GSA. In this case, there is an upstream/production venture that may have different participants or different participation interests than the downstream/liquefaction venture. A variation occurs when two or more upstream ventures supply gas independently to a liquefaction plant. In either case, a comprehensive GSA is needed to establish the parameters of the gas supply deliveries. However, even though there are separate upstream and downstream ventures, the parties engaged in the upstream are most likely to be the participants in the downstream liquefaction venture.

When there are different participants in the upstream and downstream segments of the LNG project, it is important that the GSA clearly define the commitments and obligations of the upstream gas suppliers and the liquefaction venture as the gas user.

Fully integrated LNG ventures

There are only a few fully integrated LNG ventures, where the participants have the same ownership interest in natural gas production facilities, liquefaction, and LNG production and sales. When there is equal ownership in all phases of the LNG project, the participants' interests are aligned. Therefore, all the parties face similar risks and rewards for the overall performance of the project, and a formal GSA may not be required. Instead, gas supply arrangements and gas utilization issues can be addressed in shareholder participation agreements and/or project operating agreements. Notwithstanding the common interests of the participants, establishing the transfer price of natural gas from the production phase to the liquefaction phase may require a formalized GSA to satisfy the host government's requirements.

With a fully integrated structure, the natural gas transfer price—the price at which gas is transferred from the upstream venture to the liquefaction venture—does not arise from bargaining among independent interests that characterize an arm's-length transaction. Therefore, the tax authorities will carefully review this price and the method used to establish it. The government may require that LNG revenues be "netted back" to the production venture, paying the equivalent of tolling fees for pipeline transportation to the plant and for liquefaction. This requires an explicit GSA to formalize these terms, and the government may well be a party to or an approving authority for the terms of the GSA.

Natural gas tolling ventures

Natural gas tolling agreements are based on the premise that liquefaction plants and ships are tools for monetizing natural gas assets and transporting gas to markets. The upstream and downstream participants must agree on the formulation of the tolling fees, including rates of return most appropriate for liquefaction and shipping investments. The LNG sales revenue is then allocated to the upstream/production venture(s) after the agreed payments are made for the liquefaction tolling fees and ship charters. In this business structure, the gas supplier owns/controls the LNG at the tailgate of the LNG plant and is responsible for marketing the LNG. This can raise another set of issues in the case where the LNG is sold FOB to the upstream gas suppliers. If they have established back-to-back LNG sales to third parties (whether ex-ship/DAP or FOB), then transfer pricing should not be an issue. But if they are acquiring the LNG for portfolio supply, without such sales, there may be challenges from the host government on the pricing of the FOB purchases.

The LNG tolling structure limits the scope of the GSA because the party selling LNG is also responsible for supplying gas to the LNG plant. Nevertheless, some form of gas supply and processing agreement is required, and the scope of an individual agreement depends on the specific nature of each arrangement. The issues covered in the agreement will be somewhat different than in the agreements described previously and may also have more of the characteristics of a processing agreement, including the following:

- Obligations of the operator of the liquefaction facility, including LNG daily and annual production capacity, scheduled beginning and duration of operations, and feed gas requirements provided by the downstream venture.

- The gas suppliers' rights to utilize LNG production capacity, storage, and loading facilities, including accommodating the rights of other gas suppliers.

- Scheduling the delivery of natural gas into the liquefaction plant and the production and loading of LNG, as well as balancing arrangements to accommodate timing differences between gas deliveries, LNG offtake, and variations in agreed production rates between different gas suppliers.

- Payment obligations for the liquefaction capacity, which may include a capacity charge for the allocated capacity and a volume-based processing fee for its use.

- Feed gas and LNG quality specifications and liabilities for supplying off-specification gas or LNG.

- Rights or obligations of the unaffected suppliers to supply gas in the event of a failure of one of the gas suppliers.

Beyond these issues, the tolling agreements can quickly become much more complex, providing terms and conditions for, among other issues: rights and obligations under future plant expansions or debottlenecking of the existing plant; rights to spare capacity and excess cargoes; rights to use common facilities shared by subsequent expansions; adjustments to the tolling fees for changes in operating expenses or additional capital expenditures; obligations of the operator; and mechanisms to replace a nonperforming operator.

The US LNG export projects represent a sharp departure from the earlier tolling arrangements because they do not rely on a dedicated natural gas resource or on a traditional GSA. Typically, the liquefaction plant is owned primarily by third parties that have avoided or limited both price and commodity risks by locking in a firm payment from the capacity holders. The capacity holders, who may take stakes in the liquefaction venture, are responsible for acquiring feed gas, primarily buying it on the market from gas producers or other wholesale gas suppliers at prevailing market prices and then marketing the LNG at prices that they hope will generate a satisfactory profit. Typically, the acquisition of gas supply will be subject to forms of agreements that prevail in the market for any large buyer of natural gas.

Typical Terms in a GSA

Quantity

Perhaps the most critical part of the GSA is the dedication of proven natural gas reserves and production capacity. The ability of the project to meet its long-term LNG delivery obligations depends on the reserves committed to the project. One million tons (MMt) of LNG is equal to about 47 to 50 Bcf (1.3 to 1.4 Bcm) of natural gas (depending on gas composition). An additional 8%–10% of the gas is needed as fuel for the liquefaction plant. Thus, a 20-year sale of 1 MMt/y (1.4 Bcm/y) of LNG will consume about 1.05–1.15 Tcf (30 to 33 Bcm) of gas over the life of the contract. These volumes exclude any 'shrinkage' that occurs in processing due to LPG, acid, and inert gas removal.

Since liquefaction plants typically operate about 340 days per year, the production of 1 MMt/y of LNG requires a feed gas rate of 150–160 MMcf/d (1.5 to 1.7 Bcm/y); this rate has to be sustained from the first day of full commercial operations through the last day of the GSA. The reserve base required to support this production rate throughout the term of the contract is about 20% larger than the natural gas consumed, allowing for the supply of natural gas during plant commissioning and testing, and the natural decline in reservoir performance at the end of the primary production period. Since the typical liquefaction plant will produce above its design rate and will remain productive beyond the initial contract term, the GSA will also spell out procedures for supplying gas above the volumes needed for the firm LNG SPA volumes, as well as rights or obligations to supply gas during SPA extension periods. To the extent the LNG buyer(s) incur take-or-pay and earn makeup rights, the GSA must also address the shortfall (if any) for gas deliveries during the LNG customers' failure to accept LNG as well as provide for the supply of gas for this makeup LNG.

Producers dedicate specific natural gas reserves certified by an internationally recognized third party, such as DeGolyer and MacNaughton, Gaffney Cline, or Netherland-Sewell. Moreover, the producers commit to construct and maintain production facilities and to produce and deliver natural gas at specified production rates over the life of the project. Hourly, daily, and annual production rates must be agreed on, as well as procedures to coordinate maintenance periods. If there are multiple suppliers, the specific quantities delivered by each supplier must be agreed on, as well as procedures for suppliers to substitute for one another and balance uneven deliveries.

The volumes are usually dedicated under a take-or-pay arrangement, closely mirroring the take-or-pay arrangements between LNG suppliers and LNG buyers, and, in turn, the GSA may provide for claims arising against the supplier in the event of an unexcused failure to deliver gas as required. The LNG SPAs will contain obligations to deliver LNG in specific quantities within a specified schedule and with tightly defined quality conditions. The SPAs will also establish the commercial terms of an LNG sale, including damages for an unexcused failure to perform (that is, to deliver LNG). In addition, if the LNG SPAs provide for volume flexibility on the part of the buyers (for example, downward quantity adjustments or rights to excess quantities), the GSA must in turn reflect these provisions to match the supply commitment with that of the LNG sales.

The GSA also spells out the expected schedule for the development and commissioning of the project facilities (production, pipeline, and liquefaction) and provides mechanisms to coordinate these efforts (as well as rights and obligations if one of the parties fails to complete their required facilities on time). Also important are the obligations to supply gas during the plant's commissioning and start-up period (often described as the initial supply periods) when the LNG production process will be somewhat uncertain, and the LNG buyers may not be obligated to purchase all the output. Generally, the full obligation for purchase and any corresponding take-or-pay obligations do not commence until the liquefaction plant passes all its performance tests and is declared commercial by the operator.

Quality

An LNG SPA contains obligations to deliver LNG in specific quantities within a specified schedule and with tightly defined quality conditions (see chapter 14). The LNG SPA contains specific composition and heating value ranges with which the delivered LNG must comply; it also imposes upper limits on impurities permitted in the LNG. The liquefaction plant is designed to process feed gas with a specific range of hydrocarbon constituents, and it is important that the feed gas conform to this range. In addition, impurities in the delivered feed gas must be limited to the maximum levels which the LNG plant is designed to accommodate in the front-end gas processing, where impurities are removed.

Delivery point

It is important to establish which party is responsible for building, operating, and maintaining the pipeline system (which may in fact be developed and operated independently of both the upstream and the liquefaction ventures) that delivers feed gas to the liquefaction plant and to establish at what point the title of the gas is transferred. In some cases, the upstream gas-producing venture is responsible for the pipeline and must deliver the feed gas to the liquefaction plant. In other cases, the liquefaction venture is responsible for the pipeline and purchases the gas at its inlet. Occasionally, a third party who may or may not be a sponsor of the liquefaction plant may develop the pipeline, normally under contract to the producer. The delivery point is also the point at which the feed gas volumes and composition are measured through metering and sampling.

Measurement and delivery conditions

To determine the volume, heating value, and composition of natural gas at the delivery point, one party must install and operate measurement devices. The parties must agree on the following: who will install, operate, and maintain the measurement equipment (usually the liquefaction plant); the accuracy of the equipment; how and by whom will the measurement be conducted; and who will set procedures for periodic verification of the measurement equipment. It is also necessary to agree upon procedures to adjust past measurements in the event that a measuring device is found to be inaccurate.

Delivery pressure plays an important role in the efficient operation of the liquefaction plant, as all of the gas cooling and refrigeration occurs at high pressure. The higher the feed gas delivery pressure, the higher temperature at which gas liquefaction takes place, reducing the required refrigeration power and improving the process efficiency. The LNG flow is finally flashed to near-atmospheric pressure by passing the flow through a throttling valve or an expander after all the refrigeration has been applied. Specifying the minimum delivery pressure is a key component of the GSA and is a significant consideration in the design and operation of the liquefaction plant. Where the upstream gas field operates at naturally high pressures, this may not be an issue, but where this is not the case, there will be potential trade-offs as to which party is responsible for compressing the gas to optimize pressures at the liquefaction plant. As the gas fields are produced, the natural pressure will decline and may require the installation of compression by the producer during the later stages of the field life.

Natural gas sales price

This provision establishes how LNG revenues are allocated to the segments of the LNG supply chain—specifically between the upstream/production and downstream/liquefaction (including shipping in the case of CIF or DAP sales) ventures. Equally important, this allocation of revenue impacts the taxes and other payments to the government resulting from the LNG sales. The natural gas sales price can be set in a variety of ways and is often the subject of significant debate between the LNG project and the host government.

Governments tend to view natural gas as having great intrinsic value and perhaps as the most valuable portion of the LNG supply chain. The liquefaction venture, usually led by a multinational oil and gas company, often takes the view that its technological and commercial know-how, market access, and financial support for the LNG project bring significant value, allowing monetization of an otherwise stranded natural gas resource, which it may have discovered. Both views are valid. Clearly, there can be no LNG project without large gas resources to be dedicated to the liquefaction venture. Likewise, stranded gas could remain undeveloped without the technical, commercial, and financial support of the international sponsors.

The goal of negotiation between the government and commercial parties is to establish a commercial basis to develop an LNG project. To accomplish this, the project's commercial agreements must balance the rewards the international sponsor receives for the skills and financing that they bring to the venture and the risks they will likely take, with the host government's views of the value of their natural gas resources. Many of these terms may be incorporated in the original concession agreements under which the gas exploration and discovery was carried out, but these agreements rarely include all the provisions necessary to define the allocations of revenues and costs (although the fully integrated project model based on production-sharing contracts comes close, since all the costs, including liquefaction, are generally treated as development costs). Since all gas monetization options are unique in their costs and revenues, PSCs and concession agreements generally do not contain specific gas terms, and the negotiation of these terms can be a source of project delays.

Some GSAs set the feed gas price as a fixed or formulaic percentage of the price of the LNG, usually on an FOB basis. This method implicitly values each segment of the LNG project, assigns a relative value to each, and allocates revenue accordingly. This method is used in Nigeria LNG, Malaysian LNG, Brunei LNG, and the first train of Atlantic LNG.

Other agreements set a floor price for the feed gas and establish percentage sharing of LNG price whenever the resulting gas price exceeds the floor. Examples of this are the Qatargas and RasGas projects. In Oman, Oman LNG purchases feed gas from the Sultanate of Oman at a fixed price that is adjusted periodically. As a variation on this approach, these price mechanisms may also provide that, when the percentage sharing yields a price below the floor price, the difference in revenue from the formula share is notionally "banked," and the gas buyer is then allowed to recoup the overpayment if the price rises above the floor by paying a price below the formula. Once the bank balance reaches zero, the regular formula applies. This approach provides a more stable flow of revenue to the gas producer and, in turn, to the government through royalties and taxes.

For tolling facilities and some integrated ventures, the feed gas price is essentially a netback from the LNG sales price. That is, producers deduct largely fixed fees for shipping and liquefaction from the LNG revenue, and the remainder is allocated to upstream gas. This method is used in the Indonesia projects, Atlantic LNG Trains 2, 3, and 4, Kenai (Alaska), Egyptian LNG (Idku), and Australia's Darwin LNG venture. At US tolling plants, the feed gas price is determined by the market.

Allocation of the value of by-products

Valuable by-products, such as natural gas liquids (NGLs) and condensate, are often recovered during the production and treating of natural gas before liquefaction. They may be recovered by the upstream venture when natural gas is treated in the field before being delivered into the pipeline and/or at the LNG plant when gas is treated before liquefaction. In the latter case, an explicit allocation of the value derived from those by-products may be included in the GSA or the government's concession or PSA terms. It may not be appropriate to value NGLs and condensate on the same $/MMBtu basis that is used to value feed gas, and a separate measure tied to the market value of the by-products may be developed. This is particularly relevant when different suppliers are providing gas streams of different qualities and are expecting to receive a different share of the NGL sales proceeds. If this is the case, the GSAs will also need to include separate measurement, pricing, and other mechanisms to determine each producer's share of the NGL proceeds.

Liabilities

If the plant does not receive its properly nominated feed gas supply, the liquefaction venture will not be able to meet its contractual delivery obligations and may be subject to a claim for damages from its customers. Producers may claim damages where the liquefaction plant or LNG ships do not perform. On the other side, the liquefaction venture may be entitled to receive take-or-pay compensation from its customers or damages from its EPC contractors or other contractors if they fail to perform. In both cases, the GSA must allocate risks, liabilities, and recoveries between the upstream and downstream. The allocation and treatment of payments received by the liquefaction venture for take-or-pay obligations (or in the alternative, amounts paid by the liquefaction venture for an unexcused delivery failure) are the most complex to address, as issues of alternative sales, makeup, and other provisions of the LNG SPA must be, in turn, reflected in the GSA in a balanced fashion.

Force Majeure

Force majeure is an issue closely negotiated in both the GSA and the LNG SPA. Force majeure excuses a party from performing its obligations, wholly or partially, if events beyond its control prevent it from doing so. The right of the LNG seller to be excused from its GSA obligations under force majeure for a downstream failure (a problem affecting the liquefaction plant, shipping, or the customers' facilities) is an important issue. A topic of ongoing debate is whether damage to natural gas reservoirs or loss of reserves that result in the upstream not being able to perform should constitute a force majeure event that must be resolved in the GSA. As the industry has matured, with more examples of failures or potential failures in performance, so too have the force majeure provisions become more complicated.

Conclusion

The GSA, a critical agreement in the LNG supply chain, should be given as much attention and effort as the LNG SPA. The host government will be particularly interested in transfer price issues, as this has a major impact on the government's tax realization from the LNG project.

It is difficult to characterize the specific terms that must be covered in each GSA of each LNG supply chain. The nature of preexisting agreements between the gas suppliers and the host government will set the tone for the GSA. The LNG SPAs will set the terms by which LNG is purchased (volume, price, and flexibility in offtake), which in turn will flow back to the upstream producers via the GSA. However, the specific business model selected for the LNG supply chain will dictate the types of terms needed in the GSA to clarify rights, obligations, and liabilities and to allocate the risks and rewards of the producing and liquefaction segments of the supply chain.

14

LNG Sale and Purchase Agreements

Introduction

Sale and purchase agreements (SPAs) are the commercial cornerstone of an LNG project. The SPA is the key agreement that apportions risks and rewards along the LNG chain. It sits at the intersection of the upstream (where production and liquefaction remain very closely integrated in most projects, except those in the United States), and the downstream, with the meeting point being either the loading port, in the case of a FOB contract, or the unloading port, in the case of a CIF or DAP contract. The SPA incorporates provisions governing pricing, volume commitments, LNG specifications, shipping arrangements, measurement and testing procedures, payment terms, scheduling (initially of the project and ultimately of the cargo loadings or discharges), governing law, dispute resolution, and force majeure provisions.

The arrangements for shipping the cargoes are usually specified in summary terms in the SPA, but the shipping itself will often be the subject of separate agreements, including LNG tanker charters, which along with similar agreements are covered in chapter 15.

Historical Context

The LNG industry has traditionally been characterized by long-term contractual relationships between buyers and sellers. When the industry developed these long-term contracts, two major elements emerged almost immediately. The first element reflected an adaptation from the then-prevailing practices in the pipeline industry, namely, the take-or-pay gas sales contract. In the 1960s and 1970s, virtually all pipeline gas contracts involved a take-or-pay commitment on the part of the buyer. (Most of these contracts were found in the US and Canadian markets, where the natural gas industry was more developed and more diversified than in other parts of the world.) Under take-or-pay provisions, the buyer is obligated to lift a specified volume of natural gas or LNG each year or pay for the shortfall at the contractual price. The rationale behind this approach was to secure the financing of the capital-intensive natural gas supply chain through a predictable revenue stream.

The second element was the adoption of pricing mechanisms that tied the price of LNG to oil prices. This was not the first LNG pricing model: the first models (early Alaska and Algeria) treated the liquefaction project almost as a utility undertaking that tied the price of LNG (either CIF or FOB) to a fixed price with escalator mechanisms, often including elements of labor and steel pricing as a proxy for the underlying costs of building and operating the projects and recovering an appropriate return on investment. This was not dissimilar from the pricing mechanisms that had evolved for US interstate pipeline tariffs, where the rates for transportation services were set by reference to the recovery of "just and reasonable" costs of and returns on the pipeline investments.

This pricing structure broke down rapidly under the combined (and, to some degree, linked) impacts of rapidly increasing inflation and oil prices in the mid-1970s. Following the oil shocks and the energy crises of the 1970s, oil exporters (many were also members of the Organization of the Petroleum Exporting Countries [OPEC]) who were also LNG exporters argued for a linkage between oil and natural gas prices. This move had the added benefit of providing a way to enhance revenues and avert the collapse of several LNG projects that had found themselves in financial difficulties. The early contracts were adjusted to reflect new terms at the insistence of the sellers, with a few notable exceptions; for example, in the case of El Paso's contract with Algeria, the parties were unable to reach a final agreement, and the contract then collapsed.

By 1980, virtually all LNG sales contracts included take-or-pay provisions and pricing formulas directly linked to crude oil, frequently including an S-curve formulation, which protected the buyers by dampening the linkage to high oil prices and sellers by dampening the linkage to low oil prices. The establishment of this pricing structure was aided by the position of the buyers who almost without exception tended to be utilities and, in the case of non-US companies, utilities that were either owned directly by the host government or strongly influenced in their energy purchase policies by these host governments. In all cases (the US buyers included), the utilities' cost structure and financial drivers did not mitigate against this behavior. Under the rules generally in place at that time, utilities' earnings were a function of their invested capital base, and provided the costs for LNG supply acquisition were deemed reasonable, the cost of LNG supplies would be treated as a direct pass-through to consumers. As discussed in chapter 3, this did not reflect competition in the end-use market (since utilities at that time were thought to be natural monopolies and not susceptible to market forces), nor did it lead to particularly hard bargaining between buyers and sellers. Buyers were more concerned about the security of supply for what was generally believed to be a finite and depleting commodity than about the price.

Reinforcing this tendency was a belief that LNG represented a diversification of national energy supply (particularly for Japan) away from a dependence on crude oil (dominated then by supplies from the Persian Gulf and OPEC), or the securing of gas supplies where (in the case of the United States and Europe) domestic supplies were seen as being limited. Finally, because many of the importing countries had a high degree of dependency on crude oil and oil products for much of their energy mix, especially in power generation, natural gas (whether in the form of LNG or otherwise) could be seen as being in direct competition with oil and therefore could be reasonably priced on the same basis. Beginning in 1964 and continuing uninterrupted through 1976, the LNG industry experienced rapid and sustained growth almost independent of the pricing of LNG supplies.

However, beginning with a series of legislative and regulatory changes in the United States, this model came under increasing pressure. In 1978, the United States passed the Natural Gas Policy

Act, which set the domestic producing industry on a path toward wellhead price decontrol by 1985. Gas consumption in the United States had been rising steadily until the early 1980s, but higher prices, coupled with a collapse in industrial demand following the severe recessions that occurred during the same time period, began to take their toll. A sharp fall in gas demand resulted, just as supply was increasing under the stimulus of higher wellhead prices, and the infamous gas "bubble" emerged in the United States (later characterized more as a gas "sausage," reflecting its extended duration).

In 1984, the FERC issued Order 380, which permitted the customers of the gas pipeline companies to refuse to purchase gas, which they had been otherwise contractually obligated to buy. However, Order 380 did not afford the same opportunity to the pipeline companies, which remained bound to their agreements with their suppliers. The result was a severe financial crisis that saw the major pipeline companies doing everything in their power to escape their purchase obligations, leading producers to assert major claims under the take-or-pay provisions of these contracts. Several pipelines entered bankruptcy (some more than once), and all suffered serious financial consequences in settling take-or-pay claims. The sole LNG supplier to the United States, Algerian state-owned company Sonatrach, also saw each of its contracts collapse under the same pressures, and, between 1979 and 1985, the United States went from being the largest to one of the smallest LNG importers in the world.

Europe was less affected, since the number of suppliers and buyers was much smaller and the market was less well developed than in the United States. However, the effect that oil pricing had on LNG contracts—and the loss of the US market—led European buyers to exert much stronger downward price pressure on their LNG suppliers (essentially Sonatrach). When oil prices collapsed in 1986, LNG prices fell sharply. In many cases, price indexation was renegotiated and tied to oil product pricing (distillate and heavy fuel oil) rather than crude oil pricing, since products were seen as both less volatile than crude oil and more reflective of the alternative prices competing against natural gas (gasoline, diesel, and jet fuel were excluded from the price mix since natural gas did not compete directly against these products). Asia (for all practical purposes, Japan) initially continued to experience steady economic growth and did not seek the same form of price restructuring as seen in Europe and the United States. However, the rate of growth slowed as costs in the LNG chain escalated and global economic growth slowed. Many of the contractual terms today reflect experiences from these decades, as well as the realities of natural gas deregulation in major LNG markets, including electricity.

LNG contracts are lengthy documents and have only become lengthier with the passage of time. To a degree, the technical provisions of these contracts have become standardized over time. The procedures for measuring and testing LNG volumes and quality are one example. In other respects, though, these contracts remain highly specific to the individual trade being covered, and this aspect is unlikely to change while LNG is supplied into disparate markets through dedicated chains.

Long-Term Contracts

Nearly three-quarters of the LNG traded in 2014 was still bought and sold under long-term contracts. In the most simplistic view, the value of the trade is established by the quantities, prices, and duration of the LNG sales under the SPA. The other fundamental commercial consideration in the SPA are the terms of sale—FOB, CIF or DAP—and the resulting obligation to provide shipping.

Quantities

Most long-term LNG contracts contain very specific provisions regarding the quantities of LNG to be bought and sold. Most are subject to take-or-pay provisions. Typically, the contract will provide for an annual contract quantity (ACQ), which is the quantity of LNG, usually expressed in MMBtus, that the buyer must purchase and pay for if not taken. However, the ACQ is subject to significant requirements and adjustments that must also be taken into account.

Most contracts covering new trade arrangements contain provisions that govern the buildup of deliveries during the initial period of the contract. These buildup provisions are designed to accommodate both buyer and seller.

- On the seller's side, very often the start-up of a liquefaction plant will require a period of reduced production in order to fully test the plant's technical systems, accommodate any delays in plant completion, and address the delivery of LNG tankers to be used in the trade. As the technology of liquefaction plants has improved, the technical window needed for plant testing has narrowed, reducing the need for buildup on the seller's side.

- On the buyer's side, the buildup period is designed to address the need to grow the downstream market in order to handle the full ACQ.

Most contracts provide for the supply and purchase of LNG during the start-up period, when the volume and timing of LNG that can be produced at the liquefaction plant is uncertain. Generally, LNG sales and purchases during the start-up period are not subject to take-or-pay requirements but are instead governed by reasonable efforts to deliver and take LNG on both sides. There are frequently LNG volumes available for spot sales if actual production volumes exceed the planned production ramp up.

Once the buildup period has ended, the contract generally enters a plateau period when the ACQ remains constant over an extended period. As described earlier, the ACQ is normally expressed as a fixed quantity (in MMBtu) to be delivered on an annual basis. However, the ACQ is subject to a variety of adjustments. Among those adjustments are:

- **Volume flexibility, or downward quantity adjustment.** This may permit the buyer to reduce its ACQ obligation by a fixed amount. Generally, this is a small quantity, normally around 5% of the ACQ, and some contracts limit the number of adjustments, or the aggregate adjustment that the buyer can make over the term of the contract.

- **Round-up/round-down provisions.** These provisions are designed to address the reality that while the ACQ may be expressed in a fixed quantity (in MMBtu), such a fixed quantity cannot be delivered precisely in cargo parcels during each contract year. The variations in the actual quantities delivered can arise through scheduling provisions of the contract. For example, a cargo which loads or delivers the day after the contract year ends may actually be attributed to the delivery obligation of the previous contract year, and vice versa. Other factors relate to variations in cargo sizes (either because different sized vessels are used or because the MMBtu content of the LNG varies slightly during the period) and to permit reductions in ACQ obligations to allow for scheduled maintenance of the facilities (including the LNG tankers).

- **Excess quantities.** These arise when the plant performs better than expected or when a buyer in a multiple-buyer project exercises a downward quantity adjustment. Most LNG plants are designed with a margin of technical conservatism, which results in the plant being able to produce more LNG than the aggregate of the ACQs. The treatment of excess quantities in SPAs can vary and can be the focus of hard bargaining during the SPA negotiations. Normally, these excess quantities are highly profitable for the seller, and may also be profitable for the buyer, especially for buyers who can now earn a profit margin on each LNG delivery. The SPA must determine whether the buyer has preferential rights to these excess quantities, and if so, whether these rights are subject to the same terms and conditions as the ACQ.

- **Incorporation into the ACQ.** Another consideration is whether, once confirmed, these excess quantities become incorporated into the ACQ and covered by take-or-pay obligations. The treatment of excess quantities is also driven by the control of shipping in the SPA. An FOB buyer may be in a much stronger position to argue for preferential rights to the excess, as the buyer could have the shipping to lift these quantities. For a DAP or CIF contract, the leverage may be with the seller, who is in a better position to control excess shipping. Excess quantities can have different characteristics. Predictable improvements in plant performance result in predictable excess quantities, while other excess quantities may be less predictable (arising from a buyer's force majeure, for example, or seasonal variations in plant performance). The former are often treated as part of the ACQ (and some contracts even provide mechanisms for increasing the ACQ to formally incorporate this), whereas the latter require different treatment owing to the inability of either the seller or buyer to predict them.

- **Makeup quantities.** These constitute a class unto themselves. In the circumstance of a take-or-pay contract where the buyer is unable to or declines to receive the ACQ (a shortfall) and is not otherwise excused from performing (as a result of a force majeure incident, for example), the buyer is nonetheless obligated to pay for the shortfall at the price specified in the contract (which may not be the full contract price). In return, the buyer is usually afforded the opportunity to take delivery of the equivalent (makeup) quantities at a later date. The makeup quantities are normally given scheduling priority over all other quantities (such as excess quantities) after the firm delivery obligations. The makeup quantities can be taken in subsequent years, including a period after the primary term of the contract. In some instances, there may be a limit (five years following the incurrence of the take-or-pay is typical) during which time the makeup quantities must be taken or else be forfeited. When the makeup quantities are taken, they may also be subject to a price adjustment (either to make up any price discount given when the shortfall was incurred or to take account of any price difference between the time of the shortfall and the time of delivery).

- **Force majeure makeup quantities.** These are not always found in LNG contracts. These are quantities that have not been delivered following a force majeure incident, but that the parties agree can be delivered at a later date. In this respect, they are treated much as take-or-pay makeup quantities, although they usually have a lower priority in scheduling.

From this description, it should be clear that the quantity provisions could become very complex, especially when there are multiple buyers and/or sellers in the project. This is particularly true in the case of the interaction between the quantity provisions and scheduling provisions (which establish the cargo delivery programs), which will require addressing the priority rights between different types of quantities and flexibilities, as well as tracking the quantities attributable to each buyer by category.

Another element of the quantities section of the contract is the treatment of expansion quantities. Generally, in contracts that are negotiated for the start of new projects, the parties may reasonably expect that the liquefaction plant be expanded at a subsequent point in time. For buyers, access to these expansion quantities can be a valuable right, which they will naturally seek to negotiate in the contract. There is no standard treatment of expansion quantities. They may be quantities taken at the buyer's option (subject to demonstration that the buyer has adequate terminal capacity to take them); they may be quantities which are treated as obligations of the buyer and the seller; they may be treated as quantities added to the basic quantities of the contract; or they may be subject to the negotiation of an entirely new contract (which may or may not incorporate many of the provisions of the original contract).

Sellers may be less inclined to offer rights over the expansion quantities, as they wish to be free to market these on their own terms. The contract generally spells out the buyers' and sellers' rights with respect to any expansion, along with notice periods, negotiating periods, and any conditions precedent, which must be satisfied before the commencement of negotiations. Buyers were generally the parties seeking rights over expansion volumes. The sellers were generally resistant, because they hoped to obtain more favorable sales terms in their expansion sales contracts. This dynamic shifted with the advent of a buyers' market in the late 1990s, and the issue of expansion quantities became less important and less controversial (and ultimately disappeared from many SPAs). By 2004, the pendulum began to swing in the opposite direction, once again in favor of sellers. Clearly, the treatment of these provisions becomes a function of the buyers' and sellers' relative bargaining leverage.

Pricing

While quantity provisions in most long-term SPAs generally share common characteristics, the same cannot be said of the pricing provisions. The pricing formulas tend to be tied to specific markets. They also tend to be set by reference to prices of other fuels (including natural gas). These generally reflect perceived market conditions in the three largest LNG market regions: Asia-Pacific, Europe, and the United States (though the United States is no longer so important). The influence of the industry's history, as described in the opening of this chapter, also factors into the evolution of pricing terms in LNG SPAs.

While prices in long-term SPAs have typically been linked to prices in downstream markets, the emergence of US LNG exports from 2015 tends to introduce a new pricing dynamic with a direct link to US hub prices, primarily Henry Hub in Louisiana, which is considered the North American benchmark. International buyers have been pursuing these supplies, betting that they will be available at a discount to LNG supplies that are primarily linked to crude oil (at least that was the case when oil prices were near or above $100/bbl). For many buyers, this represents an attractive diversification of pricing in their portfolio of LNG supplies.

Asia-Pacific. Generally, LNG prices in the Asia-Pacific region are set with reference to crude oil prices. Because this LNG is sold into markets where there are no competing supplies of natural gas, this mechanism has been well established for many years. In Japan, LNG is pegged to crude oil prices by a formula indexed to the Japan Customs Cleared (JCC) price for crude oil imports (also called the Japan Crude Cocktail). Indonesian LNG had been priced against Indonesian crude oil, but recent contracts have shifted to JCC as well. Meticulously compiled by the government, the JCC consists of up to 200 types of crudes from about 30 oil-producing countries depending on the sources for any particular month. This combination makes it highly immune to manipulation. It also captures the cost required to ship fuel supplies to the region. The original concept tied the landed cost of LNG in Japan quite closely to the landed price of crude oil imported into Japan. At the time, many power stations in Japan burned crude oil. This price would be adjusted downward if the LNG were purchased FOB to take into account the cost of LNG shipping assumed by the buyer.

The basic pricing structure of Asia-Pacific region pricing can be represented as:

$$P_{LNG} \ (\$/MMBtu) = \text{\ss} \times P_{CRUDE} \ (\$/bbl) + \alpha$$

where:

- ß is 0.1485 for early Indonesian sales. In subsequent contracts, it has been negotiated higher in a tight market or lower in a weak one.

- P_{CRUDE} is the average JCC price. Different contracts specify different periods over which prices may be averaged and subsequently lagged.

- α represents non–oil price related terms that may include: a constant, shipping cost indices (in CIF or DAP contracts), and inflation indices. Some contracts contain minimum price provisions.

As figure 14–1 shows, Japan's LNG prices closely follow the JCC's movements. Any time lag is the result of the delay in Japan's customs authorities publishing the data (up to two months after the end of the month) and the time taken for LNG to be shipped, invoiced, and paid for. This delays the immediate impact of sudden changes in the oil price on the LNG market.

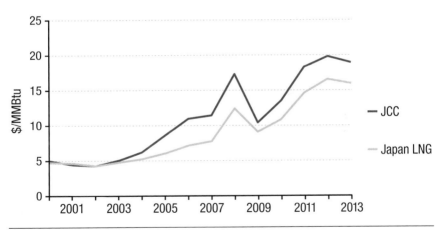

Fig. 14–1. LNG pricing in Japan: JCC vs. weighted average LNG price. (*Source:* Poten & Partners.)

The increased volatility of oil prices led to modifications to the original formulas to reduce the associated volatility in LNG prices, which neither buyer nor seller saw as beneficial. Price formulas began incorporating dampening mechanisms, the best known of which is the S-curve (fig. 14–2). In this formulation, the rate of increase (or decrease) in the LNG price is reduced in relation to the rate of increase (or decrease) in the crude oil price above (or below) a certain price level, often with built-in floors and ceilings. This gives the seller downside price protection and the buyer upside price protection. From the seller's perspective, this also makes the upstream project more readily financeable, as the lenders (more focused on the downside) can count on a more predictable flow of revenues. South Korea and Taiwan have followed the Japanese approach to the pricing formula.

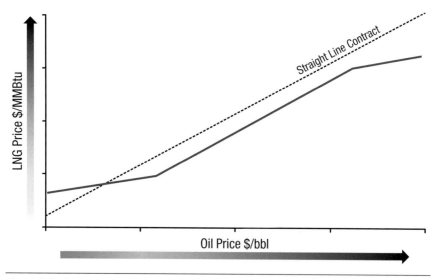

Fig. 14–2. S-curve moderation of price increases and decreases. (*Source:* Poten & Partners.)

In the buyers' market of the early 2000s, these pricing mechanisms came under pressure as the rate of growth in the traditional markets slowed down, just as LNG supply costs were falling. New customers in China, India, and South Korea aggressively sought lower prices. China held competitive bidding, which resulted in LNG prices significantly lower than prices paid by Japanese buyers (depending on the crude oil price assumptions). India also sought to import LNG at lower prices. Initial Indian contract prices included price mechanisms, largely fixed over the initial life of the contract or subject to very minor adjustment—an ironic shift back to the pricing formulas of the earliest days of the industry. India differs somewhat from the other Asian markets, because it has access to domestic gas production and significant industrial gas demand. More recently, as their appetites for LNG grew and the market moved in favor of the sellers, buyers in China and India agreed to contract prices in new term purchases consistent with those paid by other Asian buyers.

Europe. European gas markets outside the United Kingdom and the Netherlands do not have significant domestic production. As in the Asia-Pacific, LNG (and pipeline gas) prices are set by reference to oil, but in this case to oil products as well as crude. The pricing structures tend to follow the pattern of a base FOB price indexed to a single crude oil (for example, Brent), or a basket of crudes, or a basket of products that compete with natural gas in the end-use market, including low-sulfur light and heavy fuel oils.

Most European-oriented contracts do not have S-curves, but many have floor prices. The use of a floor price in any contract raises the issue as to whether the contract should also include a ceiling price or, alternatively, a recoupment mechanism that applies when the floor price payments exceed the formula price payment that would have been otherwise applied. This recoupment provision would allow the buyer to reduce the contract price to recover the overpayment once the contract price moved above the floor price.

The sample price formula below is typical of a traditional European purchase contract. It contains references to several fuel alternatives with different weightings that reflect a particular market's proportion of alternative fuel uses. Moreover, instead of pegging the LNG price solely to current prices, the formula compares the changes in the various alternative fuels from a base period.

$$P_n = P_0 \times (W_1 \times F1_n/F1_0 + W_2 \times F2_n/F2_0)$$

where:

- P_0 is original negotiated price (at time 0). P_n is the price at time n.
- W_1 and W_2 are weighting factors/percentages of alternate fuels.
- F1 and F2 are alternate fuels' prices published by third parties. Gas oil, low-/high-sulfur fuel oil, and coal are common alternatives. $F1_0$ and $F2_0$ are the prices for F1 and F2 at time 0, and $F1_n$ and $F2_n$ are the prices at time n.
- An inflation component may be added.

Alternative pricing mechanisms

In the Atlantic Basin market, gas increasingly provides the fuel for power plants, which need to compete in increasingly open electricity markets. As such, the electricity produced from gas must compete on a dispatchable basis with electricity generated from other fuels, such as heavy oil, coal, and even nuclear. To ensure a reasonable level of dispatch, the pricing in power-oriented gas contracts may also include factors indexed to high-sulfur fuel oils, coal, inflation, and even power prices. These indexation factors tend to produce a less volatile gas price and in turn permit the power generator to dispatch the plant(s) with a reasonably high load factor. The advantage of these indexation factors from the seller's perspective is that they provide a more stable cash flow. Note that such indexation does not necessarily translate into lower gas prices, since often the base price can be higher than in a conventionally indexed contract, given the expected high efficiency of modern combined cycle gas turbine power plants.

Hub-based pricing (imports). *North America.* Given the more financially liquid gas markets in North America, LNG pricing mechanisms are generally set with reference to price indices in those markets. For LNG terminals in the Gulf of Mexico, the natural price point is set by reference to Henry Hub pricing or Houston Ship Channel pricing, both of which are viewed as liquid and transparent reference price points. In 2005–06, it appeared that CIF or DAP sales were being made at prices between 85% and 93% of Henry Hub, irrespective of the discharge terminal. On the East Coast, the pricing may be more complex since the basis differentials between this region and the Gulf of Mexico could be taken into account in the price formulas. Historically, East Coast prices have been at a premium relative to the Gulf of Mexico, reflecting, among other factors, the cost of transporting natural gas from the producing region to the end market. However, with the surge in Marcellus Basin's shale gas production, this has changed, as natural gas supply from this prolific basin has resulted in lower prices in Northeast US markets during much of the year.

Europe. The United Kingdom, similar to the United States, is a liquid natural gas market. Natural gas prices in the UK market, as well as pipeline and LNG imports, reflect pricing at the National

Balancing Point (NBP). NBP is a virtual or notional trading location for the sale, purchase, and exchange of UK natural gas. It is the pricing and delivery point for the International Exchange (ICE) natural gas futures contract. It is the most liquid gas trading point in Europe. Natural gas at NBP trades in pence per therm (a dekatherm [10 therms] equals 1 MMBtu). The NBP is similar in concept to Henry Hub but is not an actual physical location, so one cannot effect the physical delivery of natural gas under a futures contract as is occasionally done at Henry Hub.

As trading hubs in northwest Europe become increasingly liquid, they are influencing the market price for natural gas in this regional market. Sellers of pipeline natural gas, such as Norway and Russia, are now revisiting their term sales contracts to include a natural gas hub price component at the buyers' insistence. These buyers were losing sales to rivals with access to hub-based natural gas and to cheap US coal for power generation. Market participants are showing greater confidence in the continental hubs (TTF, Zeebrugge, NCG, and GASPOOL), where traded volumes have greatly increased in recent years (fig. 14–3).

Oil-indexation is still prevalent in southern and eastern Europe. The lack of market liquidity in the southern hubs and lack of supply competition given limited connections to the European grid have limited the penetration of hub pricing in these markets. The dominant position of Russia and Algeria in supply for many southern and eastern European countries gives these sellers a strong hand in maintaining oil-indexation in their supply contracts to these countries. In the longer term, markets in southern and eastern Europe will likely follow the lead of northwest Europe, and there promises to be an erosion of oil indexation, with prices becoming increasingly linked to traded natural gas hubs. European energy policy is increasingly focused on eliminating barriers to free trade in natural gas, through the requirement that all pipelines be open access, and the promotion of cross-border interconnector pipelines.

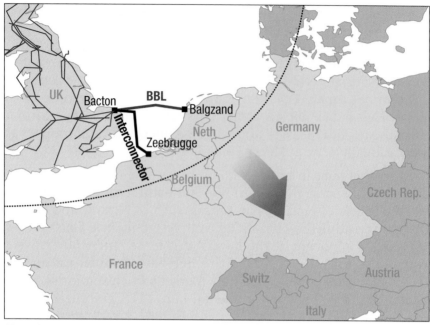

Fig. 14–3. Front line between gas-on-gas and oil indexation moving south in Europe. (*Source:* Poten & Partners.)

Hub-based pricing (exports). The emergence of a US liquefaction industry with LNG exports linked to hub-based prices introduces a new paradigm into global LNG pricing. The first Henry Hub–based LNG will likely reach the market in late 2015, when Cheniere Energy's Sabine Pass Liquefaction project is commissioned. Within 10 years, these exports could reach 60 MMt/y (82 Bcm/y), representing a large, new component in global LNG trades. Buyers, attracted by the opportunity to diversify their LNG purchase portfolio away from purely oil-linked pricing, have apparently also been attracted by the potential for trading profits in these purchases.

A typical FOB Henry Hub–linked price formula is:

$$P_{USGC} = \text{Henry Hub} \times 115\% + \text{liquefaction fee}$$

where:

- P_{USGC} is the FOB price US Gulf Coast.
- Henry Hub is the benchmark trading price in Louisiana.

- The 15% premium to Henry Hub is intended to cover pipeline and basis costs.
- Liquefaction fee is typically $2.50 to $3.50/MMBtu, with very small indexation factors.

In 2013, Henry Hub averaged $3.71/MMBtu, which would yield an FOB price of approximately $7.75/MMBtu under this formulation. Adding a trading margin of approximately 10%, or $0.75/MMBtu, and a transportation charge of approximately $2.50/MMBtu yields a price of $11.00/MMBtu ex-ship Japan. In 2013, the average LNG import price in Japan was $16.17/MMBtu, underscoring the interest of Japanese and South Korean buyers in LNG from the US projects. By contrast, the UK NBP averaged $10.13/MMBtu in 2013, yielding a far narrower margin even though the transportation cost is approximately half of that to Asia. The margin of less than a dollar for European sales would quickly disappear if oil prices weakened (as happened during the second half of 2014 into 2015) and/or if Henry Hub prices strengthened. By early 2015, with oil trading at or less than $50 per barrel, the attractiveness of US LNG in either Asian or European markets had disappeared, even with a Henry Hub price falling to less than $3.00/MMBtu.

Generic price provisions. There are certain provisions that are often (but not universally) applicable across pricing mechanisms, regardless of the specific trade. For example, many contracts allocate the responsibilities for payment of taxes, fees, and customs duties on LNG sales between buyers and sellers. Generally, the sellers assume the obligation for these payments in the exporting countries, and buyers assume them in the importing countries. Assessments on LNG tankers depend on whether the terms of trade are FOB or DAP. Port and harbor dues are treated as transportation costs and generally are not reflected in LNG pricing terms.

Irrespective of the expected LNG sales destination, many LNG contracts provide mechanisms for the periodic renegotiation of the price terms to reflect changes in the end-use markets. These renegotiation mechanisms can be open-ended or may contain specific guidelines that limit the range of price adjustments that can be adopted. They also provide for the ability to refer price reopeners to third parties (including arbitration provisions) when the parties are unable to reach agreement on their own. Generally, LNG SPAs provide specific intervals during which prices can be reopened; however, if these occur frequently, it is conceivable that one reopener provision may still be in negotiation/arbitration by the time the next scheduled reopener comes along. These provisions are less prevalent where gas markets are liquid and there is little dispute over how to determine the market value of natural gas (for example, in the United States and United Kingdom). Even in Europe, the rapid spread of gas trading and hub-based pricing has led to the resetting of LNG SPAs (often through arbitration) to reflect hub pricing.

Contract duration

Most traditional LNG contracts tend to have durations of 20 years or longer, reflecting a balance between the gas reserves required to be proven up front to support the sales commitment and the need for a period long enough to ensure that the appropriate investment return can be achieved and term financing secured. The primary terms of the contracts do not include the start-up and buildup periods, which are treated separately (as described earlier). As discussed in the "Quantities" section, the primary term of the contract may be extended to provide for the buyer to recoup makeup LNG quantities, whether arising out of a take-or-pay obligation or force majeure. Other mechanisms can also result in an extension of the primary term. For example, the contract may provide for an extension mechanism automatically or at either the buyer's or seller's option for a certain period (for example, five years). This extension may be on the same commercial terms in the initial contract, may provide for limited reopening of key commercial terms (price or quantity), or may simply offer one or the other party the exclusive right to a limited period of negotiation for the extension quantities. Normally, such extension provisions are conditional to the buyer and the seller being in compliance with the contract at the time the extension right is exercised.

Other Terms and Conditions

Unlike the elements discussed above, other terms of LNG SPAs have more in common across different markets. These provisions include gas quality, transportation, scheduling, invoicing and payment, measurement, transfer of risk and title, force majeure, choice of law and dispute resolution, allocation of liabilities, and termination rights.

Gas quality

Where gas quality was once almost a formality, it has now become more of an issue as the LNG trade has expanded and diversified. Traditionally, the Pacific customers look for LNG that has a richer or heavier blend of constituents (a higher proportion of ethane, propane, and higher hydrocarbons) than is sought by other buyers. For most buyers without domestic gas supplies, the issues of interchangeability and blending are of less concern. As a practical matter, heavier LNG means that more MMBtus are transported for each cubic meter or metric ton of LNG, thereby reducing the transportation costs per unit of energy, as well as making life simpler for the LNG producer, who would otherwise have to deal with these constituents as by-products with limited markets. However, when buyers are considering LNG of markedly different qualities than they traditionally accepted (as is the case with eastern Australian and US LNG supplies, which are very lean in contrast to most Asian supplies), the handling of the product can give rise to concerns over rollover in storage, as well as emission issues on finely adjusted gas turbines. The result has been the adoption of provisions, often through price adjustment, to compensate the buyers for the costs of handling atypical supplies.

Conversely, the North American market looks for much leaner product that typically reflects the composition of domestic pipeline gas, which is stripped of components for petrochemical feedstock. The European market falls somewhere between the two extremes. This issue, known as gas interchangeability, continues to evolve, especially in the United States and the United Kingdom. Regulatory authorities are attempting to find a balance between facilitating new LNG supplies and assigning cost responsibility for ensuring gas interchangeability between the suppliers and the customers.

At the same time, the introduction of advanced gas turbines and tighter emissions standards in the power generation sector has placed an increasing emphasis on LNG composition, since these power plants are designed to run within a very limited range of gas quality variances. As the trade has become more globalized, balancing market requirements has become more of an issue for LNG sellers and buyers. SPAs now reflect tighter gas quality specifications, and the risks associated with dealing with off-quality or off-specification (off-spec) LNG have increased. Typically, the contract remedy for a cargo of off-quality LNG has been to give the buyer the right to reject the cargo, but this remedy is so draconian that it has never been used. Rather, contracts incorporate provisions that require the buyer and the seller to cooperate to deal with off-quality LNG, usually by mandating that the seller reimburse the buyer for the cost of treating the LNG to bring it into the desired specification range. For rich LNG entering a lean market, this can be achieved by blending nitrogen into the regasified LNG, and for lean LNG entering a rich market, this can be achieved by spiking the LNG with heavier hydrocarbons. Major LNG sellers are now designing and modifying their liquefaction plants to produce differentiated streams of LNG, often with segregated storage, to enable them to customize LNG blends for different markets.

What has not changed is a limited tolerance for impurities entrained within the LNG, which in any event is much less of a concern, since liquefaction plants typically remove these impurities as they would interfere with the liquefaction process.

Transportation

In the context of an SPA, transportation is, in its simplest terms, the obligation to provide sufficient shipping of an agreed specification to deliver the ACQ. On the basis of the earlier discussion, it is clear that transportation becomes a more important issue in the flexible trading world that is evolving. The responsibility for transportation is very different between an FOB contract (buyer provides transportation from the loading port) and a CIF or DAP contract (seller provides transportation to the discharge port). Contracts of affreightment occasionally appear in the LNG trade, but for practical purposes these are similar to DAP contracts. Generally, LNG contracts afford the party not providing the transportation the right to inspect or vet the ships.

In consideration of other aspects of the transportation provisions, the following section is written from the perspective of an FOB contract. In the case of a CIF or DAP contract, the obligations and undertakings are essentially the same but the roles of seller, buyer, load port, and discharge port are effectively reversed. In an FOB contract, the seller is obliged to provide a safe berth, which the LNG tanker can access and leave from under all normal conditions. The buyer is obligated to ensure that the tankers are compatible with the berth, meet the requirements for loading times, have up-to-date measurement equipment (since quantities are measured by gauges in the cargo tanks), be of the proper size (or size range), and satisfy the operating standards and quality specified in the SPA. The buyer is responsible for compensating the seller if the LNG tanker remains on the berth for longer than the allowed period as a result of a problem with the LNG tanker (excess berth occupancy time). The seller is responsible for compensating the buyer in the event the tanker's loading is delayed owing to a problem with the seller's facilities (demurrage), including the failure of an earlier-arriving tanker to have completed loading and departed on schedule. The buyer is also responsible for ensuring that the tanker notifies the seller in advance of its anticipated arrival time. When the tanker arrives at the loading port, it will serve a notice of readiness to the seller confirming that it has cleared all customs and immigration checks and is ready to proceed to the berth and load the cargo. In a CIF or DAP contract, these obligations and undertakings are essentially transferred from the seller to the buyer and from the loading to the discharge port.

A unique aspect of FOB transportation relates to the temperature of the tanker's cargo tanks when it arrives at the loading port. If the LNG tanker has failed to retain sufficient heel on board since its last cargo discharge (and almost certainly if it has just left a shipyard where it has been gas freed), then the cargo tanks must be gradually cooled to cryogenic temperatures before the tanker can begin accepting cargo at full loading rates. The cooldown process not only extends the normal loading time but also requires the liquefaction plant to provide quantities of LNG, which is returned as vapor, requiring further liquefaction, as a result of being warmed in the cooling process. Typically, the buyer must compensate the seller for the extra time and the cost of the LNG used in cooling down the cargo tanks. In some cases, the seller may delay loading or even refuse to load a ship not already cool.

Other miscellaneous provisions include responsibility for provision of liquid nitrogen or bunkers for the tankers, and/or facilities to load ship's stores at the berth.

Scheduling

In the traditional LNG project model, scheduling was a relatively straightforward process, given the inherent spare capacity in almost every element of the LNG chain. In a typical Asia-Pacific contract, schedules were agreed between the buyers and sellers months in advance of each contract year and varied little, if at all. If they did, it would only be for unusual circumstances, such as a typhoon disrupting the loading, transportation, or discharge of a limited number of cargoes.

In the trading world, scheduling takes on more critical importance. A failure to optimize schedules can result in idle liquefaction, shipping, or terminal capacity and the loss of an associated financial opportunity; but given the potentially different incentives and objectives of buyers and sellers, one party's opportunity may come at another party's cost. No one wants to leave idle capacity anywhere in the chain, and as a result, spare capacity is squeezed out of the system. Most contracts still provide for annual scheduling, but of greater importance are the provisions that allow for the annual schedules to be updated on a rolling basis. It is not unusual for schedules to be reset every 30 days to cover the upcoming 90-day period.

Scheduling provisions often call for mutual agreement, but the issue of what happens if the parties fail to agree takes on much greater importance, and with take-or-pay provisions hanging over the buyers, schedules leading to shortfalls in ACQ obligations can carry significant financial penalties. The situation becomes far more complicated when there are multiple buyers and sellers using the same facilities (whether liquefaction plants or import terminals). In some cases, plant operators have appointed independent experts or referees to sort out scheduling conflicts between the parties. The SPA scheduling provisions interact heavily with the ACQ obligations and become even more complex when addressing makeup quantities. The SPA scheduling provisions must further take into account scheduling of deliveries to the designated import terminals, and vice versa.

Invoicing and payment

When LNG buyers were typically large, stable, creditworthy utilities, this aspect of the contracts was less of a concern. Typical terms called for payment for a cargo of LNG within 10 to 30 days of title transfer (differing for FOB, DAP, and CIF contracts). With more diversified and often less creditworthy or special purpose buyers (often subsidiaries of major companies but without the full faith and credit of the parent behind them) emerging, the payment obligations of buyers are now increasingly underwritten by letters of credit from a financial institution or separate guarantees. These provisions often require separate negotiation, which can be complex: for example, when and how LNG sellers are allowed to draw down letters of credit (especially irrevocable standby letters of credit) or claim against guarantees, and what consequences arise if the drawdown is subsequently judged to have been unjustified. Increasingly, such guarantees carry limits (driven by accounting rules that require disclosure of such contingent commitments on the issuer's balance sheet), raising further questions as to what will happen if and when the limits of these guarantees are reached.

Measurement

LNG cargo measurement has become increasingly standardized. Generally, measurement has two components. The volume is measured by gauges in the tanker's cargo tanks, by determining the level in the tanks before and after loading at the liquefaction plant and discharge at the import terminal. The composition of the LNG is established by taking frequent or continuous samples of the product as it is being loaded or discharged, and measuring the density and constituent elements of the LNG. The sampling information can then be used in conjunction with the volume to calculate the quantity of energy loaded or discharged (usually measured in MMBtus). Sampling also permits the buyer and seller to confirm that the LNG meets SPA quality specifications. The International Group of Liquefied Natural Gas Importers (GIIGNL) has developed an LNG measurement protocol, which is now standard in most SPAs.

Title transfer

In an FOB agreement, title to the LNG transfers to the buyer as the product passes onto the tanker at the flange, where the tanker's manifold is connected to the loading arms. At that time, the buyer assumes the risks associated with the cargo, usually covered by cargo insurance. In a DAP agreement, the reverse situation occurs, with the title and risk transferring to the buyer as the cargo is unloaded from the tanker.

CIF contracts can be more complicated. In a CIF contract, the seller is responsible for securing shipping and cargo insurance, but the costs of both are passed on to the buyer in the LNG selling price. However, there is no absolute rule as to where title and risk transfer. In several agreements, title and risk transfer on the high seas outside the territorial waters of both seller and buyer. This is often done to ensure that the buyer is not subject to taxation or liability in the exporting country, with the same being true for the seller in the importing country. While the seller has obtained the insurance for a CIF cargo, once title transfers, the benefit of the insurance in the event of a cargo loss has also to be transferred to the buyer, who will be liable for paying for the cargo once title transfers, regardless of whether the cargo is delivered or not. Often the agreement will call for the buyer to assign the insurance proceeds to the seller in the event of a cargo loss and be then absolved of making payment for that cargo. In a CIF contract, there is no requirement that title and risk be transferred simultaneously.

Force majeure

Force majeure provisions are included in SPAs to address the situation where one of the parties fails to perform for reasons beyond their control. Most force majeure provisions describe examples of circumstances beyond the parties' control, such as storms, floods, hurricanes, wars, failure to complete facilities on time because of uncontrollable events, inability to obtain permits or licenses for the construction or operation of the facilities, or damage to any of the facilities involved in the delivery chain (from gas field to receiving terminal).

These provisions have become increasingly complicated as the trade has become more complex and as the world has changed. Most force majeure examples today would cite terrorism as an event. Strikes and industrial disturbances are usually classified as events of force majeure, and the party being subject to a strike is under no obligation to settle the strike and bring the force majeure situation to an end (to avoid giving undue leverage to the strikers in an industrial dispute). In the more complex traded world of LNG, physical events affecting third parties downstream of the import terminal may also be specified as events of force majeure if they prevent the buyer from moving the LNG or gas to its customers, or sometimes if the customers are unable to pay.

Somewhat more controversial are provisions that state that acts of governments, agencies, or changes in rules and laws can also be events of force majeure. Very often the effects of these types of actions may be unclear, and the parties may disagree as to whether they prevent one party from performing or simply result in much higher costs or even losses to that party. In the case where a party is a state energy company, there is always a question as to whether the actions of the state can be seen as truly separate from the energy company, and thus whether the state company can claim force majeure as a result of an action of its shareholder.

Not classified as force majeure are events that render the agreement uneconomical for one of the parties, no matter how these arise. Such events are often described as "price majeure" events, giving rise to economic losses by one of the parties as a result of changed circumstances. Force majeure fundamentally means that a party is physically prevented from performing, not economically prevented from performing.

Force majeure provisions allow the affected party to notify the other party as to the extent and expected duration of the event and require the affected party to take measures to remedy the problem (the one exception being industrial disputes mentioned earlier). Where an event of force majeure reduces the capacity of the facility (liquefaction plant or terminal), the provisions also cover allocation of any available product or capacity between the various buyers and sellers who may be affected. During a force majeure event, the unaffected party is usually permitted to sell LNG to or buy LNG from third parties, so as to mitigate any damages the seller or buyer respectively would otherwise incur. Finally, if the force majeure persists for an extended period of time, the contract may contain provisions allowing the unaffected party to terminate the agreement.

Choice of law

The choice of law is not always as straightforward as it might seem. Earlier LNG contracts tended to be governed either by English or New York law. A party rarely accepted governing law from the home country of another party. Understanding the implication of the choice of law is important for understanding the standards of performance under a given provision of an SPA. For example, under English law, the term "best efforts" means "leave no stone unturned" in correcting a problem encountered in the commercial relationship, and cost is no object. US law tends to interpret this phrase as meaning commercially reasonable efforts. Parties often negotiate a standard of commercially reasonable efforts in an attempt to further define this issue.

Although there are many examples and court precedents that can be used to understand how take-or-pay is interpreted in the US legal system, almost no take-or-pay claims have ever been adjudicated in the English courts, most having been addressed in sealed arbitration decisions. The parties to an agreement also have to understand, when they choose a legal regime, whether they are also incorporating other standards into the agreement, for example, the Uniform Commercial Code in the United States or the English Sale of Goods Act. Rather than risk incorporating these provisions by default, SPAs often include provisions specifically excluding their applicability to the agreement. This subject requires the careful attention of legal advisers during contract negotiation and drafting. This is particularly true of legal systems based upon codes (for example, Napoleonic Law) where there may be onerous default provisions in the law that neither party intended. Choosing an unfamiliar governing law adds costs, as every party to the deal needs to undertake their own due diligence on the impact. This may also make the project difficult to finance.

Dispute resolution

Disputes are rarely left to be resolved in the courts, as most parties prefer more confidential handling of their differences and are not comfortable with going to courts outside their own country. As a result, most legal disputes are handled by international confidential arbitration, with several different standards available to address the mechanics of the arbitration process. Arbitration can be brought using the rules of the UN Commission on International Trade Law (UNCITRAL), the International Chamber of Commerce (ICC), the Latin American Arbitration Association, the American Arbitration Association (AAA), to name but a few authorities. There are also provisions to establish an appointing authority or a party who will appoint the third arbitrator (the first two being appointed by the parties to the dispute) if the parties cannot agree to the appointment. Again, this can be a body such as the ICC or AAA. Finally, the locus of the arbitration must be designated—New York, London, Paris, and Geneva being among the common choices.

For disputes of a purely technical nature, the SPA will often provide for the appointment of a single technical expert who will settle such disputes. Examples of technical disputes include measurement errors, invoicing errors, and LNG quality issues.

In both cases, the cost of the dispute resolution must be determined. Usually, the agreement calls for the parties to split the costs of arbitration or expert determination (apart from their own legal and witness costs), but it is also possible to use the English system and allow the arbitrator to allocate costs, usually against the loser. The enforceability and collection of an arbitration award must also be given consideration and may require that the agreement reflect the right of the winning party to go into the courts of the losing party's host country to enforce and collect an arbitration award. Arbitration proceedings tend to be lengthy and expensive, and most contract disputes are normally resolved through negotiated settlement.

Allocation of liabilities, liquidated damages, termination

As is evident from history, some companies in the LNG industry have incurred very large financial costs associated with disputes and contract failures. In response, a growing trend is the use of caps and other limits on liabilities and/or financial guarantees. When one party reaches the cap on liability as a result of a failure to perform, the other party then often has the right to suspend or terminate the SPA. In the case of a suspension, the SPA can be resumed once the nonperforming party makes good on its liabilities. In the case of termination, the contract ends and the nonperforming party loses any further rights under the SPA, with the other party then being free to do business with other buyers or sellers.

As well as liability limits, SPAs also may introduce provisions for liquidated damages for the failure of one party to perform. These are more usually applicable to the seller for failure to deliver, while the buyer's liability is normally take-or-pay; however, liability arising from the buyer's failure to take may also be resolved by the payment of liquidated damages. These damages are often calculated following a formula in the contract (for example, the product of the quantity not delivered by the seller and the additional costs incurred by the buyer to replace these quantities), thereby removing the uncertainties associated with leaving damage calculations to the arbitration process.

In the most extreme instances, the resolution of a dispute may be to give the performing party the right to terminate the SPA unless the other party cures the nonperformance within a certain time. However, this gives rise to further issues, such as when the nonperformance arises out of another underlying dispute. If one party terminates the contract but the other party is later determined to have acted correctly, the consequences may be unclear but could be very painful. Even provisions that would appear to be straight-forward, such as termination, turn out not to be quite so simple in practice.

Spot contracts

As mentioned earlier, spot transactions are increasingly being carried out under so-called master agreements. Master agreements contain many of the same provisions as an SPA, although they tend to be less elaborate, covering only a limited number of cargo deliveries and limited time periods. The typical master agreement will include standardized provisions covering measurement and testing, invoicing, payment and associated security requirements, marine transportation and shore-based facilities, force majeure, choice of law, and dispute resolution. The key commercial terms—including quantities, price, delivery schedule, LNG quality specifications, and the identification of the loading and discharge ports—are handled in short schedules or memoranda of agreement, which are agreed to and executed as each LNG cargo or group of cargoes is committed for sale and purchase.

With the growth of spot market arbitrage opportunities, upside sharing between buyers and sellers has become a significant issue. Buyers with FOB contracts have discovered that they can make money by diverting their cargoes to premium markets. Sellers can restrict these arbitrage plays by including strict destination restrictions in the SPA with the buyer. Alternatively, they can negotiate upside sharing terms in the SPA to the benefit of both parties. In a separate but related development, cargo reloads have also become more commonplace. Sellers again are concerned that they are being shut out of the arbitrage profit when, after being delivered into an LNG import terminal, their LNG is reloaded and resold for arbitrage. Sellers, anxious to secure some of these trading profits, are seeking to fashion SPA terms to ensure they again get a share of the arbitrage upside from cargo reloads, though this is complex since it is often unclear whose molecules are being reloaded.

Given the critical nature of the LNG SPA in determining the commercial underpinnings of a relationship that may last for decades and involve the delivery, sale, and purchase of billions of dollars of LNG, there is no substitute for obtaining good legal advice on both sides of the negotiations.

15

LNG Tanker Contracts

Introduction

There are two types of standard contracts covering the carriage of goods by sea: the bill of lading and the charterparty. The bill of lading covers the sale and ownership of the cargo or goods being shipped. It is a document of title to the goods and serves as a receipt for the cargo, stating its quantity and condition. It is also evidence of the contract of carriage. The charterparty contract addresses the transportation service. It is a contract between one party, the charterer (or shipper), and the vessel's owner by which they agree to the terms and conditions under which the charterer has the right to use the owner's tanker.

Charterparty Overview

In the past, charterparties were drawn up by notaries to cover each voyage or series of voyages of a particular vessel. However, with the expansion of the world's merchant marine, many charterparties are now negotiated around standard forms published by various shipping organizations, especially major oil and natural gas companies in the context of chartering oil and oil product tankers. They are certainly not drawn up by notaries. Each type of charterparty incorporates a standard set of clauses. Any divergence (additional clauses and/or amendments to the standard forms) is dictated by the physical nature of the commodity and the negotiations between the contracting parties.

The charterparty is, in essence, the framework for a long-established set of rules, rights, obligations, and remedies that leaves ample room for the negotiation of terms and conditions between the contracting parties. The emphasis on standardization reflects the unique features of admiralty law that have been developed over centuries of merchant trading on the high seas. Unlike other assets involved in the LNG supply chain, the vessels are mobile and pass through many legal jurisdictions during their lifetime and even on individual voyages. Many aspects of the marine transportation industry are covered by international legal conventions to which most countries subscribe. In turn, these conventions spell out many of the legal requirements that must be recognized and addressed within the charterparty. The wording and intent of charterparty clauses themselves have usually undergone intense scrutiny and have become strictly defined through maritime case law.

Figure 15–1 shows typical charterparty contracts in the ocean transportation link of the LNG project chain. A variety of transportation contracts or charterparties are available to the charterer and shipowner. To determine the most appropriate type of charterparty, the project participants must evaluate their LNG supply and purchase obligations and the operational requirements that these obligations place on the project's shipping requirements. Until recently, the LNG tanker industry has used variations of standard oil tanker charterparty contracts amended to meet the LNG industry's technical and commercial requirements and standards. However, as the commercial model has changed and the industry has developed more spot and short-term trading, the contractual terms appear to be moving toward more standard formats characteristic of the oil tanker industry, which are typically of a shorter duration.

The main charterparties used in LNG shipping are bareboat charters, time charters, trip-time charters, single-voyage charters, and contracts of affreightment. The obligations, liabilities, and remedies of the contracting parties will vary depending on the charterparty employed. The key commercial terms are described later.

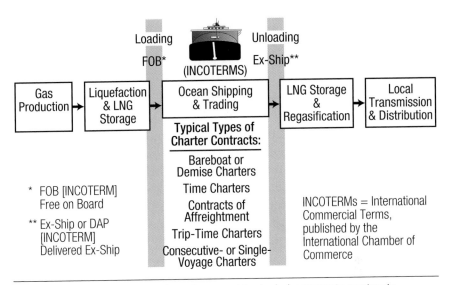

Fig. 15–1. Ocean transportation phase and typical charterparty contracts. (*Source:* Poten & Partners.)

Evolution of LNG Charterparties

Unlike oil and most other bulk shipping segments, there was virtually no spot charter market (single voyages) for LNG tankers because of the technical and commercial structure of the business. The LNG trade was based on long-term SPAs, which in turn required that the shipping capacity be locked in under long-term time charters for the duration of the SPA (typically 20 years). Although long-term charters remain dominant, there have been some changes owing to the availability of short-term LNG supply and import capacity, and a growing number of LNG tankers have been built and delivered on a speculative basis, unassociated with any long-term trade.

LNG charterparties in the traditional shipping model

Traditionally, LNG tankers were dedicated to specific LNG projects, because the long-term SPAs required the provision of a fleet of tankers of sufficient capacity to meet the sale and purchase obligations for the duration of the SPA. The high capital cost and specialized nature of LNG tankers offered project participants little maneuverability in the SPA terms and very limited flexibility in shipping, which was fully integrated into the project. The LNG tankers were perceived as a moving pipeline. Tankers always called at the same terminals, and their loading and discharge schedules were programmed well in advance with great precision. This is still true for many older, ongoing contracts.

The most commonly used charterparty template in the traditional LNG shipping model is Shell LNG Time 1. (This was largely based on the Shell Time 3, 4, and 5 oil tanker charterparties with modifications to accommodate LNG-specific clauses.) The industry incorporated a lengthy set of standard clauses found in these oil tanker charterparties and amended them to fit the LNG trade. However, in the early projects, shipping was such an integral component that, in many instances, the ships were owned, one way or another, indirectly by the project itself. Under such an arrangement, there were no true charterparties as such; rather, the shipping conditions were implicit in the LNG SPA, which contained FOB, CIF, or DAP delivery terms. An example of this type of arrangement is the fleet of Brunei LNG (started in 1972).

Notwithstanding the integrated nature of these projects and the shared ownership of the tankers, the fleets themselves were often organized into affiliated companies, and charterparties were still established between the affiliates just as if they were independent third parties. This reflected the unique taxation and liability regimes covering shipping and served to insulate the other segments of the LNG project from the legal risk of an incident involving an LNG tanker. The existence of affiliated charterparties also helped to establish appropriate transfer pricing mechanisms within a given project to ensure the agreed sharing of revenues, costs, and risks. When the shipping company is an independent entity, the fleet is chartered in under long-term charterparties (time or bareboat charters) to the project.[1]

The time charterparty. This is a contract under which the charterer hires the LNG tanker from the shipowner for a specified length of time, ranging from months to years. The charterer enjoys full control of the ship's movements for the contracted period. The shipowner is responsible for keeping the vessel in a seaworthy condition, compliant with international regulations and conventions; ensuring that it is manned by a competent and experienced crew; assuming the risks of operations; and bearing all operating costs other than those reserved in the charterparty for the charterer's account. The money paid by the charterer to the shipowner, called the hire (daily rate payable on a monthly basis, usually in advance), is divided into two main components: the capital expenses (CAPEX), which represent the reimbursement to the shipowner for the capital cost of the vessel, and the operating expenses (OPEX), consisting of the daily cost of the crew, stores, management, maintenance, insurance, and all other agreed expenses. The charterer must also pay the voyage costs (including fuel oil [bunkers and diesel], port charges, canal dues, and nitrogen). Boil-off incurred during the voyage is also a cost borne by the charterer, though the boil-off gas may be made available at no additional cost to the shipowner for use in the ship's boiler or engine.

CAPEX costs are generally fixed for the term of the charterparty, although they can be revisited in the event that, for example, new regulations or a change in loading or discharge port terminal or berth configurations require that significant modifications be made to the tanker. OPEX costs generally vary over the contract term to reflect changes in the underlying costs, although charterers and shipowners have begun agreeing to formulaic approaches to OPEX, rather than direct cost pass-through. One exception tends to be insurance, where the costs are established by the market and may, under certain circumstances, vary dramatically (for example, additional premiums are assessed for war risks in areas subject to armed conflict and high-risk areas where piracy may be prevalent). Formulaic approaches to OPEX may include linking operating cost elements to one or more consumer price indices or allowing revision of certain cost elements (for example, manning) if these costs vary by a certain percentage greater than the relevant price index.

The charterparty also spells out the obligations of the parties, with most of the obligations being assumed by the shipowner, who is the service provider. The shipowner will be required to crew and maintain the vessel to high standards and to guarantee the tanker's performance in areas such as the boil-off rate, service speed, fuel consumption, and rate of cargo pumping, all of which can fundamentally affect the costs of the LNG transportation. Failure to maintain these standards can result in the assessment of financial penalties against the shipowner by the charterer and, in extreme cases, can lead to the termination of the charterparty. The charterer may also have the right to terminate the charterparty in the event that the tanker suffers excessive off-hire. Off-hire is any period when the tanker is unable to perform the service required by the charterer, during which time the charterer is under no obligation to pay charter hire (for a breakdown, for example).

The charterparty will spell out the provisions for dry-docking the vessel (intervals, time allowed, etc.). The charterparty also specifies the locations where the vessel is to be delivered to the charterer and redelivered to the owner at the start and end, respectively, of the charterparty term. The charterparty can further specify the right of the charterer to subcharter the tanker to a third party. In this instance, the original charterer now takes on the role of the owner vis-a-vis the new charterer, yet the rights and responsibilities toward the shipowner remain unchanged. Choice of law and arbitration is often England, since the English legal system has extensive experience and precedent in handling maritime disputes, usually through arbitration. Charterparties can also go into more arcane (and often sensitive) areas such as the naming of the tanker, the flag state, the ship registry, and even the paint scheme.

Shell issued their Shell LNG Time 1 charterparty in November 2005. It is mainly based on the Shell Time 4 oil tanker charterparty, adapted specifically for LNG tankers. The main differences from Shell Time 4 relate to the unique nature of the cryogenic cargo in LNG tankers. Among the important issues are cargo boil-off rates, forced vaporization, spray cooling of tanks, and heel retention (to keep the tanks cold). For instance, Clause 26 (g) and (h), titled "Key Vessel Performance Criteria," specifies the vessel's guaranteed maximum daily boil-off rate for both the laden and ballast voyage. Shell LNG Time 1 represents a meaningful step toward the development of a standard LNG charterparty across the industry for both the short- and long-term market.

The bareboat (or demise) charterparty. Under this type of contract, the shipowner hires out the tanker (that is, the physical ship itself) and bears no operating or maintenance responsibilities. It is the charterer's duty to fully man, operate, maintain, and insure the tanker; obtain the appropriate certifications and flag registry; and pay the associated administrative costs. For the duration of the charter, the vessel is in the full possession of—and under the absolute technical, operational, and commercial control of—the charterer. The charterer may opt, under a bareboat charter, to repaint the vessel in his own house colors, change the name, or otherwise represent the vessel as part of his existing fleet. There is often a tanker purchase option on the expiration of the bareboat charter. In this case, the hire paid to the shipowner is the equivalent of the CAPEX of the time charter, as the charterer assumes all the operating and voyage costs. At the end of the charterparty, the contract spells out the obligation of the charterer to return the tanker to the owner in a specified condition and at a specified location.

LNG transportation contracts in the new era

Short-term LNG trading activity is giving rise to different forms of transportation contracts that are new to the LNG industry, albeit utilized in other merchant marine sectors. The following charterparties are increasingly being considered both for strategic purposes and for adjusting transportation costs and risks.

Trip-time charterparty. This contract runs for a short period of time, generally for one trip, and most companies now rely on Shell LNG Time 1. The money paid to the shipowner is based on a daily hire rate that incorporates voyage costs but excludes fuel and port costs, which are for the charterer's account. There are no demurrage claims, since the charterer continues making daily hire payments until the cargo has been discharged.

Voyage charterparty. This is a shorter contract than a trip-time charterparty, although the physical movement of the ship is the same. It usually does not include clauses related to the long-term operations or management of the ship, such as dry-dock provisions or maintenance. The money paid to the shipowner is normally based on a currency amount per ton of cargo loaded, which incorporates all the voyage costs. There are also consecutive-voyage charters. Voyage charters are not common in LNG trades due to disputes over the valuation of cargo boil-off used by the shipowner as fuel as a means of managing cargo temperatures and pressures. GIIGNL is working with Baltic and International Maritime Council (BIMCO) to develop a voyage charterparty to replace its existing format.

Contract of affreightment (COA). This is a voyage charter variation that offers flexibility when a shipper (the charterer) agrees with the shipowner to transport a specified quantity of LNG on a ship-or-pay basis. No vessel is specifically designated, and the time interval between cargo loadings is not necessarily linked with round-trip voyage time, since the shipper usually has multiple destination options. A certain volume of cargo is contracted for loading and delivery, usually at regular intervals, for a specific time period. The shipowner can nominate any capable and available tanker from the fleet or charter in a vessel for this purpose, provided that the tanker is approved by the charterer. If the shipowner has access to vessels of differing cargo capacities, then their utilization and impacts on shipping schedules must be accounted for in this agreement. The payment to the owner takes the same form as for voyage charters. The use of COAs is fairly limited in LNG trades owing to the dominance of long-term contracts and the lack of consistency in the market that would support a COA structure.

Controlling LNG Tankers: Adaptation to an Emerging Model

Historically, LNG suppliers looked to control the LNG tankers, through outright ownership or charterparties, to have consistency in newbuilding designs, operations, and maintenance. LNG supply projects, which had the greatest capital exposure in the chain and were perceived to have more limited options in the event of a shipping failure, generally sought to control their own LNG tankers.

This trend is changing. The reliability of LNG tankers has been well established (subject to proper maintenance practices), and the perception that control equates to reliability appears to be less of a concern. Buyers have become more interested in controlling shipping, and in some instances, sellers have been more willing to cede this control, although there is still a lingering concern by sellers that a buyer could limit LNG production rates if vessels are late and shore storage reaches a "tank tops" situation. Reasons for this shift include the following:

- Improved control over tanker expenses and residual values (if the vessel is owned and not on charter) that may lead to a reduction in transportation costs
- Greater flexibility to meet operational requirements, as buyers contract with multiple sellers and their downstream markets become less predictable
- Destination flexibility to deliver at multiple locations
- Capability of sourcing alternative LNG supplies, if required
- Capability of securing extra cargoes
- Concerns on the part of the sellers over legal and fiscal liabilities arising from tankers underway in foreign jurisdictions (for example, the United States is notoriously litigious and has onerous antipollution laws)

In addition, buyers with higher credit ratings can generally secure lower charter rates (through reduced finance costs for shipowners) than suppliers, who may be project financed and have lower credit ratings.

The traditional LNG business model committed tankers to specific trades and to carrying predictable quantities. Given the very high capital costs of LNG tankers, only a limited number were available on short notice and for limited duration to meet sellers' and buyers' opportunistic or seasonal requirements. Most parties felt that more assured access to these opportunities was best achieved through control of tankers. Control of shipping became a major issue in many LNG sales negotiations.

With a growing fleet of uncommitted tankers (both new builds and vessels coming off long-term charters), the actual spot charter rates during any given period may be higher or lower than the long-term charter rate, and spot chartering avoids the need to make long-term commitments. History has shown that the spot market can be very volatile based on seasonality and natural and man-made disasters. For transient business opportunities, not controlling shipping and incurring its attendant fixed costs may in fact be an advantage. The commercial allocation of the risks and rewards, including the factors mentioned previously, associated with a long-term SPA between buyers and sellers will probably be more reflective of the relative bargaining power of the parties during the negotiation process than the particulars of the one who controls the transportation link in the chain. The only compelling reason to seek shipping control is if one party is concerned that the other party will default, in which case the party controlling the shipping may have more ability to mitigate its risks than if the shipping control rested with the defaulting party. Of course, one might ask why any party would enter an LNG SPA where a potential for default by the counterparty was a primary consideration. Regardless of the type of LNG sale, one of the parties to the SPA will be obligated to provide the necessary oceangoing transportation and thus be faced with the decision of whether to own or charter the tankers.

Owning LNG tankers

Under this option, the future owner (buyer or seller, not independent shipowner) contracts for a new LNG tanker to be built by a preselected shipyard; funds the vessel with equity, debt, or both; and contracts with an independent manager to operate and maintain the vessel and provide technical management services (unless the owner has an in-house ship management department). Even though the owner may not necessarily operate the ship, he or she would need to establish a shipping organization to monitor the manager's performance.

The following are some of the main considerations in owning an LNG tanker:

Financial. Ownership exposes the company and its balance sheet to the capital cost of the tanker (including finance and owner's costs during construction). For a nonshipping company, tankers may not be regarded as core assets, and balance sheet exposure to the tanker's capital expenditure may be an issue. However, the financial and balance sheet implications of a tanker can be partially mitigated through project financing, leasing, or tax advantaged sale/lease back arrangements. If structured carefully, they may require only a footnote to the balance sheet and have a lower impact on the company's leverage.

Liability. Tanker ownership exposes the party to liabilities associated with shipping. Tanker owners carry substantial protection and indemnity (P&I) insurance for third-party liability claims and insure the value of the ship through what is termed "hull and machinery" insurance. Nevertheless, one of the most serious risks that cannot be insured is that of damage to the company's reputation and name in the event of a serious casualty or a significant pollution incident (although LNG tankers are much less risky in this regard than oil tankers). In addition, liabilities can arise with a failure to meet the shipping obligation under an SPA if the tanker does not perform as expected.

Residual value. This has taken on increasing importance in light of the longer-than-expected trading lives of LNG tankers, which are now approaching 40 years. Even as recent advances in technology are extending the trading lives of LNG carriers, it is raising fears of obsolescence for ships employing steam and first-generation DFDE propulsion systems. Ownership provides full rights to the residual value at the end of an LNG SPA or charterparty, unless some sharing arrangement is agreed upon in the initial charterparty. The owner benefits if, at the end of the initial employment period, the tanker continues in service at market rates, is chartered to a third party, or is sold. It is usually in the owner's best interests to ensure that the vessel is properly maintained throughout its life, rather than run down toward the end of the fixed charter period. Longevity studies and extensive life-extension work in dry dock can significantly improve the prospects of older vessels gaining further useful employment.

The scrap value of LNG carriers is always difficult to determine, partly because so few LNG vessels have been scrapped. Scrap value is determined on a US$/lightweight ton basis. However, LNG carriers have a higher than usual proportion of "exotic" metals, such as nickel, stainless steel, and/or aluminium, the prices of which are significantly higher than steel. Another valuable element of old LNG vessels can be the equipment—both equipment on board and depot spares usually held for a fleet of vessels. These can sometimes have considerable value to another shipowner who may have a similar vessel out of service due to a damaged component or system which can be replaced from a vessel in layup or going for scrap.

Control. Ownership ensures control of the operating and maintenance standards of the tanker. A company without in-house ship-operating capability would usually employ a ship manager to be responsible for the operations and maintenance of the LNG tanker to agreed standards (which are usually in excess of regulatory minimums). The shipowner has the right to replace the manager if these standards are not met. Since most ship managers are responsible for a large number of vessels, they usually have access to a larger crew pool, have more leverage with shipyards and insurance providers, and have in place all the systems and controls required to run a large fleet. As such, they can usually perform ship management services at a lower cost than an owner's in-house operation, unless the owner also operates a large fleet of ships.

Technical and market risks. Owning LNG ships exposes the company to all the technical risks and their associated responsibilities, such as failures or breakdowns. Similarly, technical obsolescence of the ship is the owner's problem and could result in early withdrawal of the ship from service and a reduced residual value. Mitigation of this risk comes from shipyard guarantees. At delivery there are liquidated damages if the ship does not meet the owner's specifications. The shipyard guarantees the tanker for the first one or two years, and breakdowns and failures are repaired at the yard's expense. However, the shipbuilding contract typically does not allow consequential damages, thus lost cargoes are the owner's responsibility (although these losses can be insured).

The owner of a fleet of similar LNG vessels will normally invest in depot spares which are maintained in a central location in climate-controlled storage conditions. The purpose of depot spares is to have spares of all long-lead-time items that may result in a vessel going out of service for an extended period of time. The spares are usually crated and ready for immediate transportation to where they may be needed. The range of spares will vary between steam turbine and DFDE vessels and may include reduction gears and bull wheel for steam turbine gearboxes, tail shaft, propeller, generator crankshaft, turbine high- and low-pressure rotors, and other equipment which may have a 6- to 18-month lead time. As soon as a depot spare is removed for use, a replacement is ordered, unless it is possible to refurbish the defective part to Classification Society approval, in which case it is then returned to storage. Depot spares are likely to cost about $4 million for a DFDE vessel and $12 million for a steam turbine vessel.

An owner takes little market risk after signing a shipbuilding contract with a shipyard if the owner intends to utilize the vessel in his or her own long-term service. The price of the tanker is locked in, and the only remaining item is the management cost. This introduces an element of market risk as crewing and shipyard maintenance costs rise and fall with the demand for LNG-related services. However, this is only a small part of the total shipping cost and is seldom avoided if chartering is selected as an alternative. The owner is sheltered from short-term market conditions and charter rate cycles during the primary term of the vessel's utilization and needs to consider these aspects only if the LNG tanker is released from the SPA commitment before the end of its useful life.

Chartering LNG tankers

An alternative to direct ownership is the sellers or buyers chartering LNG tankers from a third-party owner for the duration of an LNG SPA. In this case, the tanker owner is responsible for contracting for the construction of the tanker, management, technical operations, maintenance, and insurance. Even though there is a third-party owner in the equation, it is still necessary for the charterer to establish a shipping organization (although it will be a smaller one than for ownership and does not need to meet the requirements of certification under the International Safety Management Code) to monitor the owner's performance.

The main considerations in choosing to charter LNG tankers are as follows:

Financial. The charterer is exposed to a long-term financial obligation to pay monthly charter hire to the owner for the duration of the contract. The commitment from the charterer underpins the financing of the tanker(s); the owner is responsible for securing the financing. Because many shipping companies are private, they are often willing to assume more debt than might be the case for a public company and consequently have a lower cost of capital than the charterer might have. As a result, the cost of chartering, when considered solely on financial grounds, may be lower than the cost of ownership.

Liability. Chartering tankers places the liability on the owner of the vessel. While the owner will provide certain performance guarantees for the tankers with liquidated damages for underperformance, this may not be sufficient to offset liabilities associated with breaching the loading obligations of an LNG SPA. Furthermore, the charterer's control over the LNG tanker management is weaker than an owner's, and the charterparty usually constrains the charterer's ability to force improvements in the tanker's operation. Therefore, the charterer must carefully consider the choice of the owner, the owner's operating record and practices, and the incentives (and penalties) built into the charterparty and their influence on the owner's behavior. It is normal for a charterer to take out charterer's liability insurance to cover any risks the charterer is exposed to, including loss of use of the vessel, any claims arising from the cargo, charterer's war risk insurance, etc.

Residual value. In typical time charter arrangements, the owner benefits from any residual value of a tanker at the end of the charter period. However, the charterer can negotiate to receive a portion of this value at the end of the charter period, and most recent charterparties reflect such potential value-sharing mechanisms.

Control. Chartering the vessel puts management in the hands of the owner. The shipowner generally has control of the operating and maintenance standards employed on the tanker. However, the charterer may be able to negotiate limited rights to require a change in tanker management in the event that the operating and maintenance practices are not satisfactory. Another tool to address the potential operating and maintenance concerns is to include a tanker purchase option or conversion to bareboat charter at a predetermined price at specific points in the charterparty. This gives the charterer the ability to take control of the tanker if necessary.

Technical and market risks. The owner is responsible for any failure or breakdown of the tanker and must pay for the repairs. The tanker would be off-hire (the charterer makes no payment of hire) during the repair period, and this would be the only mitigation the charterer receives. The charterer's claims arising as a result of lost charter hire or lost cargo deliveries would be solely compensated by the off-hire provision. Some of the consequential risks from off-hire, such as the one of take-or-pay, may be mitigated by force majeure provisions in the SPA or the purchase of charterer's off-hire insurance or business interruption insurance. Obsolescence of technical onboard machinery is usually the owner's concern, as the charterer will have no risk of economic and technological obsolescence at the end of the charterparty. However, for mandated changes to an LNG tanker (for example, those arising from regulatory measures or from the need to fit new port conditions), the costs are either shared or passed through to the charterer, in whole or in part, by means of an increase in the CAPEX component of the charter hire. Table 15–1 defines key commercial terms that have a financial impact on either party to the charter.

Table 15–1. Key commercial terms in an LNG charterparty

Term	Description
Duty to maintain	Requires the owner of the vessel to provide the charterer with a plan to ensure operational reliability of equipment that affects a ship's performance, safety, and availability of vessel. This includes an understanding of the owner's plans and budget for maintaining the ship throughout the charter service. The owner must maintain the vessel in a seaworthy and cargo-worthy condition at all times during the charter period.
Period/ trading limits	Details the duration of the charter arrangement, the location of the vessel's redelivery, the nomination process of the vessel's redelivery, and the geographic limitations of where the vessel is able to trade. The clause also details the procedures to be taken should the vessel not be authorized to enter a terminal because of size.
Delivery/ canceling	Details the start of the charter, the condition of the vessel at delivery, and the charterer's right to terminate the contract if the vessel does not meet the conditions. With respect to newbuildings, the shipowner would request that the canceling details of the charterparty mirror that of the newbuilding contract. The vessel's delivery positioning is important, as the charterer would seek to minimize any ballast time paid before the first loading.
Owners to provide	Determines the owner's responsibilities and financial obligations through the duration of the charter period.
Charterers to provide	Determines the charterer's responsibilities and financial obligations through the duration of the charter period.
Rate of hire	Details the daily rate of hire that the charterer agrees to pay the owner of the vessel for use of the vessel subject to the terms and conditions of the charter agreement.

Table 15–1. (cont.)

Term	Description
Bunkers at delivery/ redelivery	States which party is responsible for the cost of bunkers at the time of the vessel's delivery and redelivery and stipulates the minimum bunkers to remain on board.
Subcharter	Spells out the conditions under which the charterer can subcharter the vessel, i.e., charter the ship to a third party.
Ship-to-ship transfer	Some charterparties will not permit ship-to-ship transfers taking place other than in an emergency situation. More recent charters have adopted a more conciliatory approach to LNG ship-to-ship transfers, allowing them to be carried out in the event that loading or discharge alongside another vessel (including FSRUs) becomes subject to international regulation and an accepted practice in the LNG industry and provided that the operation complies with international standards, international regulations, and accepted practice. Any additional insurance costs for ship-to-ship transfers are for the charterer's account.
Off-hire	This clause details the conditions under which the charterer is absolved from paying the hire rate. Conditions that trigger an off-hire are generally related to occasions that result in the vessel's interruption of service in any manner, apart from periodic dry dockings. An off-hire will tend to have a negative knock-on effect along the LNG chain, resulting in a delay in loading under a tight schedule and a delay in LNG delivery.
Periodic dry docking	Details the procedures for the vessel's dry-docking obligations. The clause provides a program nominating the dates of dry dockings and describes exactly when the vessel becomes off-hire during the actual dry-dock period. Dry dockings for LNG ships are typically completed over a 25-day period every 5 years, with in-water surveys at 2½ years. There is, however, some flexibility in the timing. An owner will use best endeavors to time the dry dockings when gas sales are not critical (during the low season or annual plant maintenance). As the number of dockyards capable of handling LNG tankers are limited (particularly in the Atlantic Basin), the vessel's deviation and time to reach the yard can be costly. Thus the timing of the off-hire is an important element generally covered in negotiations.
Vessel's performance	Details the owner's guaranteed operational parameters under which the vessel will operate given specific circumstances and set parameters. The main components include the guarantees on vessel's speed, bunker (fuel-oil) consumption, daily boil-off rate, and cargo loading/ discharge rates.
Laying up	Covers the circumstance under which the vessel can be laid up and the various terms and conditions of doing so. Laying up is extremely rare in LNG shipping and is a sign of a very poor market.

Table 15–1. (cont.)

Term	Description
Financial responsibility for pollution	Details the owner's responsibility to ensure that the vessel is legally covered and compliant with all pollution regulations from the start of the charter and throughout the charter period. For example, a vessel navigating in US waters will need to have a Certificate Of Financial Responsibility (COFR), which is a record attesting that the owner has the financial ability to pay for any (oil) pollution damages.
Cooldown	Defines the temperature the vessel's storage tanks must be at when the vessel is ready for cargo loading. This clause takes effect at the beginning of a charter, after an off-hire period or a dry docking. There is a cost associated with cooling down the tanks (purchase of LNG and time), which is negotiated between the parties. It is common for the expenses to be debited to the owner's account.
Boil-off	Determines the maximum amount of LNG that boils off from the cargo or heel and highlights the method of boil-off measurement for each laden and ballast voyage.
Bareboat charter and purchase options	These two clauses are backups for a charterer. If, for any reason, the owner defaults from his duties or obligations (e.g., loss of licenses, change in classification status), the charterer is entitled to terminate the charter or to exercise his or her option to gain full control and management of the vessel. Similarly, should a charterer be unsatisfied with the owner's management of the ship, the charterer has an option to purchase the ship. This option is generally exercisable at the end of an original charter period or any other option period.
Substitution	Details the condition under which the owner or charterer may provide or request an alternative vessel to fulfill the charter obligations.
Insurance	There are two distinct and obligatory types of marine insurances: hull and machinery (H&M) and protection and indemnity (P&I). H&M insurance covers physical damage to the vessel itself (e.g., to the hull and machinery such as engine, boilers) under the Institute Time Clauses. The P&I Club coverage (offered by mutual nonprofit associations of shipowners) provides third-party coverage for human injuries or fatalities, collisions, groundings, pollution, wreck removal, and towage operations. Insurance is generally a cost pass-through to the charterer in an LNG charterparty. There are other optional insurances, such as loss of hire, which protects the shipowner from potential loss of earnings from a vessel as a result of a casualty.
Quiet enjoyment	Stipulates the right of charterer to continue to use the ship in the event of owner default. This is provided by the owner's financiers.

(*Source:* Poten & Partners.)

Note

1. For further reading on charterparties, please see Michael D. Tusiani, *The Petroleum Shipping Industry* (Tulsa, OK: PennWell Corp., 1996).

16

Shipping Conventions and Regulations

Introduction

Unlike land-based facilities, which are clearly based in and regulated by their host countries, ships have less obvious domiciles. They may have ownership based in one country, be registered (or "flagged") in another, and operate in the open ocean or territorial waters of any country. To participate in the movement of goods and commodities in international trade, LNG tankers, like other vessel types, must comply with a large number of international rules and regulations, as well as with those of their home nations. These rules and regulations are embodied for the most part in international laws. International conventions governing shipping are negotiated and passed through the International Maritime Organization (IMO), an arm of the United Nations.

The IMO itself does not have enforcement powers. Every signatory nation to an IMO convention must pass an equivalent domestic law, regulation, or standard to give the convention force within its territorial waters. For example, in the United States, an international convention must first be passed as a law by Congress. The US Coast Guard, now under the Department of Homeland Security, is charged with enforcing a domestic law reflecting the international convention. Generally, the domestic legislation mirrors the wording of the international convention. The United States and other nations are free to pass maritime legislation that may differ from an international convention or be unique in nature.

Although the IMO membership comprises government delegations from every country, much of the work and technical input to IMO codes and conventions comes from nongovernmental organizations (NGOs), the more significant of which have "consultative status" at IMO. This means that they can participate in IMO meetings, take part in working groups, and submit technical analysis on matters being discussed, but are unable to vote or propose changes in IMO codes or conventions. In some cases, NGOs that may not have consultative status, but nevertheless have important contributions to make on a specific subject, can be coopted into a national delegation in order to allow their input to be heard. Some examples of NGOs with consultative status are given below.

The International Chamber of Shipping (ICS) is not a regulatory body but, as the international trade association for merchant ship operators, represents the collective views of the international industry from different nations, sectors, and trades. ICS has consultative status with a number of intergovernmental bodies, including the IMO, which is a major focus of ICS activity. The stated objective of the ICS is to maintain a global regulatory environment in which well-run ships can operate safely and efficiently. It is committed to the principle that maritime regulation should be formulated at an international level. The International Association of Ports and Harbors (IAPH) also holds IMO consultative status and has a major focus on safety and security issues in port areas.

The main NGOs in the IMO are:

- The Society of International Gas Tanker and Terminal Operators (SIGTTO), which represents the interests of the LNG and LPG shipping and terminaling industries. SIGTTO produces industry best practice guidelines and recommendations regarding vessel and terminal operations, which feature prominently in any LNG or LPG carrier new building specification.

- The Oil Companies International Marine Forum (OCIMF) promotes high standards of operation in the oil shipping and terminaling industry, and has produced extensive best practice guidelines and recommendations across a broad spectrum of shipping concerns from mooring equipment and manifold standards to ship vetting. The principles and guidance behind much of OCIMF's output are applicable to all shipping sectors, not just oil.

- PIANC—originally the Permanent International Association of Navigation Congresses but now known as the World Association for Waterborne Transport Infrastructure—produces guidance and recommendations on port marine civil engineering design issues. Of particular relevance to LNG shipping are their recommendations on channel width and sizing of turning areas.

- The International Association of classification societies (IACS) is responsible for coordinating the views of the various classification societies. One of IACS's main duties is sharing agreed interpretations of IMO codes (particularly those related to vessel design and construction) so that each classification society adopts a common approach to code wording. This is of particular relevance as the different country administrations in the IMO have now delegated their ship survey obligations to the classification societies.

The Rise of Modern Shipping Regulations

Establishment and structure of the IMO

The growth of international trade in the 19th century created a need for an organization that would institute a standard set of laws for international maritime commerce. The development of international trade would be impeded if each nation developed and enforced its own standards. The first international maritime organization, the Comité Maritime International (CMI), was established in 1897 as a common center for national maritime law associations. The CMI has remained the main coordinator for harmonizing maritime law and still performs this function today.

In 1948, a UN conference in Geneva voted in favor of creating a body for the sole purpose of governing the shipping industry. The convention was adopted in 1954. Four years later, the Inter-Governmental Maritime Consultative Organization (IMCO) was created, and in 1982 was renamed the International Maritime Organization (IMO). Today the IMO, headquartered in London, is recognized as the global authority for setting maritime law and standards. Other organizations, including the CMI, continue to play an important role in drafting maritime conventions and cooperate with the IMO to achieve their mission.

The purpose of the IMO is to facilitate cooperation among governments in all shipping matters and to encourage the adoption of the highest practicable standards of maritime safety, navigation, and prevention and control of marine pollution. The organization consists of five main committees:

1. **Maritime Safety Committee (MSC).** This committee deals with all safety-related matters including navigational safety, search and rescue, lifesaving, standards of training and watchkeeping, carriage of dangerous goods, fire protection, and radio communications.

2. **Marine Environment Protection Committee (MEPC).** Established in 1973, this committee is responsible for the prevention and control of all forms of pollution of the marine environment by vessels.

3. **Legal Committee.** Established to deal with the legal issues raised by the world's first major tanker disaster, the 1967 sinking of the *Torrey Canyon*, this committee has responsibility for international conventions dealing with liability and compensation.

4. **Technical Cooperation Committee.** This committee assists governments, primarily of developing nations, which lack the technical knowledge needed to operate a shipping industry, to improve their ability to comply with international standards.

5. **Facilitation Committee.** The purpose of this committee is to address and simplify the formalities and bureaucratic documentation, which can delay vessels entering or leaving ports and terminals.

The five main committees are supported by specialized subcommittees, which deal with technical and legal matters and prepare resolutions and recommendations. The governing body of the IMO is the Assembly, which meets every two years to make policy and adopt resolutions, although if required, special meetings may be called at any time. Over the years, a set of comprehensive international maritime conventions has been developed, prompted at times by large-scale disasters that focused attention on certain aspects of vessel design or operation. These conventions continue to evolve.

The creation of the IMO coincided with a period of very important change in world shipping, and the organization has been kept busy from the start, developing new conventions and ensuring that existing conventions keep pace with changes in shipping technology. Now responsible for more than 40 international conventions and agreements, it has adopted numerous protocols and amendments. Currently the IMO's emphasis is on trying to ensure that these conventions are properly implemented by the countries that have accepted them. To do so, the IMO works with the flag state (the state of the vessels' registry) and port state (countries where vessels trade to and from) control agencies that have the direct responsibility to regulate shipping and promote safety. The basic idea is that every vessel must comply with the same set of standards of vessel design and operation regardless of where it is built, its ownership, the nationality or nationalities of its crew, or its nation of registry.

To pass an international convention, a series of hearings is held with interested parties after which the IMO prepares draft legislation that meets the requirements of maritime nations. The IMO then proposes an international convention that is again reviewed by the world maritime community. A final version is produced and presented to each nation for signature. Most international conventions circulate until a prerequisite number of signatures are obtained, typically from two-thirds of the member nations. However, some conventions are passed after they are signed by contracting governments or signatory nations that represent a percentage of the world fleet tonnage affected. Once an international convention enters into force, domestic legislation is needed for its enforcement. The process can at times appear to be painfully slow, although improved procedures have ensured that changes following specific incidents can now be enacted more quickly.

Major IMO conventions and codes applicable to LNG tankers

A detailed list of international conventions, codes, and regulations with which liquefied gas carriers must comply is given at the end of this chapter. The following six have the greatest impact on the LNG shipping industry, although only the IGC Code is specific to LNG tankers.

1. The International Convention for the Safety of Life at Sea (SOLAS)

2. The International Code for the Construction and Equipment of Ships Carrying Liquefied Gases in Bulk (IGC Code), which is most relevant to the LNG industry

3. The International Convention on Standards of Training, Certification and Watchkeeping for Seafarers (STCW Convention)

4. The International Convention for the Prevention of Pollution From Ships, 1973 as modified by the Protocol of 1978 (MARPOL 73/78)

5. The International Safety Management Code (ISM Code)

6. The International Ship and Port Facility Security Code (ISPS Code)

The STCW is a stand-alone convention, whereas the IGC, ISM, and ISPS Codes are all amendments to the SOLAS Convention. Though not the only regulations in existence, they are indicative of the nature of international controls placed on ship design, management, and operation.

SOLAS. The IMO's first achievement was the passage of a new version of the International Convention for the Safety of Life at Sea (SOLAS). The initial SOLAS Convention dates back to 1914 in response to the sinking of the *Titanic* two years earlier. This was an early example of the influence of the public in shaping international conventions. The loss of this vessel resulted in the deaths of more than 1,500 people in the icy waters of the Atlantic. In light of this highly publicized disaster, maritime nations met and established SOLAS, a set of requirements related to safety concerns associated with the *Titanic* sinking. SOLAS was brought under the auspices of the IMO in 1960 and has subsequently been revised several times.

The original set of SOLAS requirements were a direct result of a number of problems highlighted by the *Titanic* disaster, even though the ship was compliant with the British regulations in force at the time. The shipbuilder had installed lifeboats for fewer than half of the passengers. Once the ship began sinking, confusion reigned among the crew members as to how to lower the lifeboats. There was no loudspeaker system to notify all the passengers of the danger they faced, nor were the crew trained to evacuate passengers from the vessel. The *Titanic* was the first vessel to send out the then newly introduced SOS distress signal; however, at the time, vessels in the area were not required to monitor radio receivers to pick up distress calls.

The initial SOLAS convention addressed each of these concerns. Ships now carry sufficient lifeboats and life rafts for all passengers and crew, and the crew must be trained for and conduct drills covering a wide range of emergency situations. Lookouts must be posted at all times, and ships must continuously monitor radio frequencies dedicated to distress calls.

SOLAS also stipulated that passenger ships must be subdivided into watertight compartments so that the vessel remains afloat and stable, assuming a certain degree of damage to the ship's hull. This requirement has been extended to all types of ships, including LNG tankers, whose design must ensure stability in normal and emergency conditions; this is known as intact stability and damage stability. Machinery such as steering gear and electrical installation is required to remain in service under stipulated emergency conditions. Other areas covered by SOLAS include fire protection, detection, and extinction, design and sufficiency of lifesaving appliances, and conditions associated with the carriage of dangerous goods.

Of particular significance to LNG ships is Chapter V, Regulation 22, "Navigation Bridge Visibility." It sets requirements for achieving good visibility from the bridge of a ship, including specifications for the field of vision from various positions. The elevated level of LNG cargo tanks, especially Moss-type spherical tanks, together with their associated relief valve vent stacks and risers, poses a challenge to naval architects in complying with the visibility regulations. The situation is exacerbated where LNG ships are required to trade in ports having relatively low bridges spanning the access channel. To comply with SOLAS regulations and provide good visibility, the navigating bridge needs to be raised well above the cargo deck, which can conflict with the air draft restrictions imposed by low bridges. The net result is that larger LNG ships, designed to comply with SOLAS, cannot access certain ports where channels are crossed by low bridges.

The actions required by signatories to the SOLAS Convention are not just applicable to ships that fly their flag or enter their ports. Governments are required to establish and maintain aids to navigation such as lighthouses, buoys, and radio aids. They are also required to ensure that necessary arrangements be made for the rescue of persons in distress off their coasts. This requirement was supplemented in 1979 by the International Convention on Maritime Search and Rescue.

The IGC Code. Of the six IMO international standards discussed here, the most specifically relevant to LNG ships is the IGC Code. It forms Chapter VII of the SOLAS Convention. The latest amendments are contained in the 1993 edition, which applies to ships, regardless of their size, engaged in the carriage of liquefied gases and certain other hazardous substances.

Unusually, the IMO agreed to allow the gas shipping industry to work on revising the IGC Code. This was the first time that an IMO code had been updated outside the direct control of the IMO and national administrations. SIGTTO took the lead role in this activity, with the support of the United Kingdom, and assembled a team of about 140 gas shipping experts from around the world to update different sections of the code. Most of the revisions were aimed at aligning IGC Code requirements with recent changes in other similar and complementary codes. Other revisions were necessary due to changes in technology, ship design, and operating practices.

The working groups involved included representatives from classification societies, national administrations, and others connected to IMO delegations. In this way, the delegations, which were to approve the revised Code, could understand and have input to the Code during its revision. After a five-year effort, the revised Code has been approved by the IMO and will come into effect on January 1, 2016 and will apply to ships contracted or having keels laid on or after July 1, 2016. Existing ships will not require modification to comply with the new Code but will be "grandfathered."

The Code provides an international standard for the safe carriage of liquefied gases at sea by prescribing the design and construction standards of ships and the equipment they carry, all aimed at minimizing the risk to the ship, its crew, and the environment. The Code takes into account present knowledge and technology, is under a constant state of review, and is supplemented by amendments incorporating the latest experience and technological developments. The original 1976 Code served to approve the LNG fleet as it existed at that time through retroactive application, rather than require significant modifications to ships already in service. The fact that many of the ships predating the 1976 Code remain in safe and reliable service almost 40 years later is a testament to the competence and conservatism of those who designed, constructed, and operated the first generation of LNG ships.

The Code prescribes the safe location of cargo tanks, survival standards under various conditions of damage and flooding, and requirements for ship stability and freeboard. It is especially concerned with the various cargo containment systems, the design of the tanks, the filling limits of the tanks, their construction materials, welding, and testing. It establishes requirements for cargo tank pressure control and relief, environmental controls, instrumentation, and monitoring systems. Fire protection and extinguishment, personnel protection, and operational procedures, together with other special requirements, are all comprehensively addressed.

Ships that comply with the IGC Code are issued an International Certificate of Fitness for the Carriage of Liquefied Gases in Bulk, commonly known by its abbreviated name, Certificate of Fitness. The certificate is issued by a classification society under the authority of the government of the ship's country of registry and is valid for a maximum period of five years. The Certificate of Fitness stipulates the cargoes that the ship is certified to carry and any temperature/pressure restrictions associated with the carriage of that cargo. It is mandatory for a ship in service to undergo annual surveys and an intermediate survey (performed by the classification society) at which times the Certificate of Fitness must be endorsed in order to retain its validity. Renewal of the certificate requires a full survey of the ship, and this is usually carried out coincident with the ship's five-yearly special survey cycle. For an LNG ship, a Certificate of Fitness issued under the IGC Code is the primary document attesting that the ship has been constructed and maintained according to accepted international standards. Without a Certificate of Fitness, the ship cannot trade.

STCW Convention. The 1978 STCW Convention established minimum standards with regard to training, certification, and watchkeeping for seafarers on an international level. Prior to its passage, each individual nation had its own set of standards, usually set without reference to practices in other countries. As a result, standards and procedures varied widely. The original STCW Convention entered into force in 1984 and recognized that safety lies with the crew, not just with the ship.

An unusual and significant feature of the Convention is that it applies to ships of nations not party to the Convention when visiting ports of nations that are party to it. This ensures that ships of nonparty nations cannot avoid the control measures applied under the Convention. The potential difficulties that could arise for ships of a nonparty nation are among the reasons why the Convention, by December 2000, had been adopted by nations representing more than 97% of the world's shipping tonnage.

The STCW Convention has individual chapters addressing the requirements for masters and the deck department, the engine department, the radio department, and special requirements for tankers. Within this latter chapter, there are special requirements for liquefied gas tankers. These cover training related to the cargo and cargo handling equipment, as well as the requirement for officers to have completed an appropriate shore-based firefighting course and to have had a specific period of shipboard service or other approved familiarization commensurate with rank.

The STCW Convention was subjected to a major revision in 1995 (STCW 95). This revision incorporated many lessons learned as a consequence of the 1994 sinking of the *Estonia* in the Baltic Sea. More than 850 lives were lost when this trans-Baltic ferry sank after its bow doors failed during a storm. The amendments corrected weaknesses in the original Convention, particularly by ensuring proper training of crewmembers. The revised STCW Convention places greater obligations on shipping companies to demonstrate that seafarers are competent and qualified to carry out their assigned duties.

STCW 95 requires the following of every vessel-operating company:

- Each seafarer must hold an appropriate certificate.
- Ships must be manned at safe levels.

- Necessary documentation on seafarers is maintained and readily accessible.
- Seafarers are properly trained to perform their specific duties.
- Seafarers are trained to handle routine and emergency situations.
- Communications are in a language that the crew members understand. (This can be a problem on vessels where the officers and the crew are of different nationalities.)
- Minimum rest periods are required to ensure a safe and alert crew.

A significant feature of the revised STCW Convention is the establishment of uniform standards of competence, minimum seagoing time, and experience for specified shipboard functions. This is now required for all types of vessels, not just for tankers. This brings the regulations for all vessel types into line with requirements which have applied to liquefied gas tankers, including LNG ships, since the 1978 introduction of the STCW Convention.

STCW 95 also recognizes the contribution of technical innovations such as simulators for training and assessment purposes. Ship handling, cargo handling, and engine room control simulators have all become important tools in the training of officers who serve on LNG ships. Ship maneuvering simulations are now an essential item in the submissions required by the US Coast Guard in carrying out a waterway suitability assessment (WSA) for any new US LNG project.

In an ever-changing world, amendments to the STCW Convention in 2006 added mandatory training and certification requirements for designated ship security officers. This is yet another regulatory measure of significance to LNG ships, which have become a focus of attention as potential targets of terrorism.

MARPOL 73/78. MARPOL is the main international convention aimed at the prevention of pollution of the marine environment by ships. The 1973 Convention focused on pollution by oil, chemicals, harmful substances in packaged form, sewage, and garbage, and was prompted by the loss of the *Torrey Canyon*, which went aground in the western approaches to the United Kingdom, spilling 120,000 tons of crude oil into the sea.

The 1973 Convention was never ratified. A number of significant tanker accidents in 1976–77 caused the IMO to reconsider tanker safety and pollution prevention and measures affecting tanker design and operation. This became the 1978 Protocol to MARPOL 73.

The combined MARPOL 73/78 entered into force in October 1983. MARPOL 73/78 now comprises the main convention and a number of annexes (Annexes I and II are compulsory, but the other annexes are voluntary):

- Annex I   Prevention of Pollution by Oil
- Annex II   Control of Pollution by Noxious Liquid Substances
- Annex III Prevention of Pollution by Harmful Substances in Packaged Form
- Annex IV   Prevention of Pollution by Sewage from Ships
- Annex V   Prevention of Pollution by Garbage from Ships
- Annex VI   Prevention of Air Pollution by Ships

The main focus of MARPOL 73/78 has been the prevention of air pollution by ships, which is covered by Annex VI. This annex set limits for sulfur oxide and nitrogen oxide emissions from ships' exhausts and prohibited the deliberate release of ozone-depleting substances. More stringent sulfur oxide emission restrictions can be applied in special areas under this Annex, requiring fuel oil with a sulfur content of less than 1.5% to be used. In 2008, further amendment to Annex VI brought in a phased reduction of the global sulfur cap, further reduction of the sulfur content in fuel oil for sulfur emission control areas (SECAs) and introduction of wider emission control areas (ECAs). 2011 amendments to Annex VI identified the requirement for an Energy Efficiency Design Index (EEDI) for new ships and a Ship Energy Efficiency Management Plan (SEEMP) for all ships.

International Safety Management Code (ISM Code). The ISM Code was adopted in 1993 as a new chapter (Chapter IX) to the SOLAS Convention in order to provide better guidelines for ship management procedures both ashore and on board. It was passed after a number of shipping accidents in the 1980s and early 1990s were deemed to have resulted from shortcomings in management which facilitated human errors. One was the 1987 capsize of the

Herald of Free Enterprise, a roll-on, roll-off car and passenger ferry sailing between Belgium and England. The main cause of the tragedy was the failure to close the bow doors before leaving port. Nearly 200 people died in this accident. The subsequent investigation revealed that no management plan was in place that required securing the ship's bow doors, no one had been assigned to ensure the doors were closed, and there was no bridge checklist or indicator showing or recording the status of the doors.

The ISM Code requires a safety management system (SMS) to be established by the company, which is defined as the shipowner or any person, such as a manager, who has assumed responsibility for operating the ship. It entered into force for passenger vessels and tankers, including LNG ships, in 1998 and for other types of cargo ships and mobile drilling units in 2002. The ISM Code requires every shipping company to establish an SMS that:

- Initiates a safety and environmental protection policy
- Provides instructions and procedures to ensure safe operation of ships and protection of the environment
- Defines levels of authority and lines of communications between shore and shipboard personnel
- Establishes procedures for reporting accidents and nonconformance with the ISM Code
- Sets forth means to prepare for and respond to emergency situations
- Ensures compliance by means of internal audits and management reviews

The shipping company must provide the necessary resources and shore-based support to the ship, and is required to appoint a designated person ashore (DPA) who is contactable at all times in case of an accident or other ISM Code issue. The DPA must also have direct access to the highest levels of onshore management.

International Ship and Port Facility Security Code (ISPS Code). In response to the perceived threat to ships and port facilities in the aftermath of the terrorist attacks in the United States in September 2001, the IMO introduced the ISPS Code as an amendment to SOLAS. It addresses the security of ships and port facilities as a risk management activity and requires that an assessment of risks be made to determine which security measures are appropriate in each particular case. Under the code, shipping companies are required to appoint a company security officer and each ship has to have a ship security officer and a ship security plan. Similar security officer and plan requirements apply to all ports facilities.

Flag state regulation. The UN Convention on the Law of the Sea (UNCLOS) is an umbrella convention concerned with many aspects of the sea and its uses, including the granting of registration of a ship by a state. A ship must be registered with a nation which is a signatory to IMO conventions and carry that nation's flag in order to trade internationally. Ship registry administrations charge a registration fee and collect taxes based on ship tonnage. Under UNCLOS, once a ship is registered, the state is responsible for inspecting it and for ensuring it complies with international laws. The state is responsible for issuing certificates that allow ships to trade internationally and for withholding or withdrawing certificates from any of their ships found to be noncompliant.

Some ship registry nations carry on their own inspections, an example being the United States, where the US Coast Guard performs inspections on all US flag ships. Other nations do not have the necessary resources and delegate their authority to a classification society. The classification society, utilizing their worldwide network of surveyors, can do the work more efficiently with less potential delay to the ship by carrying out the inspections coincident with its own routine surveys.

Some nations provide registry to ships that are owned in another country and have come to be known as "flags of convenience" or FOCs. They offer shipowners such benefits as reduced registration costs, reduced or no taxation on profits, and greater latitude on the choice of the nationality of the crew. The leading FOC nations are Panama, Liberia, Malta, Cyprus, the Bahamas, and the Marshall Islands, but others such as (landlocked) Mongolia and Bolivia also exist. The main reason for choosing a particular registry is more likely to be associated with the financing and tax arrangements of owners than any intent to operate below international standards.

Port state regulation. No vessel can enter or leave a port without the permission of the local port authority. In this way, maritime nations exercise regulation over the movements of shipping, which is known as port state control. In the United States, for example, the US Coast Guard enforces port state control by allowing or denying entry to ships arriving at US ports. Should the Coast Guard find a deficiency on a ship already in a US port, it can detain the ship and prevent it from leaving until the deficiency is corrected. This is especially important if the deficiency is safety-related and would expose the ship or its crew to undue risk should the ship proceed to sea. In Europe, the local port authorities exercise similar regulations over shipping movements.

Historically, port authorities operated without knowledge of what other port authorities were doing. Port authorities are now sharing information so that there is much greater transparency about which vessels are granted or denied permission to enter ports and the nature of any deficiencies found. The US Coast Guard places the results of ship inspections on its website for all to see. Various intergovernmental memoranda of understanding (MOUs) harmonize inspection procedures designed to target substandard ships with the main objective being their eventual elimination. Each country participating in an MOU agrees to a common set of inspection criteria and sharing of information to identify substandard vessels. Preventing substandard vessels from trading by regulating their port movements eventually results in their removal from the world fleet.

Industry self-regulation. Over the years, specific sectors of the shipping industry, including the oil, LNG, and LPG sectors, have established associations and societies through which they have sought to make their professional expertise and views available to national and intergovernmental bodies. In response to increasing public awareness of marine issues, these groups have also sought to demonstrate responsible management of their businesses, through processes of self-regulation. New international regulations originating in the IMO, often initiated by public and political reaction to a high-profile marine incident, can be heavily biased toward political expedience rather than practical solutions. It is therefore in the shipping industry's best interest to exercise sound, practical self-regulation to avoid accidents. The various associations and societies have established industry standards, produced best practice guides, and published common operating procedures, which are aimed at raising the overall performance of the industry. Although the documents they produce are all reference in nature and provide voluntary guidance, not mandatory regulation, they are widely recognized and generally acted upon.

The Oil Companies International Marine Forum (OCIMF), formed in 1970, was granted consultative status by the IMO in 1971. OCIMF has produced an extensive range of documents, the most noteworthy of which, published in cooperation with ICS and IAPH, is the International Safety Guide for Oil Tankers and Terminals (ISGOTT), now in its fifth edition. ISGOTT is recognized worldwide as the definitive reference document on tanker and terminal safety. OCIMF is also custodian of the Ship Inspection Report Exchange Programme (SIRE) database, which holds records of inspections of more than 8,000 oil and product tankers.

The SIRE inspection reports are available to OCIMF members to enable them to prescreen tankers before chartering them. It provides a snapshot view of the condition of the vessel (and any identified deficiencies along with the owner's response). It is each individual charterer's task to assess the report and determine if that vessel presents risks to its operation. Unlike the oil trade, which has always employed a large percentage of ships on short-term charters, the LNG industry traditionally employed ships under long-term charters, which are managed by experienced operators often directly involved in other segments of the trade. There was little third-party chartering activity and no requirement to have a database of ship inspection reports available to the industry. However, in light of changes brought about by the rapid expansion of the LNG industry, especially the increase in short-term trading, the SIRE database was expanded in 2002 to include inspection reports on LNG ships. The SIRE data have also been made available to governmental bodies that regulate shipping, as well as to port and canal authorities. SIRE has proved a very effective self-policing mechanism for the industry, identifying substandard ships and ultimately forcing them out of the market. Most charter parties require a positive SIRE inspection every six months.

The LNG industry's shipping association is SIGTTO, established in 1979. Its members, representing LNG and LPG ship and terminal operators, recognized the importance of ensuring that the then-fledgling LNG industry be operated in a responsible manner for the benefit of all participants. To do this, they established a body through which members could exchange technical information and experience to enhance the safety and operational reliability of gas tankers and terminals, free of any commercial considerations. By 2014, SIGTTO had grown to 133 full members, 38 associate members, and 24 noncontributory members, including more than 90% of LNG ship and terminal owners and more than 50% of the global LPG shipping and terminal business. The IMO granted SIGTTO consultative status in 1982, enabling it to participate in the framing of international regulations. SIGTTO itself has produced a comprehensive set of guides, standards, and best practices.

Following the 2002 expansion of the OCIMF SIRE program to include LNG ships, SIGTTO published *Guidelines for the Vetting of LNG Tankers* in 2004. Vetting is the process of ensuring that the overall quality of a vessel and its management meet acceptable standards for charterers, cargo owners, and terminal operators. The guidelines incorporate the lessons learned from the oil industry.

An area where SIGTTO has been extremely proactive is in addressing the shortage of qualified LNG vessel crewmembers and competency. The problem arises from time to time when there are significant peaks in newbuilding activity. With a fleet of about 400 vessels as of October 2014, the 120-plus vessels on order represent a one-third increase in fleet size, which will severely strain experienced LNG crews worldwide, as was seen when the Q-Flex/Q-Max vessels entered service.

Over the past three decades, the shipping industry as a whole has swung away from steam turbine to diesel engine propulsion, reducing the need for steam qualifications among mariners. Although dual- and tri-fueled diesel electric propulsion systems are being selected for newbuildings, there are still many steam turbine-driven LNG carriers requiring qualified steam engineers. The operational life expectancy of 40 years means that steam turbine LNG ships will be around for a long time to come and officers and crews will need to be trained to operate these ships for just as long.

Add to this the rapid expansion of the LNG fleet and the entry of many new operators who do not have their own experienced LNG officers and crew, and the result is a sharply rising demand for qualified LNG officers, crew, and steam engineers. Shipowners and operators are being forced to "poach" from the existing pool of experienced officers, with salaries rising in response. SIGTTO has responded by developing *LNG Shipping Suggested Competency Standards*. This comprehensive document may be incorporated by the IMO into the next revision of the STCW Convention. SIGTTO has also produced the *LNG and LPG Experience Matrix*, which sets out the minimum experience requirements for senior and junior deck and engine room officers. These requirements are frequently cited in LNG time charterparties.

In another project reflecting new technology around floating LNG, SIGTTO, in conjunction with the OCIMF, the ICS, and the Chemical Distribution Institute, has addressed the issue of LNG ship-to-ship (STS) cargo transfer operations. The 2013 *Ship to Ship Transfer Guide for Petroleum, Chemicals and Liquefied Gases* now provides broadly accepted industry guidance.

The Role of Classification Societies

Ship classification societies date back to the 17th century when shipowners and cargo shippers met in London coffee houses to discuss business and read press clippings. The owner of Lloyd's Coffee House, which opened in 1687, started a newsletter that became *Lloyd's List*, published weekly from 1734 and daily (except Sundays) since July 1, 1837 to cover happenings in the shipping business. Ship and cargo underwriters also met at the same coffee house to arrange insurance for vessels and their cargos against damage or loss at sea. However, the ship or hull underwriters had no knowledge of the underlying physical condition of the vessels they were insuring nor was there a system in place to carry out ship inspection in a uniform manner.

In 1764, a group of insurance underwriters formed the Register Society and began to publish *Lloyd's Register Book*, a loose system of classing ships by equipment, hull, and owner. That system was based on a rating for hull condition (A, E, I, O, U) and a rating for rigging (now covering mooring and anchoring equipment), which was initially G, M, B (Good, Middling, Bad) and later became 1, 2, or 3. Thus a vessel classed as A1 was of the highest standard. The "100" notation (as in 100A1) signified that the vessel was suitable for seagoing service. Over time, the guidelines became better defined, leading to the 1834 publication of *Lloyd's Register of British and Foreign Shipping*, which set forth the condition of individual vessels along with rules for constructing and surveying ships. Thus, Lloyd's Register was formed and the first ship classification rules were issued. Other classification societies were soon organized in other maritime nations: the American Bureau of Shipping was established in 1862 and Det Norske Veritas in Norway in 1864. These three ship classification societies remain the largest and cooperate as the "LAN" group to promote unified standards.

Classification societies establish and apply technical standards for the design, construction, and survey of ships. These standards are issued as published rules, and a ship designed and constructed in accordance with the rules of a society may receive a certificate of classification from that society following completion of a satisfactory survey. The owner must subsequently submit the ship to periodic surveys to verify that the ship continues to meet the relevant rules. Classification societies also undertake inspections and surveys delegated by the ship's flag state administration. If a ship is found to be out of compliance with class standards, the classification may be withdrawn, and a ship that is not classed cannot be insured or carry cargoes.

First-generation LNG ships were all classed by the classification societies of the countries (US and European) where the industry started and where the ships were designed and constructed. With the spread of the LNG industry to Japan and South Korea, both were quick to build and class their own LNG ships. In certain circumstances where national ambition is particularly strong, LNG carriers may be dual-classed. For example, ships built in China have come into service with both ABS and China Classification Society (CCS) class as part of a technology transfer arrangement. Ships to be built for the Yamal LNG project will be dual-classed by Bureau Veritas (BV) and the Russian Maritime Register of Shipping (RS).

Rules and regulations applicable to LNG ship construction

The following list is an extract from a typical LNG ship newbuilding technical specification, detailing a number of relevant rules and regulations with which the design and construction of the ship must comply. The list (which should not be considered exhaustive) covers construction, but not operational, requirements. However, it illustrates how shipping in general, and LNG in particular, is a highly regulated business.

Classification society. The vessel, including hull, machinery, equipment, and outfittings, shall be constructed in accordance with the rules and regulations of the nominated classification society and under survey of the society's surveyors.

Rules and regulations. The vessel shall comply with the following rules, regulations, and requirements of the regulatory authorities in force at the date of contract and in accordance with any rules and/or regulations officially ratified but not in effect at the date of contract, provided they are scheduled to come into effect for the vessel prior to the delivery date and IMO amendments coming into effect and becoming compulsory to this type of vessel:

- Maritime regulations of the country of registry
- International Convention on Load Lines 1966, as modified by the 1988 Protocol, amendments up to and including the 2006 amendments and later amendments
- International Convention for Safety of Life at Sea (SOLAS), consolidated 2004, and later amendments
- International Code for the Construction and Equipment of Ships carrying Liquefied Gases in Bulk (IGC Code)
- International Convention for Prevention of Collision at Sea 1972 and later amendments, including IMO Resolution A.464 (XII)
- International Convention for Prevention of Pollution from Ships 1973 (Annex I, IV, V, and VI), as modified by the Protocol of 1978 relating thereto (MARPOL 73/78) and later amendments
- International Telecommunication Union's Radio Regulations 2004 and SOLAS Chapter IV, as amended
- International Convention on Tonnage Measurement of Ships 1969, as amended by IMO resolutions and later amendments

- IMO Resolution A.468 (XII): Code of Noise Levels on Board Ships
- US Coast Guard rules for foreign-flag LNG tankers for LNG ship operating in US navigable waters and involves compliance with USCG CFR 33 Parts 155, 156, 159, and 164, and CFR 46 Part 154 excluding Alaskan waters and without certificate or inspection
- International Ship and Port Facility Security Code (ISPS Code), 2005
- OCIMF, Guidelines and Recommendations for the Safe Mooring of Large Ships at Piers and Sea Islands, 1994
- ICS Guide to Helicopter/Ship Operations, 2008
- CAP 437 UK Civil Aviation Authority (if helideck fitted)
- SIGTTO: Guidelines for the Alleviation of Excessive Surge Pressures on ESD, 1987
- International Electro-Technical Commission (IEC) Publication 60533: Electrical and Electronic Installations in Ships—Electromagnetic Compatibility
- ISO 4406:1999 Hydraulic Fluid Power—Fluids—Method for Coding the Level of Contamination by Solid Particles
- ISO 10816 Parts 1 and 3: Evaluation of Machine Vibration by Measurements on Nonrotating Parts
- IMO Resolution A.272 (VIII) and A.330 (IX): Safe Access to and Working in Large Cargo Tanks and Ballast Spaces
- IMO: Code on Alarms and Indicators, 1995
- IMO Resolution A.601 (XV): Provision and Display of Manoeuvering Information on Board Ships
- IMO Resolution A.708 (XVII): Navigation Bridge Visibility and Functions
- IMO Resolution A.751 (XVIII): Interim Standards for Ship Maneuverability
- ISO 8861 Shipbuilding—Engine Room Ventilation in Diesel Engine Ships—Design Requirement and Basis of Calculation (1998[E])
- ISO 484-1: Ship Screw Propellers—Manufacturing Tolerances

- OCIMF, Guidelines on the Use of High-Modulus Synthetic Ropes as Mooring Lines on Large Tankers, 2002 (if fitted)
- Suez Canal navigation regulations and tonnage measurement of ships
- ILO Convention 92: Concerning Crews Accommodation (1949)
- ILO Convention 133: Concerning Crew Accommodation on Board Ship (1970)
- OCIMF, Mooring Equipment Guidelines 2008
- OCIMF, Recommendations on Equipment for the Towing of Disabled Tankers, 1981
- SIGTTO Manifold Recommendations for Liquefied Gas Carriers, 2011
- CDI/ICS/OCIMF/SIGTTO Ship-to-Ship Transfer Guide for Petroleum, Chemicals and Liquefied Gas, 2013
- SIGTTO, ESD Arrangements & Linked Ship/Shore Systems for Liquefied Gas Carriers 2009
- IEC Publication 60092: Electrical Installations in Ships
- ISO 6954:1984 Guideline for the Overall Evaluation of Vibration in Merchant Ships
- IMO Resolution A.330 (IX): Safe Access to and Working in Ballast Spaces
- IMO Resolution MSC.137 (76): Standards for Ship Maneuverability
- OCIMF Recommendations for Ships' Lifting Equipment, 2005

17

LNG Terminal Siting and Regulatory Issues

Introduction

Governments, energy consumers, and policy makers around the world value safe and secure access to reasonably priced, clean, and abundant supplies of energy. However, this does not necessarily mean that they embrace the infrastructure through which energy is delivered. Ideally, energy transportation and infrastructure should be "close, but not too close" to customers—that is, close enough so that transportation and distribution costs remain moderate, but not so close as to be on top of them. However, over the past few decades, the agreement as to what constitutes acceptable closeness has been evolving. Increasingly, local communities have embraced the concept of "not in my backyard" (NIMBY), indicating opposition to anything perceived as "too close" for comfort. The NIMBY concept has been further expanded to include BANANA (build absolutely nothing anywhere next to anything), and NOPE (nowhere on planet Earth).

When it comes to the siting and construction of unsightly and potentially hazardous energy-producing or delivery infrastructure, the policy dilemma is that the negative impacts from the infrastructure are felt most acutely by the host community, while the benefits are spread over a much wider spectrum of the population. Opposition to these projects, increasingly the norm as opposed to the exception, is not simply limited to a few environmentally sensitive and/or wealthy areas. Recently, local communities in southern

Algeria have vocally protested the potential development of shale gas and the introduction of fracking in the southern Sahara. Aboriginal tribes have protested and at times blocked the development of LNG projects in Northern Australia and British Columbia.

Even so, the public's attitude toward the construction of LNG import terminals in the vicinity of their communities is only one of the concerns that terminal sponsors must keep in mind when considering locations. While it is difficult to build an LNG terminal in a designated location without the support of the local community, it is impossible to do so if the site does not have advantageous physical and economic characteristics. This narrows dramatically the list of suitable sites. The onus is on terminal developers to identify locations with the appropriate physical characteristics that are located in a high-value market (assuring the developer that a profit can be made selling natural gas in the region), and, perhaps most importantly, that minimize the opposition of local residents and municipal, state, and federal authorities.

LNG Import Terminal Design Basis

Codes and regulations

Individual countries have different standards and specifications that govern the design and construction of LNG facilities. Given the potential for a major incident arising from the improper handling and storage of large quantities of any form of energy, government authorities in every LNG importing country have enacted codes and regulations governing the design, construction, and operation of LNG facilities and the tankers that serve them. Countries that are considering the introduction of LNG into their national energy mix usually borrow regulations from countries with established codes of practice. For some countries, there are advanced codes and regulations governing the siting of export facilities as well.

- In Japan, provisions for LNG facilities are detailed in the High Pressure Gas Control Act (HPGCA), the Gas Utility Industry Act (GUIA), and the Electricity Enterprises Act (EEA). The GUIA applies if the LNG facility is used only for the gas industry; the EEA is applicable if the LNG facility is used only for the electric industry; and the HPGCA covers all cases that do not fall under the first two categories.[1]

- In the United States, the major design and operating criteria are found in the National Fire Protection Association's (NFPA) "Standard for the Production, Storage, and Handling of LNG" (NFPA 59A) and "LNG Facilities: Federal Standards" (49 Code of Federal Regulations [CFR] Part 193), which are established by the US Department of Transportation (USDOT) and administered by FERC. In addition, the United States has codes governing the waterfront facilities and LNG tanker transits administered by the US Coast Guard (33 CFR Parts 127 and 165).

- In Europe, onshore LNG import terminals are classified as an 'upper tier' hazard under Directive 2012/18/EU of the European Parliament and of the Council dated July 4, 2012, on the control of major accident hazards involving dangerous substances (Seveso Directive). The primary technical standard is EN 1473, "Installation and Equipment for Liquefied Natural Gas—Design of Onshore Installations." EN 1473 covers many of the same aspects of LNG terminal siting that NFPA 59A does in the United States. However, the European standards do not generally provide for prescriptive exclusion zones for public protection, but rather require that the developer undertake a site-specific hazard analysis in support of the final design and siting. Under the umbrella of EN 1473, individual countries may have their own additional codes and requirements.

Although host countries have unique siting and permitting regulations, they incorporate many common requirements. Each country mandates that LNG receiving terminals operate according to the highest safety, environmental, and security standards. A prospective site must meet or exceed these regulatory criteria while providing a technically and commercially viable location on which to construct an LNG import terminal. The focus of the regulatory review can vary among jurisdictions. For example, in the United States, primary siting authority for interstate facilities (which covers all LNG import and export terminals) is vested in FERC to ensure uniform national siting standards. However, in the United Kingdom and other European countries, the siting authority usually lies with local county council planning boards, which will consider national or transnational siting standards and will also turn to other national agencies (for example, the United Kingdom's National Health and Safety Executive) for technical input.

Location

Sufficient available land—that can be used for industrial purposes and is close to a deepwater port that can accommodate LNG tankers— is an absolute prerequisite for short-listing a location for an LNG terminal. The site must be able to accommodate the planned LNG storage tanks and sendout equipment (including future expansions) and have reasonably close access to the pipeline grid. Usually, LNG terminal developers identify multiple sites as potential locations and then enter into a process of elimination based on certain criteria to determine the most suitable location.

Land area. LNG import terminals comprise tanker berthing facilities, storage tanks, vaporization and sendout equipment, pipelines, and miscellaneous facilities (for example, power generation and gas metering equipment). Even though these installations take up ample space, it is difficult to pinpoint the minimum amount of land needed because that is contingent on several details, including:

- The number and size of the LNG storage tanks, the type of containment system, and the size of the resulting tank footprint
- The host country's thermal radiation and vapor dispersion exclusion zone requirements or other hazard assessment protocols
- The adjacent land uses

The terminal developer must also be able to identify and secure a corridor for the pipeline spur that connects the regasification facility to the regional pipeline grid.

Seismic. It is important to determine whether the site under consideration will be affected by seismic activity. Fault lines are often found to be far more numerous and extensive than originally believed, but few are active. In some parts of the world, especially

Japan, some existing terminals are located in earthquake-prone areas. The plant can be designed to counter the effect of seismic shocks or peak ground accelerations, although these adaptations can translate into expensive civil and structural designs requiring elaborate foundations under the critical equipment and tankage.

The other danger associated with seismic activity is that of tsunamis, which are waves generated by earthquake activity or by slumping of the seabed, possibly on the edge of the continental shelf. In open waters, tsunami waves pass virtually unnoticed, but as they reach the shore, they change their shape while still possessing the same kinetic energy. An LNG terminal should be built with sufficient protection from any tsunami wave impact.

Significant weather events. In addition to the possible effects of tsunami waves, the coastal location of LNG plants gives rise to other potential impacts related to weather. Many LNG facilities (both import and export) are located in tropical areas prone to the effects of tropical revolving storms (TRS), known variously as cyclones, hurricanes, and typhoons, dependent on the location. The very high winds generated by these storms can, in themselves, do significant damage, either directly or by airborne debris smashing into equipment, but they also affect the sea. The high winds generate huge storm surges that can inundate large areas of low-lying coastline, causing flood damage, water ingress, and scouring. US Gulf Coast experience has shown that protection against storm surges may require sites to be raised up to almost 9 meters above sea level in some locations, particularly those areas where there are large expanses of relatively shallow water. Needless to say, operations at marine terminals are untenable in such conditions, and LNG vessels will be instructed to proceed to sea or delay their arrival long before the landfall of a TRS. However, if suitable time permits, the LNG carrier will want to ensure that it is in a safe condition to depart from the terminal and that membrane tanks will not be damaged because of sloshing loads caused by heavy rolling and pitching that will be experienced in the open sea.

For a floating LNG facility in a port that can be affected by a TRS and storm surge, the option of putting to sea to ride out the storm may not be feasible. Barge-mounted facilities, which will most likely not be self-powered, will need to have a mooring system and berth that can withstand the effects of very high winds (for example, 150 knots) and storm surges. Floating storage and regasification units (FSRUs) usually have suitable propulsion systems to allow them to proceed to sea and ride out the storm. Internal cargo transfer may be required on membrane containment FSRUs to ensure that there is no damage to the tanks from sloshing.

Safety. LNG is considered a hazardous material because of its cryogenic properties, dispersion qualities, and flammability.[2] According to a staff report by the California Energy Commission:

> *The extreme cold of LNG can directly cause injury or damage. Although momentary contact on the skin can be harmless, extended contact will cause severe freeze burns. On contact with certain metals, such as ship decks, LNG can cause immediate cracking. Although not poisonous, exposure to the center of a vapor cloud could cause asphyxiation due to the absence of oxygen. LNG vapor clouds can ignite within the portion of the cloud where the concentration of natural gas is between a 5% and 15% (by volume) mixture with air. To catch fire, however, this portion of the vapor cloud must encounter an ignition source. Otherwise, the LNG vapor cloud will simply dissipate into the atmosphere.[3]*

On the ocean, LNG disperses and evaporates at a much faster rate than on land because the water acts as a heat source for the liquefied gas and also because a spill cannot be contained. The potential impacts from a marine spill can be more widespread, though of relatively shorter duration, than a landside spill. On land, terminals employ a series of impoundments and other designs to contain spills and limit their effects on other equipment—especially to reduce or eliminate the possibility that the spill effects will travel outside site boundaries. Spills on land evaporate much more slowly and travel over shorter distances as heat transfer is inhibited by the freezing of the ground beneath the spill.

The design must ensure that adequate distances exist between the following components of an LNG terminal: the storage tanks, the berthing and unloading facilities, the vaporization process area, and the other parts of the facility. LNG facilities may also be required to have a prescribed exclusion zone—that is, an area surrounding a facility in which an operator legally controls all activities. Exclusion zones assure that public activities and structures outside the immediate LNG facility boundary are not at risk in the event of an LNG fire or a release of a flammable vapor cloud.[4]

There are two types of exclusion zones: thermal radiation protection (from LNG fires) and flammable vapor-dispersion protection (from LNG clouds that have not ignited but could migrate to an ignition source). Thermal radiation exclusion distances vary by country. In the United States, for example, they are determined by applying the NFPA standard for the production, storage, and handling of LNG and by using computer models that account for facility- and site-specific factors, including wind speeds, ambient temperature, and relative humidity. For example, the required distances assure that heat from an LNG fire inside the storage tanks' bunds would not be severe enough at the property line to cause serious public injury. Computer models that consider average gas concentration in air, weather conditions, and terrain roughness are used to determine safe distances that ensure that vapors created by an LNG spill will be diluted below flammable levels before leaving the site.[5]

LNG import terminal developers always prepare an LNG spill prevention, detection, containment, and countermeasure plan that includes spill prevention practices, spill handling and emergency notification procedures, and staff-training requirements. All of a plant's safety and security manuals are submitted to the host country's energy regulatory authorities for review and comment before the relevant regulatory body gives permission for terminal construction to go ahead. Notably, the safety and security parameters of the terminal sponsor are likely to be even more stringent than those prescribed by energy regulatory bodies. No multinational or state-owned petroleum company or gas and power utility would be willing to risk its reputation, endanger life and property, or risk the financial consequence of being held accountable in the event of an accident by designing a terminal with less than the highest safety and operating standards.

LNG receiving terminals have excellent safety records and consistently meet the stringent safety and environmental requirements that are imposed. However, as M.W. Kellogg noted in a feasibility study for a proposed LNG receiving terminal in Turkey, "One should not overlook the probability of opposition by local and national groups that may not fully understand the nature of the process."[6] Usually, LNG import terminal sponsors will consider possible objections from such groups and attempt to establish lines of communication with the local community and other interest groups very early on in the development process in order to demonstrate that the impact of the proposal on the local environment and community has been carefully considered. Without question, LNG safety is hard to explain to the public as it involves elaborate technical analysis and design; the opposition to LNG terminals is rarely concerned with factual discussion and can readily frighten the local population with nightmarish scenarios of conflagration and destruction. A general distrust of government agencies and energy companies can make siting LNG facilities a challenging experience, notwithstanding the presence of highly protective regulatory requirements.

Port access. To be considered for an LNG terminal, a prospective site must be near a deep shipping channel. Because LNG tankers are very expensive to operate, the proposed location should be relatively unaffected by or protected from extreme weather conditions and severe oceanographic conditions that could delay tanker transit and unloading operations for more than a few days each year.[7]

Although it is possible to dredge both the ship channel and the unloading berth, dredging can be extremely expensive, hard to permit, and may add significantly to the project's schedule and construction costs. It can also be very difficult to find a nearby disposal site for the dredge spoils (particularly if it is contaminated with toxic material). Other environmental considerations include air quality (from gas-fired vaporizer and tugboat and LNG tanker emissions), water quality (from open rack vaporizer discharges and ship and tugboat operations), and the environmental impacts associated with the construction of the pipelines connecting the terminal to the natural gas transmission grid.

Regulatory Issues

Terminal-licensing procedures

The first step taken by a company planning an LNG import terminal at a designated location is to sign a basic engineering and design contract and employ an environmental consultant. The contractors will work with the developer to determine whether the designated site is suitable by conducting technical and environmental assessments of the site and by comparing and contrasting it with other nearby locations. After a site is chosen, the contractor will determine the extent of the civil work required in order to prepare the site and prepare preliminary designs of the terminal's berthing, storage, and regasification facilities. The developer will also design the pipeline that connects the terminal to the natural gas grid, assessing alternative routes from an environmental and safety perspective, unless arrangements are made with the grid operator to perform this work. At this point, the import terminal sponsor will generally also open an informal dialogue with the regulatory agencies.

The developer will usually establish a line of communication with local political officials to address and allay their concerns as early as possible. In addition, the developer will initiate contact with other interested political, community, and/or environmental advocacy groups. Organized interest groups often view an LNG import terminal in their community as inimical to the interests of the local population. If they feel their concerns are not being adequately addressed by the company and local elected political officials, a project can easily be delayed or derailed.

Once the design scope and environmental assessments are in hand, the developer usually submits an application to the national energy regulatory body or appropriate government entity for a license to build the terminal. This application will provide a detailed description of the project, including its safety and security features and its environmental impacts, especially effects on water use and quality, air quality, vegetation and wildlife, cultural resources, geological resources, soils, land use, recreation, aesthetics, and noise. This assessment will also look at the impacts associated with the terminal's construction activities. It may discuss alternatives, including alternate means of supplying gas, other sites considered and the reason for choosing the specific site, alternate layout considerations within the chosen site, and alternate pipeline routes. The regulatory authority generally takes an applicant's environmental report as the starting point for preparing the government's own environmental review. The relevant national regulatory body will solicit additional information and/or clarification from the applicant as needed and will accept public comments on the application.

Provided that all the safety, security, and environmental criteria of the host country are met, the energy regulator will usually issue a final authorization approving the construction of the project. This authorization will generally incorporate conditions with which the developer must comply during both the construction and operational phase of the project. These conditions may cover design, construction, other permits, and so forth; they must be satisfied or accepted by the owner before the terminal developer will be permitted to begin construction and start operations. Particular concerns usually include the proposed location of the terminal, dredge disposal arrangements, LNG storage and containment provisions, and the attendant thermal radiation and vapor dispersion exclusion zone calculations. These authorizations can also be the subject of legal challenges by aggrieved parties; once seen as limited to the United States, such legal challenges have spread rapidly to other host countries, notably in Europe.

Terminal access

Historically, LNG import terminal sponsors had proprietary access to the receiving facilities they built. Proprietary access to LNG receiving terminals was—and in many cases still is—considered to be essential by participants in the LNG chain. Exporters are understandably reluctant to sign a 20-year take-or-pay LNG purchase contract with an LNG importer without a guaranteed market outlet. Traditionally, Asian city gas and power utility companies, either publicly owned (for example, Taipower in Taiwan) or privately owned (all utilities in Japan), built LNG terminals to ensure a steady supply of gas to their own customers.

Proprietary access to LNG receiving terminals has been one of the underpinnings for financing the entire project. As described in chapter 2, the regulatory bargain granted proprietary or monopolistic rights to a utility in exchange for being subject to government regulation of the prices or tariffs charged to its customers, including the profits that the utility could earn from the infrastructure it developed.

The opening of gas and power markets to competition has effected dramatic changes in LNG import terminal access in Europe, where this is governed by Directive 2009/73/EC of the European Parliament and Council of July 13, 2009, "Concerning Common Rules for the

Internal Market for Natural Gas (3rd Gas Directive) and in North America—and even in the more conservative Asian markets. Pipeline gas and LNG imports are no longer the sole province of the state gas and power monopolies. Other firms are claiming the right to import natural gas and/or generate electricity for sale into the market, rather than purchasing gas or power from a state monopoly. In Europe and even in South Korea, private companies have built their own terminals to secure the gas supplies or market access they need. These companies generally have the blessings of their host governments, which are anxious to promote the private construction of natural gas delivery infrastructure to meet gas demand by consumers.

Nevertheless, market liberalization is a double-edged sword for LNG import terminal sponsors. Gas market liberalization is generally characterized by the requirement that the transmission—and often the distribution—pipeline operators offer open access or third-party access on a nondiscriminatory basis to any qualified participant, often under terms and conditions, including prices, regulated by the government. While host governments appreciate the need for new LNG import infrastructure, they are also aware that few companies will bear the risk of investing in a liberalizing gas market, especially in an LNG import terminal, unless they are guaranteed some access to the facility or are allowed to charge access fees that reflect the risks they have taken. Some companies will not go through the laborious process of siting and licensing a receiving facility and spend hundreds of millions of dollars to build it if they are required by regulations to permit other companies—in many cases their competitors—to use it. Other terminal developers have positioned themselves as infrastructure operators, willing to build terminals for third parties on the premise that the terms of access be freely negotiated and allow the developer to profit from his or her efforts.

To that end, governments and regulatory bodies worldwide are looking for ways to create a regulatory climate that is not only conducive to the construction of new LNG terminals but also respectful of the spirit of gas market competition. In Europe, for example, some governments appear amenable to the idea of an import terminal sponsor using all or most of its terminal capacity on a proprietary basis. However, the remainder must be made available to third parties on a nondiscriminatory basis.[8] Some European governments have granted exceptions to third-party access on a use-it-or-lose-it basis; the terminal operator can use the terminal for its exclusive importation, but, if the terminal capacity is not fully utilized, the regulator reserves the right to revisit the terms and conditions of the authorization and may require the granting of third-party access.

The United States had a stronger history of ensuring third-party access to LNG-receiving infrastructure at the time when additional sources of gas were required to meet demand. Of course, that demand has now turned to excess through the exploitation of shale gas. But authorities had to revisit this policy and in 2002, the FERC exempted LNG terminal sponsors from having to provide open-access service. In addition, the US Congress passed legislation in 2002 that granted owners of offshore LNG import terminals exclusive access to their facilities. An onshore terminal open-access exemption was codified in the US Energy Policy Act of 2005.

Offshore terminals

One of the drivers for the development of offshore terminals was that they avoided many of the more controversial aspects of onshore facilities, especially when it comes to overcoming concerns about safety and security that factor heavily into onshore terminal siting. However, while these concerns may be lessened, they are not ignored by the relevant agencies. Most offshore terminal proposals are subject to rigorous safety reviews, although the approach used is more akin to that applied to other offshore energy infrastructure, such as producing platforms, as opposed to onshore terminals. Public concerns are not always assuaged by offshore siting, although such terminals are subject to safety and security measures similar to those applied to LNG tankers transiting to onshore terminals.

Offshore terminals may also incorporate novel technologies, including ship-to-ship or similar cargo transfer methods, that require separate certification and approval before they can be deployed. In contrast, onshore terminals tend to use relatively well-developed technologies that do not require additional certification.

Finally, offshore terminals face unique environmental issues that do not apply to onshore terminals. Offshore pipelines, connecting the terminals to the mainland, raise their own sets of concerns and issues that must be addressed, though this is not a concern for dockside regasification terminals. The selection of vaporization technology has proved controversial—ORVs are simply not used in the United States because of the perceived environmental impacts associated with permitting once-through water cooling. As a result, most offshore terminals use STV vaporizers (see chapter 11).

Adding liquefaction

The US gas market has been fundamentally changed by the recent increase in shale gas production, which has reduced the requirement for LNG imports except in pipeline gas–constrained markets such as New England during winter months. As a result, many of the existing LNG import terminals have filed with the FERC to add liquefaction facilities in order to facilitate LNG exports. In most cases, having an approved import facility eases the burden of review since existing infrastructure siting requirements have been met (storage tanks, marine facilities, ship traffic, etc.) and site characteristics have already been approved. As with LNG import facilities, the US FERC will assess the same environmental issues as required under the National Environmental Policy Act (NEPA) for the addition of liquefaction trains. Exceptions may apply where a terminal sponsor will have to expand to previously undisturbed acreage requiring reviews of environmental factors such as wetlands, vegetation, and wildlife. However, the FERC, in conjunction with the Pipeline and Hazardous Materials Safety Administration (PHMSA) of USDOT, has introduced another layer of complexity in US LNG projects, by now requiring developers to perform detailed three-dimensional computational fluid dynamic (CFD) models on potential spills, ruptures, and other postulated accident scenarios. The result has been a lengthier and costlier approval process.

Notes

1. There are several cases in which the GUIA applies even though LNG is used both as a residential and public utility fuel source for electric power generation. In addition, requirements in the Factory Location Act and the Petroleum Industry Complex Casualty Prevention Act are enforced for all LNG plants and facilities. The Labor Safety and Health Law (for boiler, pressure vessels, etc.) and Fire Prevention Law (for facilities handling hazardous materials such as petroleum) are also enforced to ensure safety of all LNG facilities. The Japan Gas Association (Standards for Safety of Facilities and

Recommended Practice for LNG Inground Storage), Japanese Industrial Standards, Japanese Petroleum Institute, and Steel Structure Calculation Standards are also specific codes and standards that supplement these laws.

2. California Energy Commission, "Liquefied Natural Gas in California: History, Risks, and Siting," Staff paper, (2003), 2.

3. Ibid.

4. Ibid.

5. Even if an LNG storage tank is breached, this does not automatically mean that a fire will result. This can happen when the concentration of LNG in the atmosphere is between 5% and 15%, and there is a source of ignition to light the vapor cloud.

6. M. W. Kellogg, "BOTAS Petroleum Pipeline Corporation New LNG Receiving Terminal Study" (in-house study, Poten & Partners, 1993), 1: 82

7. Ibid., 84.

8. See, for example, "BG Group and Enel Become Partners in Brindisi LNG," BG press release, June 24, 2003. BG Group and Enel were to share 80% reserved capacity in their 6 MMt/y receiving terminal at Brindisi, located in southeastern Italy. The remaining 20% was to be subject to regulated third-party access by Italian energy authorities. (In June 2005, BG announced that it had bought out Enel's share of the Brindisi LNG project.) This project was subsequently dropped.

18

LNG Import Terminal Use Agreements

Introduction

LNG import terminals are the final link in the LNG value chain. Traditionally, terminals were constructed and operated by utility and pipeline companies with the responsibility of acquiring and delivering gas supply to the market. Examples of companies that own and operate LNG receiving terminals include Japan's Tokyo Electric and Tokyo Gas, Fluxys of Belgium, Elengy (GDF Suez), Korea Gas (Kogas), and pipeline companies such as Trunkline (Energy Transfer Partners) in the United States, which owns the Lake Charles terminal. In all of these markets, LNG terminals were treated as regulated facilities, necessary to secure needed gas supplies for downstream markets. As gas markets matured, views on the role of pipeline companies in securing and selling gas supply evolved. So did the commercial structure of LNG regasification terminals.

The process of natural gas market deregulation began in the United States through regulatory initiatives from the FERC, which first eliminated take-or-pay contracts in interstate gas sales and, then altered the role of pipeline, storage, and gas treatment facilities to require open access, including capacity made available to market participants on a nondiscriminatory basis.

Deregulation and open access in the United States was followed by liberalization of the UK and European markets. The European Union has issued a series of gas market directives to liberalize the

European gas market in order to create an open and competitive market. In addition to open access for new import capacity, the regulators directed that some interruptible service be made available to the market. Far less progress has been made in Asian markets, except for the new LNG terminal operating in Singapore, which hopes to become a trading hub. There are discussions regarding liberalization of markets and allowing third-party access in Japan and elsewhere, but progress on these initiatives is slow.

The early LNG regasification terminals were mainly built by natural gas and electric utilities for their own exclusive use. This changed as downstream markets liberalized. Third-party terminal services became increasingly prevalent in the LNG industry. At first, authorities set the rates for service and this service was auctioned in a bidding process. But it was soon realized that regulated third-party access would not facilitate the building of new capacity.

Third-party access (TPA) agreements fall into two general categories. Where the access is governed by the regulatory agencies (regulated TPA), the access terms and conditions are usually spelled out in tariffs developed by the owner/operator and approved by the relevant regulatory authority (they can be subject to comment by interested parties) and are then published by the terminal operator for prospective customers.

In the second case, where the terminal is operated on a proprietary basis, the terms and conditions are covered by a terminal use agreement (TUA). (There is no discernible difference between a TUA and a terminal service agreement [TSA].) Such agreements are negotiated between the terminal operator and each customer (capacity user), and their terms are not usually subject to regulatory oversight.

United States

In the late 1990s and early 2000s, US terminal operators were required to offer capacity under regulated TPA as a condition of reopening shuttered US terminals, whose investment costs were largely depreciated. This policy did not threaten any underlying throughput rights and was consistent with the conditions being imposed on the natural gas pipeline business. Proprietary access came later in the regulatory scheme as the difficulties inherent in regulated TPA became apparent. (It was difficult to justify building new capacity under a regulated TPA regime). Proprietary access was applied initially only to new terminals, then subsequently to expansion of existing terminals. This allowed terminal owners, which came to include infrastructure companies, to use the capacity themselves or to negotiate TPA on commercial terms.

The decline in LNG imports, resulting from the increase of shale gas production in the United States, has led most US terminals to be severely underutilized, which reduced (and even eliminated) the value of capacity access. However, some terminal owners retained a contract obligation under their TUAs to provide import capacity, which has prompted owners and operators to negotiate a new formulation, the bidirectional terminal use agreement. As the name suggests, it allows the facility to both import and export LNG. Such an approach could raise a complex set of operational challenges, but it is unlikely to occur, except in extraordinary circumstances. Even though there has been some experience with cargo reexports from import terminals in the United States and in Europe, the first operational experience with trades under these new bidirectional agreements could occur after 2016 when the Sabine Pass liquefaction project is commissioned by Cheniere Energy.

Europe

Regulated TPA predominates in Europe. Of the 18 European terminals located in eight member states operating at year-end 2014, 13 of them are subject to regulated TPA under which terminal owners are required to share access with any third party granted access rights under transparent and nondiscriminatory conditions.[1] The remaining 5 have been granted exemptions from regulated TPA requirements: South Hook, Grain, Dragon, Gate, and Rovigo. In an exempted terminal, the owner is free to negotiate contracts directly with primary shippers, but the terminal's antihoarding mechanisms, which are monitored by regulators, must be transparent and enable secondary shippers to gain access to capacity when it is not used (the so-called "use it or lose it," or UIOLI provision).

Without a liquefaction option, European terminals have been aggressively expanding their service options beyond the unloading of cargoes with the blessing of the regulators (table 12–1). The main new services offered include:

- Ship loading or reloading for delivery to other terminals
- Truck loading for transfer of LNG into trucks for transport to small regasification plants
- Storage as an unbundled service
- Small ship loading to serve small markets
- Cooldown services for tankers without heel in the tanks
- Bunkering for LNG-fueled ships, ferries, tugs, and barges
- Transshipment of LNG from one vessel to another

Fluxys LNG's Zeebrugge terminal in Belgium is an example of a highly commercially developed LNG terminal. It is located at the heart of Europe's gas pipeline system at a major natural gas trading hub. Conceived as a means to diversify Belgium's energy portfolio and enhance the nation's security of supply, the terminal was commissioned in 1987. It was supplied with LNG under a long-term contract with Sonatrach. The terminal kept this basic role until termination of the supply contract with the Algerian exporter. In the meantime, the European Commission introduced its first energy directive aimed at liberalizing the energy market.

In 2004, Fluxys LNG concluded long-term contracts under a regulated TPA regime with three terminal users: Qatar Petroleum/ ExxonMobil, Distrigas (the Belgian gas marketing company which had been under common ownership with Fluxys), and Suez LNG Trading, making the terminal a multishipper terminal. To accommodate this new role, Fluxys LNG not only expanded the terminal's throughput capacity but also developed its marine capabilities to accommodate the smallest bunkering ships and the enormous Q-Max LNG tankers. In 2008, LNG loading services were introduced in response to requests from terminal users to be able to capitalize on commercial opportunities by reloading LNG cargoes at the terminal. This service has been extremely active, with 25 reloads in 2012 and 21 reloads in 2013 accounting for a substantial share of the terminal's activity. In 2010, the loading of LNG trucks for transport to truck fueling stations was added. In March 2015, Fluxys LNG signed a long-term contract with Yamal LNG for transshipment of up to 8 MMt/y of LNG to support year-round LNG deliveries from the Yamal Peninsula to Asia-Pacific markets. With transshipment services added to its service offering, the terminal will provide a complete range of services for large LNG volumes as well as for small-scale use for bunkering and tank truck loading virtually unmatched elsewhere.

Table 18–1. Ancillary services offered at the European LNG terminals

Member State	LNG Terminal	Ship Loading	Truck Loading	Storage as Unbundled Service	Small Ship Loading	Cooling Down and Gassing Up	Bunkering	Trans. Shipment
				Services Offered in 2013				
Belgium	Zeebrugge	×	×		×	×		×
France	Fos Tonkin		×		×	×	×	
	Montoir	×	×		×	×	×	×
	Fos Cavaou	×			×	×	×	
Greece	Revythoussa			×		×		
Italy	Panigaglia							
	Rovigo							
Portugal	Sines	×	×			×		
Spain	Barcelona		×			×		
	Cartagena	×	×		×	×		×
	Huelva	×	×		×	×		
	Bilbao					×		
	Sagunto	×	×			×		
	Mugardos	×	×			×	×	
Netherlands	Gate	×			×	×		
The United Kingdom	Grain							
	South Hook							
	Dragon							

(*Source:* Council of European Energy Regulators.)

While capacity at Zeebrugge is fully booked on a long-term basis, for short-term bookings Fluxys LNG offers a few additional slots, or shippers can go to the secondary market, which is functioning well. Shippers at Zeebrugge also have a "use it or lose it" (UIOLI) obligation. While UIOLI has not functioned well at other European terminals, it has proven effective at Zeebrugge.

Asia

Traditional proprietary terminals dominate in Asia. For example in Japan, electric and gas utilities still largely enjoy regional downstream market monopolies and operate their terminals exclusively for their own LNG supply. The Ministry of Economy, Trade, and Industry (METI) has taken tentative steps aimed at opening up the terminal business model and has established guidelines for TPA, which have not been implemented at any of the nation's terminals. In South Korea, Kogas also operates its terminals for its exclusive supply. However, private companies are allowed to build terminals for their own use. Likewise in China, three state-owned oil and gas giants (CNOOC, PetroChina, and Sinopec) own and operate virtually all import capacity for their exclusive use.

Breaking with the traditional terminal business model that predominates in Asia, Singapore LNG has set its sights on becoming an Asian LNG trading hub. The city-state is Asia's leading oil trading hub and pricing center. Now Singapore wants to repeat its oil story with LNG. Singapore has taken the first step in the process, commissioning its 6 MMt/y LNG terminal with import and export capabilities. Further development phases will increase capacity to 15 MMt/y by 2017, well beyond potential domestic needs. The terminal is owned and operated by state company SLNG with BG, an active LNG trading company, controlling the majority of the initial capacity.

Further supporting trading hub prospects, Singapore is making efforts to attract LNG portfolio players and traders (in addition to BG) through generous tax schemes with some success. Shell moved its Integrated Gas division from The Hague to Singapore in 2013. BG followed in April 2014, shifting the headquarters of its oil and LNG business from Britain to Singapore. Singapore also hosts investment banks with LNG trading arms such as Bank of America, Merrill Lynch, and Goldman Sachs. Commodity trading companies like Gunvor and Trafigura have transferred LNG traders to Singapore, and Gazprom is there as well. Elsewhere in Asia, Thailand is planning to implement TPA at its existing and proposed LNG import terminals.

South America

Proprietary access terminals dominate in South America. In Brazil, state-owned Petrobras has developed terminals for its exclusive use, although an independent power generator may be allowed to build a terminal. Petrobras' terminals are also being opened up to TPA, although the terms and conditions are, as yet, unknown. Similarly, in Argentina, state-owned Enarsa holds 100% of capacity at the nation's three LNG import facilities. Chile is proving to be the exception, with the regulator requiring that unused capacity at the nation's terminals be offered to third parties.

Regulated Third-Party Access

In many respects, tariffs governing TPA at LNG import terminals are modeled on those employed by gas transmission companies for pipeline transport and adapted for the unique attributes of LNG. Usually, these tariffs are designed to apply to throughput rights at the terminals. Many of the tariff provisions cover the same aspects of the LNG trade covered in SPAs, and often they mirror standard SPA terms and conditions—especially in technical areas, such as LNG measurement practices. The tariff provisions will cover elements such as quantity entitlements (annual and daily); fees (or rates), including the retention of LNG for vaporizer fuel; marine transportation and terminal technical specifications; scheduling, receipt, and measurement of the LNG; quality of the LNG and vapor; delivery and measurement of the vapor downstream of the terminal; force majeure; payment terms; liabilities and insurance; governing law; and dispute resolution. The tariffs are usually accessible either through the website of the terminal operators or through the regulatory bodies.

While the tariff terms and conditions do not usually pose a problem if the terminal's entire capacity is awarded to a single bidder, matters become much more complex when there are multiple firm customers. LNG terminals rarely have sufficient storage capacity to offer multiple customers dedicated storage rights (and in any event, the terminal operator usually markets throughput, not storage); therefore, the terminal activity inevitably involves the commingling of LNG from the individual customers' cargoes and their regasified LNG sendout. LNG cargoes arrive in discrete lumps, but the sendout is ideally made at an even rate between cargo deliveries (and this is one reason why the attempt to apply pipeline standards to LNG terminals can be fraught with difficulties). Inevitably, this structure results in LNG inventories being effectively swapped and loaned among the customers (whether through operation of the tariff or as a result of agreements between the customers) as the cargoes delivered by each capacity holder are scheduled according to the terms of the various SPAs. When the LNG delivered by one or more of the customers requires Btu adjustment prior to delivery to the downstream pipelines, the responsibility for the associated costs adds further complexity.

In addition to having tariffs covering firm capacity, terminal operators are typically required to post terms and conditions for interruptible terminal capacity, with maximum rates equal to the firm ones calculated at 100% load factors. Usually, the terms of the tariff provide for revenues from interruptible terminal services to be largely credited back to the firm customers. This ensures that the operator does not make an excessive return on investment but provides, at best, only a modest incentive to offer these services. Once interruptible cargo deliveries are factored into the equation, capacity scheduling becomes even more complex. This explains why so few (if any) interruptible LNG terminal transactions take place. Another hindrance to interruptible service is that the time between the posting that a cargo delivery slot is available and when the delivery is to be made may prove too short in many cases for interested parties to arrange a ship and a cargo.

These factors generally result in significant efforts to define terminal access to clarify rights and obligations of the parties among one another and with the operator, especially in the areas of cargo delivery scheduling and downstream pipeline nominations and sendout.

When allowance has to be made for force majeure events, such as bad weather that delays vessel arrival and unloading, the scheduling process takes on more serious commercial implications through the knock-on effects of scheduling delays on ship demurrage or the risk that there will be insufficient LNG in the terminal to meet the customer's downstream obligations. Consequently, the customers may have to negotiate operational agreements among themselves to manage these issues.

Proprietary Access

Given the potential problems associated with regulated TPA, it is not surprising that LNG suppliers and terminal developers have pressed regulators for different approaches to terminal services. The first move occurred in the United States in 2002 with the passage of an amendment to the Deepwater Port Act, which extended that legislation to cover gas in addition to oil ports located offshore beyond state waters (generally three miles or more). Since the Deepwater Port Act already allowed for proprietary access to oil ports (of which, to date, the Louisiana Offshore Oil Port, known as LOOP, is the only one), the same rights were automatically extended to offshore LNG terminals.

At the same time, Dynegy was developing a new LNG terminal, the Hackberry Terminal, now Cameron LNG, in Cameron Parish, Louisiana, and had requested that the FERC waive its open-access rules and permit Dynegy to negotiate terms and conditions of access with interested parties. FERC issued an authorization in the Hackberry proceeding, allowing Dynegy the right to develop a proprietary-access business model and indicated that it would permit other terminal developers to pursue the same approach, while retaining the right to reassert jurisdiction if proprietary access was found not to be in the public interest. FERC's justification of this decision was that LNG import terminals serve much the same purpose as large gas-producing fields and could therefore be allowed to compete freely in the market. The Hackberry decision was further reinforced by provisions in the Energy Policy Act of 2005, permitting onshore terminals to retain proprietary access for a period of at least 15 years.

In Europe, as noted previously, the United Kingdom adopted an approach similar to the United States, taking lessons from the US experience. However, the final decision rests with the EU regulators in Brussels, who have so far consented to the terms approved by national regulators. Variants of the UK model have also been adopted on the continent. Most LNG terminals developed in Europe over the past several years have received full exemptions to the European open-access rules, granted by the national regulators and confirmed by the Commission.

With the advent of proprietary access, there is also a need for appropriate agreements under which the operator will offer capacity to third parties. It may not be necessary to have such agreements between the operator and the capacity holder when the terminal is developed as a de facto extension of an integrated project's upstream development, as in the case of South Hook in Wales, where the operator and customer are essentially the same entity. However, other considerations may argue in favor of the development of such a formal agreement between affiliated, but legally independent, operators and customers to manage, for example, tax or other legal liability exposures in the host country of the terminal. Establishing terms and conditions, including tariffs for proprietary terminals (even if the provisions remain confidential) allows the operators to justify rates they may charge third parties under UIOLI policies, or where the open-access exemption is not for 100% of the capacity.

Common terms in TUAs generally include LNG and gas quality specifications, LNG and gas measurement, downstream nominations and dispatch (since these are generally governed by national standards, for example, the North American Energy Standards Board [NAESB] in the United States and the Network Code in the United Kingdom), technical specifications of the terminal and the LNG tankers, governing laws, and dispute resolutions.

The fee for service is a key consideration in the agreement. Generally, this negotiated rate will be high enough to allow the operator a financial return above that which could be earned in the regulated environment. The fee is usually fixed for the duration of the agreement's primary term (generally 20 years), but there may be provisions to adjust it to recover new investment obligations of the operator as a result of subsequent regulatory requirements, to cover new or increased taxes, or possibly to provide adjustments for increases in construction costs beyond the operator's control.

The fee may contain components that track operating expenses and are adjusted for general inflation or actual expense pass-through. It will also provide for vaporizer fuel retention by the operator. The fees are generally collected in advance and are subject to few, if any, reductions, even if the terminal is unavailable for force majeure reasons. In this respect, the fees for service are very similar to the payments made to a shipowner under a charterparty. However, in the case of extended force majeure affecting the terminal, the customer is often given the right to terminate the agreement. In this way, the customer has secured rate (and cost) certainty, which also benefits the operator, who is probably using the payment streams from the customers to secure highly leveraged project financing for the terminal. Lenders are as fond of commercial certainty as customers.

Cargo scheduling provisions can be tailored to match (or align closely with) the terms of the LNG SPAs. However, capacity allocation and scheduling among multiple customers are more involved, and often the TUA will anticipate the creation of a separate agreement, sometimes called an operational and coordination agreement (OCA), or another similar coordination mechanism. The OCA is negotiated among the customers, and the operator may or may not be a party to it. While TPA exemptions may not expressly prohibit it, the regulators may be skeptical as to whether such arrangements among the incumbent customers will act as a barrier to access by other parties.

It is in the allocation of spare capacity that matters become more interesting. Free of regulatory prescriptions that can limit the operator's commercial interest in providing the service, the potential value associated with spare unloading capacity in the form of incremental cargo slots can become a major point of negotiation. Does the operator retain the unilateral right to sell such slots when they become available or do the customers have any special or priority rights to these slots—and if so, at what price: the contract rate or a rate negotiated at the time? Do the customers see any financial benefit from the operator's sale of this capacity? How do the rights of a spot or short-term customer affect those of the long-term customers?

Most terminals have spare vaporization capacity, which can be made available to customers so that they can use it to vary their throughput rates over short or long periods; alternatively, spare vaporization capacity can be made available to the customers or third parties in conjunction with the delivery of extra cargoes above the aggregate ACQs. Again the issue of rates arises. Does the customer pay an extra charge for the right to utilize vaporization above the daily maximum sendout rate?

Another area fraught with complexity is the balancing of the customers' unloading and sendout rates. Clearly, the sendout rate must be balanced in some fashion with the unloading rates, as the terminal will not have unlimited or dedicated storage for each customer. Does the operator impose mandatory borrowing and lending in the TUA with commingled LNG in the tanks, or do the customers negotiate among themselves through an OCA? What rights accrue to a third party that has acquired spot or short-term unloading slots? How do these spot deliveries affect the sendout rights of the firm long-term customers?

Another advantage of negotiated access is that it avoids the regulatory requirement of mandatory capacity release postings, and the customers can generally sell their slots to third parties on negotiated terms. Extension of the TUA beyond the primary term is also a subject of negotiation without regard to regulatory policy, which generally imposes conditions on such extensions for open-access terminals.

TUAs can provide more flexible approaches to Btu adjustment, through either the stripping of LPG or the injection of nitrogen to meet individual customer requirements, rather than being limited to uniform conditions often required under nondiscrimination provisions of open-access tariffs. Other terms can be incorporated into a TUA, such as the development of a port liability agreement, addressing the respective liabilities of the terminal operator and the LNG tanker operator during shipping operations. These agreements are common in the Far East and becoming increasingly used in Europe and the United States. These agreements may be attachments to the TUA, as the shipowner is not party to the TUA.

Since TUAs normally arise in the context of developing new capacity (either by the construction of a new terminal or by the expansion of an existing one), a TUA may provide for a process of coordinating the construction and start-up schedules of the terminal with those of the upstream facilities (again, assuming that there are specific upstream projects dedicated to specific downstream terminals). The customers may gain rights to inspect the terminal plans and construction and also comment on or otherwise influence the terminal's design and operations.

Since terminals can be readily and often inexpensively expanded (at least compared to the initial cost), the rights and obligations of the parties to an expansion may be spelled out in a TUA. Do the existing customers have any rights over future expansion capacity, or can it be freely sold to third parties by the operator? How do parties to an expansion interact commercially and legally with the initial customers? In this general context, it is also important to understand whether the operator intends to enter the LNG-importing business, and if so, how conflicts between the operator's and customers' commercial interests will be resolved. While the TUA may be silent on these issues, this should be viewed as a reflection of the parties' intentions and not as a result of ignoring these aspects of the agreement.

A final consideration in the multi-customer negotiated TUA should be whether, and to what extent, the individual TUAs with each customer need to contain identical terms and conditions. A clear advantage of the TUA approach is that the TUA can be more closely matched with the upstream agreements, but most customers have different upstream agreements with different supply projects that they are trying to optimize in the downstream market. The development of a common TUA for all customers will almost certainly require compromises between them, but tailored TUAs for each customer may lead to coordination problems down the road once the terminal is in operation. There may be even more critical issues in designing these agreements to address problems that could arise from delayed upstream project start-ups.

Bidirectional Agreements

LNG exports presented existing US terminals with the potential to revitalize their business, but at the same time these terminals had to reconcile their export plans with the fact that they had long-term import TUAs with customers who expected to keep their import capacity rights. In some cases, import customers were allowed to withdraw from their TUAs for import capacity, eliminating their fixed payments for the service, and leaving the terminal owner free to develop export plans. In other cases, the terminal owner, often seeking to retain the import capacity payment revenues, had to address the possibility that the terminal might have to serve both import and export customers. To resolve this, some terminals proposed a bi-directional terminal service whereby the import customers retained full rights to utilize berth, storage, and regasification facility capacity to import LNG, while also granting LNG export customers, who signed a liquefaction tolling agreement, rights to either export or import LNG at the same time.

As with TUAs, bidirectional agreements include terms for ship scheduling, loading, unloading, regasification nominations, quality specifications and measurement, and other operational parameters. Most bidirectional agreements do require that the terminal liquefaction customer choose either import or export service for a full month in its proposed ship scheduling to allow the terminal operator sufficient time to coordinate activities among all customers. Short-term switches between import and export service are generally not permitted.

Also, bidirectional agreements need to address the complexities of storage management and regasification nominations that were required in the TUA. However, there are numerous additional complexities regarding berth lay time and windows for loading and unloading of cargoes. As with import TUAs, there are provisions for the customers to work together under an OCA with or without operator participation.

In general terms, market forces would likely rule out the possibility of imports and exports occurring at a terminal at the same time. If LNG prices are sufficiently high to support exports of US gas globally, it would likely imply that prices in the United States would be lower than overseas, thus discouraging interest in imports. However, the US gas market does incur seasonal swings in demand and price, which could lead to short periods within which both imports and exports might be of interest; and it is these events that a bidirectional agreement is intended to address.

Note

1. CEER, CEER Status Review on Monitoring Access to EU LNG Terminals in 2009–2013 (Brussels: CEER, October 22, 2014), http://www.ceer.eu/portal/page/portal/EER_HOME/EER_PUBLICATIONS/CEER_PAPERS/Gas/2014/C14-GWG-111-03_CEER_LNG_22102014.pdf.

19

Conclusion: The Future of LNG

Introduction

LNG is no longer the fuel of the future, it is the fuel of the present, with a dramatic upsurge underway in every aspect of the industry. The industry is evolving and adopting new business models, reflecting its maturity and positioning itself to accommodate shifting global demand and supply patterns. This commercial evolution is being accompanied by an onslaught of new entrants to challenge the domain of the cozy LNG club, a select few oil and gas majors and their utility customers with deep financial and technical capabilities who dominated the early years of the industry. These new entrants are helping to expand the commercial reach of the LNG industry, even as they introduce new risks and uncover new opportunities. These trends and issues have been covered in some detail in this book. As we look to the future, we feel it is appropriate to summarize our views as to where these trends may lead us and the challenges that the industry could face.

Global Demand and Supply

The world's appetite for hydrocarbon fuels continues to grow as the global economy expands. Because of its clean-burning characteristics, natural gas is the fastest growing and the preferred hydrocarbon fuel. Concurrently, the supply revolution that is North American shale gas has removed any lingering doubts as to the availability of gas over the long term. Even as economic growth shifts from mature economies to the emerging world—led by China and India—so too has growth in natural gas demand. The global shift in LNG demand is even more dramatic. LNG has been virtually pushed out of North America, where surging domestic natural gas production is satisfying growing domestic demand, and has lost significant market share in western Europe, where a prolonged economic downturn, a broken carbon trading scheme, and heavy emphasis on (and subsidies of) renewables have cut into gas demand. Still, buoyed by technological and commercial innovation, LNG is capturing new markets in South America, Southeast Asia, and the Middle East, while there are initial stirrings of interest in Africa, complementing strong growth in the new twin giants, China and India.

A robust expansion of LNG production—first in Australia and then in the United States—is virtually guaranteed into the middle of the next decade based on liquefaction projects either under construction or already sanctioned by sponsors. Yet, the path to expanded LNG production is not devoid of problems. New LNG supply projects have often experienced significant cost increases, schedule delays, or both, particularly in Australia. Moreover, FIDs for these projects were taken at oil prices closer to $100/bbl. At oil prices below $100/bbl (given that prices of most long-term LNG sales remain disproportionately linked to oil prices), the financial robustness of some, perhaps many, projects could be challenged. The pace of FIDs in Australia had slowed dramatically—in part because of rising construction costs—even before the late 2014 collapse in oil prices. With the real possibility of prolonged lower oil prices, this slowdown could spread globally, very likely delaying or even shelving proposed LNG export projects in the United States, Canada, Australia, and Africa.

Supply and demand tensions

Tensions between growing supply and uncertain demand could characterize the next decade of the LNG industry. The shift from a buyers' to a sellers' market in the wake of the March 2011 Fukushima Daiichi nuclear disaster in Japan is reversing. A potentially long LNG market was turned on its heels by the shuttering of Japanese nuclear power plants following Fukushima. This created a large and immediate incremental demand for LNG to fuel Japanese power generation even as LNG demand in China and India surged and new niche markets were opened via the fast-track installation of FSRUs, which allowed many smaller markets to switch from more expensive refined petroleum products to LNG.

It is unlikely that all of these demand drivers will remain in place. Japan is moving to restart its shuttered nuclear plants, and there are signs that the economies of China and India are slowing, even though their growth still promises to be fast compared with that in mature industrial economies. China is also experiencing a surge in new pipeline supplies as it adds imports from Turkmenistan, Myanmar, and now perhaps Russia, at competitive prices, further threatening LNG demand. The EU shows little sign of economic recovery in the near term, and the Russian standoff over eastern Ukraine creates an aura of uncertainty over European gas market prospects. Within the EU, the split between the conservative economic policies of the north and liberal economic policies of the south is complicating economic stability and market growth. At the same time, North America is transforming from a large LNG importer to an even larger exporter. Still, the boost to demand from the new markets that have been opened up by FSRUs remains in place, although these volumes are small in comparison to the traditional LNG markets in Japan and South Korea. Yet lower oil prices are undermining the rationale for LNG imports in many smaller markets where oil switching was a primary driver of LNG demand.

Technical Outlook

LNG is an industry that had largely focused on engineering and construction (or in the words of the late Malcolm Peebles, who worked in Shell's LNG business for years, "plumbing"), enabling the facilities to achieve extraordinary levels of safety, reliability, and longevity. In this regard, the industry has succeeded beyond its wildest dreams. Building upon a strong technical foundation, innovation has not halted, and the last decade has seen many advances. LNG train sizes have increased dramatically, and LNG tankers have become ever larger, producing economies of scale—although there may be a pause in scale growth as the industry seeks to optimize technology options. Liquefaction trains have been built to a capacity of 7.8 MMt/y, with LNG tankers growing to 266,000 m^3 to service these megatrains. But these sizes remain outliers and now seem more appropriate for less flexible point-to-point trades over long distances to large markets and may be challenged to accommodate the emerging, flexible commercial nature of the business. Liquefaction trains of approximately 5 MMt/y are now generally favored by the industry, while LNG carriers of approximately 175,000 m³ are preferred for general trading activity to better match market requirements and facility capacities at almost any global import facility.

At the same time, the LNG shipping industry is adopting slow-speed and dual- and tri-fuel diesel engines (propulsion systems that are found in many other merchant ships) to lower transport costs. Ironically though, for the moment, the industry shows signs of becoming a victim of its own success. Surging demand for new LNG construction has led to dramatic cost increases, most prominently in Australia, as EPC contractors and their suppliers have raised prices far faster than inflation. While there is plenty of evidence that the industry will experience continuing incremental technical improvements, the combination of falling prices and high costs currently neutralizes these gains.

Move offshore

Advances in offshore technology for both import and export facilities have been highlights of the past decade. Indeed, FSRUs for LNG import terminals are now an accepted technology, particularly where market size and siting issues make onshore terminals uneconomic or unfeasible. These quicker-to-install facilities—often characterized by lower initial capital investment costs—have helped to open up new markets, particularly those that are small in size, seasonal in nature, or simply unpredictable. FLNG production plants are also gaining traction, with one under construction for Australia, two for Malaysia, a fourth for Colombia (maybe), and a fifth that is an innovative effort to convert an existing LNG carrier to an FLNG unit. FLNG technology could unlock stranded offshore reserves by eliminating the costs of building long and expensive subsea pipelines to shore and the costs of marine facilities at an onshore liquefaction complex, as well as avoiding local opposition to an onshore plant. FLNG technology can also enable project sponsors to avoid local labor constraints, which push up construction costs. Many sponsors are adopting modular construction, with the modules built far from the site to avoid constrained local labor markets, such as Australia, which has a particularly onerous labor and construction environment.

Unconventional gas production and LNG liquefaction

While not an LNG technology, horizontal drilling and hydraulic fracturing (fracking) have unlocked vast natural gas resources with major impacts on the LNG industry—both negative and positive. During the past decade, the collapse of the North American LNG market was the result of the shale gas revolution. Once an extremely promising market for LNG, optimistic projections spawned numerous LNG import terminals and liquefaction trains to supply this market. But unconventional natural gas production (primarily shale gas) has surged, making North America not only self-sufficient in natural gas but also providing a surplus for export. Enterprising terminal owners and operators, whose import facilities were no longer needed, saw an opportunity to add liquefaction trains that benefit from existing infrastructure—storage tanks, marine facilities, etc.—at a fraction of the cost of a comparable greenfield development. Now the United States is poised to become a major LNG exporter, maybe the world's largest.

Unconventional gas production also unlocked a new LNG export province based on coalbed methane in Queensland on Australia's eastern coast. Previously, all of the country's LNG projects were developed around conventional gas fields off northwest and northern Australia. Over the next decade, with the first of these projects—both in Australia and the United States—starting production in 2014 and late 2015, LNG export projects that process unconventional natural gas resources will likely add 85 MMt/y of LNG supply to global markets. This is on a scale similar to the LNG development in Qatar over the past decade.

Commercial Trends

The last decade has seen major commercial innovation that has challenged the staid business structure of the LNG industry—one based upon back-to-back long-term contracts underpinning the historically conservative nature of the LNG delivery chain. While once considered a fringe activity of little interest to the established players, about one-third of LNG now moves under spot and short-term contracts, and new entrants as well as established LNG firms actively pursue these potentially lucrative trades. LNG sales contracts are increasingly incorporating terms that permit buyers and sellers to divert cargoes from their primary destinations to take advantage of short- and long-term commercial opportunities, including inter-regional market arbitrage.

Commercial flexibility has provided an important market-balancing mechanism and has facilitated the rerouting of large volumes no longer needed in the Atlantic Basin to Asia, where LNG was in short supply. The industry diverted large volumes to Japan on short notice when LNG was needed by electric utilities to fill the gap left by the shutdown of the nation's nuclear plants after Fukushima Daiichi. It also enables nations with large seasonal swings in demand, such as South Korea, to meet winter peaks and to dispose of surplus LNG in the shoulder months. When combined with an FSRU import terminal, spot and short-term LNG enables new markets with uncertain demand to access supplies without committing to long-term purchases. At the same time, it enables sellers to rebalance their own deliveries, as was the case of the Qatari projects that rerouted large volumes to Asia as their markets in the United States and the United Kingdom faltered. Spot and short-term trades—while not supplanting the long-term sale and purchase agreements (SPAs) that are still required for project financing (and this is not expected to change anytime soon)—provide the business with valuable flexibility while also providing savvy players with new profit opportunities.

Tolling schemes at liquefaction plants

The first tolling liquefaction model was developed at Atlantic LNG in Trinidad in the late 1990s. This business model permits more flexible and often faster upstream development as it allows companies with conflicting commercial ambitions a means to cooperate in those elements of the chain where their interests are aligned. In Trinidad, it also allowed for destination flexibility, and Trinidad LNG is now a prime source of spot and short-term LNG supplies to new markets in South America. Subsequently, BG, a partner at Trinidad LNG, adopted a tolling model for a project in Egypt (Idku LNG). As a variant on this model, but on the opposite end of the supply chain, the past decade saw many Atlantic Basin import terminals developed by third parties who, in turn, lease capacity to LNG exporters and importers.

Many US terminal owners have adopted schemes combining the two structures, which promise to be a significant feature of the next decade. They have leased liquefaction capacity to third-party tolling customers, who will arrange their own feed gas supplies and market the LNG. The customers agree to pay tolling fees to terminal owners, who then use these fees to finance construction and operation of the liquefaction train(s). The customers have complete freedom to market the LNG without destination restrictions under spot, short-term, or long-term arrangements, or a mix of all three. They also arrange their own shipping as required. US volumes will start hitting the market in late 2015, when Cheniere Energy commissions its first train at Sabine Pass in Louisiana.

Global LNG prices

Much speculation is focused around LNG pricing and whether a global LNG price will develop, as occurred in the oil market. Oil-linked pricing has dominated the LNG business for five decades. In a tight market, when sellers have the stronger negotiating position, a formula approximating oil equivalency prevails—a price ($/MMBtu) that is 0.1485 times the crude oil price ($/bbl). But when buyers are in the stronger negotiating position, they often bargain for a lower linkage—that is, a discount to oil equivalency—sometimes with the addition of ceilings and floors or S-curves to dampen the impact of high oil prices (for the buyer) or low oil prices (for the seller). The prices for cargoes sold under spot and short-term deals move above and below long-term contract prices based upon the short-term supply and demand balance. However, there is a growing influence of Atlantic Basin natural pricing hubs, particularly in Europe, which influences what European buyers will pay for spot cargoes and increasingly for term supplies as well. Buyers and sellers alike closely monitor the NBP in the United Kingdom, the most liquid European natural gas trading hub. As LNG markets in Asia loosen for a variety of reasons, the Atlantic Basin (and especially Europe) will become the clearing market for the world.

Even importers, who previously obtained very favorable LNG pricing terms after the 1997–98 Asian financial crisis, which sharply curtailed LNG demand, have come to accept higher oil-linked prices when the market tightened. China has made this transition more easily than India, though both nations are struggling with domestic natural gas pricing policies—reform of these is a prerequisite to a robust natural gas market and sustainable LNG imports.

Even as Asia-Pacific markets remain tied to oil-linked pricing for most of their LNG supplies, gas-on-gas trading hubs are increasingly prevalent in the Atlantic basin. The United Kingdom and northwest Europe benefit from increasingly liquid trading hubs, and LNG imports must meet these price markers. Long-term pipeline suppliers have been forced to recognize the pressures their buyers are under and now often include a hub-linked price component in their pipeline gas contracts. These pricing pressures are slowly moving to southern Europe as well.

Hub-linked LNG prices

A big impact on global LNG prices will result from US LNG exports with FOB prices linked to the Henry Hub in Louisiana. Buyers, primarily from Asia-Pacific, and trading firms have signed up for large volumes under these innovative pricing terms either through tolling arrangements or under long-term SPAs. Asia-Pacific buyers prize these purchases for their price diversification benefits, while trading companies are betting that these volumes provide a diversification of price risk in their LNG supply portfolio as they match varying prices in their downstream market positions. While purchases linked to Henry Hub seemed a solid win for buyers in a world where oil prices were more than $100/bbl, these benefits quickly diminished at lower oil prices, and at oil prices below $60/bbl seem destined to lose money.

While Asia-Pacific markets will continue to rely on crude oil indexation as the basis for much of their term LNG purchases, buyers and sellers worldwide will build LNG portfolios that embrace a variety of pricing indices. This is partly a response to managing LNG price risk and partly the fact that a wider range of LNG buyers and sellers will have access to market and supply outlets both east and west of Suez, and need (or want) LNG supplies that have a range of price indexation provisions. By 2020, as much as 25% of Japan's LNG supply could be priced on Henry Hub indices.

Regional LNG prices

It seems highly unlikely that there will be a world natural gas (and therefore an LNG reference) price during the next decade, probably even longer. The high cost of liquefaction, shipping, and regasification will not change, and the established LNG supply contracts and their legacy pricing will not disappear overnight. Key markets such as Japan will continue to seek supply security and thus follow the more classic utility model even as Japanese buyers seek to build supply portfolios with diversified pricing terms. While the numbers of liquefaction plants, import terminals, and tankers will grow dramatically, they are still a long way behind the breadth and depth of the world's oil market.

Henry Hub should remain the primary reference point for US LNG transactions, and NBP and the growing trading hubs on the European continent should assume a greater importance for European LNG contracts. It is unclear if and where a benchmark price would appear in Asia, reducing motivation on the part of legacy sellers or buyers to change the status quo.

The portfolio approach to LNG purchases was originally developed in order to increase physical security of supply. Because natural gas and LNG prices (and demand) have distinct regional components and do not necessarily show a high short-term correlation to oil prices, the development of a supply portfolio with diversified prices can reduce the volatility of LNG pricing for both buyers and sellers. Moreover, the development of supply portfolios has also enabled arbitrage opportunities, particularly when combined with increasingly unlimited destination options. Still, the long-term pricing trend will almost certainly remain reflective of oil prices, given the dominance of Asian utilities as LNG buyers. However, a hub-pricing component—already a part of pricing in northwest Europe—is likely to move to southern Europe as well. Term sellers to Asian buyers could be forced to lower the linkage to oil owing to competition from LNG supplies linked to US Henry Hub prices and to more market-responsive spot and short-term supplies.

Regulatory Trends

There has always been a healthy debate on whether regulations lead market developments or follow them. There are elements of truth in both arguments. Regulations, while typically implemented in response to changing market conditions, often facilitate market developments as well. Downstream gas (and electric) market liberalization has contributed to the commercialization of the LNG industry. It ended monopoly positions in major markets, opening them to new entrants and price competition. This has yielded cost savings to consumers and contributed to the commercialization of the LNG business, which is now well established and appears irreversible.

United States and Europe

The United States has, since the early 1980s, pursued energy policies based on nondiscriminatory third-party access to gas transmission infrastructure under regulated terms and conditions, including the rates charged for such services. Europe's gas directives were based on similar provisions but extended the liberalization more effectively to the distribution system and retail customers. However, this is not a straight-line process, and authorities have fine-tuned regulations based upon market developments. For example, regulatory agencies have apparently acknowledged that LNG developers are unlikely to make multimillion dollar investments in new LNG import capacity without guaranteed access to it. To that end, the regulators have largely exempted new import terminals from third-party access requirements, often on the theory that these facilities are more akin to production assets, and less similar to pipeline and storage infrastructure.

In turn, the investors in these facilities and the holders of capacity have foregone the right to assured recovery of their investments through the classical utility business model. There also appears to be an implicit or explicit recognition that different project sponsors will follow different business models, and that the last remaining barrier to entry in the terminal business is the ability to gain environmental permitting and financing. Granted, regulatory authorities have implemented certain provisions (such as UIOLI strictures in the United Kingdom) to prevent the possibility of market abuse.

Flexibility in the regulatory process contributed to the United States swinging so quickly from developing import terminals to converting them to export. There was no debate or regulatory proceedings associated with the economic or capacity rights associated with this reversal of direction, and regulatory oversight was largely limited to reviewing the environmental and safety aspects of the liquefaction plants.

Asia-Pacific

Governments and energy regulatory agencies in the Asia-Pacific region, where market liberalization has developed at a much slower pace, carefully monitor North American and European trends. Utility companies that have invested large sums in their respective nations' gas and power infrastructure cannot be expected to freely surrender market share to newcomers and allow them to erode the utilities' profit and customer bases. In any event, these monopoly companies also fulfill their respective nations' energy security objectives, which governments are reluctant to simply relinquish in the name of market liberalization. LNG provides the supply for nearly all the downstream natural gas markets in Japan, South Korea, and Taiwan, and as the global LNG trade commercializes (still far from a commodity trade), so too will pressures grow to liberalize downstream natural gas markets; indeed, it is a slow process.

By contrast, emerging economic giants China and India have become large LNG importers to complement significant domestic production. The mix of domestic production with imports—LNG in both China and India and also by pipeline in China—opens up prospects for downstream natural gas market deregulation. The challenge is to encourage domestic production and imports without undue increases in prices paid by consumers. Authorities in both nations have struggled with this conundrum for years without success. Oil prices well below $100/bbl could afford both countries an opportunity to deal at last with this issue, since the premium for LNG imports over domestic market prices would diminish, and domestic market prices could therefore move to global price levels without significantly raising prices to consumers. In the end, this would contribute to a healthier domestic market as it rationalizes demand, domestic supply, and imports (both pipeline gas and LNG).

Supply side regulation

There are regulatory issues on the supply side as well, which could significantly impact global LNG market trends. In the United States, there was an extended debate surrounding LNG exports, with some opponents arguing that significant LNG exports would raise prices to US consumers. Others opposed to gas exports also argued that it would encourage fracking, which they claim is deleterious to the environment. In the end, the authorities came down on the side of LNG exports, concluding that the benefits outweighed the costs. Still, project sponsors must proceed with a costly application and review process with the FERC in order to build export facilities, which, once approved by the FERC, must await DOE review of their applications to export LNG to nations without free trade agreements (FTAs) with the United States.

While no application to export LNG has been rejected through the end of 2014, it remains unclear what will happen if domestic market conditions change, which could happen in a low oil and gas price environment as drilling is curtailed and the domestic market tightens. The pace of development of US LNG exports and foreign buyer interest is significant. It introduces a new competitive pricing element—supplies at Henry Hub–linked prices—as well as increased flexibility into global LNG trades, which could foster additional pressures for downstream market liberalization in markets from India to Japan.

Meanwhile, countries in East Africa—Mozambique and Tanzania—are struggling to put regulatory regimes in place (including the role of domestic companies, local labor participation, royalties, taxes, and domestic market priority) to govern their natural gas industries and LNG exports. These are complicated decisions with major national implications that authorities without the requisite expertise will find challenging. While ultimately they will likely succeed, these decisions, because of the large benefits to local economies, could take time even as both countries support LNG exports and firms active in the upstream are anxious to move ahead. By way of example, Nigeria's Petroleum Industry Bill (PIB), which was originally proposed in 2008 to change the organizational structure and fiscal terms governing the oil and natural gas sectors, was still not finalized and approved as of the end of 2014. Unlike Mozambique and Tanzania, Nigeria has been extracting and exporting oil and natural gas for decades.

Small-scale applications

We have deliberately chosen to limit our book to describing and discussing the large-scale LNG industry that has developed over the past 50 years. We recognize that there is a rapidly growing segment of the industry that is operating on a much smaller scale, and as befits its small scale, is developing rapidly. This segment would include the development of LNG as a transportation fuel in truck, rail, and marine applications, as well as the increasing delivery of small LNG packages by truck or even in containers to remote sites across the world. Many of these applications are just emerging, and like their larger brethren at the global level, must survive the onslaught of a world of much cheaper oil than their promoters might have envisioned. We believe this subject may soon be ripe for treatment, perhaps in the third edition of this book when its time arrives.

Conclusions

The future of the LNG industry seems assured. Gas demand will grow, supply will expand (even with the uncertainties described above), technology will evolve, and continuing commercial innovation will characterize the next few decades. The tension between the traditional utility model and the emerging commercial trading model will be increasingly resolved in favor of commercialization. On a cautionary note, as the number of industry participants continues to grow, regulatory agencies and industry players alike will have to remain vigilant to maintain the industry's impressive safety record, which has contributed to its commercial success. These trends will create new challenges and uncertainties, but fundamentally the industry will retain much of what makes it so appealing—a size large enough to be very exciting, but small enough to be quite personal, while also operating in an international and political climate dynamic enough to keep interests alive. For those of us lucky enough to be part of it, we can look forward to challenges, opportunities, risks, and rewards aplenty.

Index

E

H

J

K

L

M

U

About the Authors

Michael D. Tusiani

Michael D. Tusiani joined Poten & Partners in 1973. From 1983 to 2016 he served as its chairman and chief executive officer. He is currently the chairman emeritus. During his career he has been active in all aspects of oil and gas trading and transportation. He has written numerous articles on energy and shipping matters and three books: *The Petroleum Shipping Industry: A Nontechnical Overview* (PennWell 1966), *The Petroleum Shipping Industry: Operations and Practices* (PennWell 1996), and *LNG: A Nontechnical Guide* (PennWell 2007, coauthor). He received a BA from Long Island University and an MA in economics from Fordham University.

Gordon Shearer

Gordon Shearer has been active in the LNG industry for more than 25 years. He is a senior adviser with Poten & Partners, which he rejoined in 2015. From 2007 until 2013 he was CEO of Hess LNG (a joint venture of Hess Corporation and Poten & Partners). Before joining Poten in 2001, he was employed by Cabot Corporation, where he was CEO of Cabot LNG Corporation, which owned and operated the LNG terminal in Boston Harbor and the LNG carrier *Matthew*. During this time, he was heavily involved in the creation of the Atlantic LNG project in Trinidad and Tobago. He received a BSc in geophysics from Edinburgh University in 1976 and an MBA from Harvard Business School in 1978. He was a member of the National Petroleum Council from 2005 until 2016.